Plant and Microbe Adaptations to Cold in a Changing World

Participants for Plant and Microbe Adaptations to Cold 2012 Hokkaido University, Sapporo Japan June 24–28, 2012

Ryozo Imai • Midori Yoshida
Naoyuki Matsumoto
Editors

Plant and Microbe Adaptations to Cold in a Changing World

Proceedings from Plant and Microbe Adaptations to Cold 2012

Springer

Editors
Ryozo Imai
Crop Breeding Research Division
NARO Hokkaido Agricultural
Research Center
Sapporo
Japan

Naoyuki Matsumoto
Silviculture & Forest Ecology Studies
Hokkaido University Graduate School
of Agriculture
Sapporo
Japan

Midori Yoshida
Crop Breeding Research Division
NARO Hokkaido Agricultural
Research Center
Sapporo
Japan

ISBN 978-1-4614-8252-9 ISBN 978-1-4614-8253-6 (eBook)
DOI 10.1007/978-1-4614-8253-6
Springer New York Heidelberg Dordrecht London

Library of Congress Control Number: 2013947056

Printed on acid-free paper

Springer is part of Springer Science+Business Media (www.springer.com)

The Conference was sponsored by the OECD Co-operative Research Programme on Biological Resource Management for Sustainable Agricultural Systems, whose financial support made it possible for most of the invited speakers to participate in the Conference.

OECD

Preface

Global warming is affecting agriculture in a wide range of the climatic zones. In contrast to the extensive debate over the effects of global warming during summer growth, the impact of rising winter temperature on agricultural production has received considerably less attention. However, climate changes can certainly affect winter conditions, and small changes in winter climate can have even more drastic impacts in many regions of the world. The tendency of winter warming is most evident in the lower latitude winter transitional areas of cool temperate regions, which include major production areas of winter wheat and forage crops. In these areas, snow cover has the greatest impact on winter crop production in both positive and negative ways: snow cover beneficially protects plants from freezing injury while also providing optimal habitat for the major winter pathogens, known as snow molds, to prevail under snow cover. Climate change affects the depth, duration, and distribution of snow cover in the cool temperate regions, resulting in increased freezing damage of crops due to reduced snow cover in some areas, as well as increased snow mold damage due to prolonged snow cover in other areas. Changes in the flora and dynamics of snow mold fungi are also being reported. Fluctuations in snow cover also affect occurrence of soil frost and freeze-thaw cycles, which result in alteration of soil physical properties, ecosystem nutrient cycling and microbial activities. Overall, there are many emerging factors that can threaten the sustainability of agricultural production.

Plant and Microbe Adaptations to Cold (PMAC) is an interdisciplinary forum for research and extension scientists working in the fields of plant pathology, plant physiology, microbiology, and crop breeding, to advance our understanding of overwintering of crops and attempt to solve the problems associated with winter damage. The first PMAC conference was held in Sapporo, Japan in 1997 and the following meetings have been held every three years in different locations around the world. The PMAC conference came back to its place of origin after 15 years, and PMAC2012 was held June 24–28, 2012 at the Conference Hall of Hokkaido University, Sapporo, Japan. The PMAC2012 conference focused on global climate change, food security, and agriculture sustainability with the subtitle "Toward risk assessment and management of sustainable agriculture in the cool and cold regions," and the entire program was organized to reflect this theme. The sessions covered a wide range of topics from soil physical properties and crop protection from frost and pathogens to current breeding strategies. In order to widen the scope of the conference and enhance interdisciplinary discussion, experts in meteorology, soil science and ecology were also invited to participate in the program. What was unique about PMAC2012 was having a special panel discussion session on global warming management. This was a great opportunity for the scientists to better understand the realities of impacts on producers and the questions that needed to be addressed through discussion with the invited panel members representing farmers, agricultural co-operators, and policy makers. PMAC2012 gathered over 100 participants from 14 countries and hosted 41 oral and 42 poster presentations.

This book is a collection of contributions from invited and selected speakers at the conference. Each contribution includes important and timely topics on plant and microbe adaptations to cold. These contributions span the topics discussed at the conference. Publication of this book has been partially suported by OECD-CRP. We also thank Hannah Smith, Melissa Higgs, and Kevin Wright of Springer for their assistance in production of this volume. Finally, we express our gratitude to all the authors and reviewers whose dedicated efforts made this publication possible.

Ryozo Imai
Midori Yoshida
Naoyuki Matsumoto

Editors; Naoyuki Matsumoto (left), Midori Yoshida (middle), Ryozo Imai (right).

Contents

Contributors

Jeffrey C. Anderson Department of Biochemistry and Bond Life Sciences Center, University of Missouri, Columbia, MO, USA

Kazuya Araki Hokkaido Research Organization Tokachi Agricultural Experiment Station, Shinsei Minami, 9-2, Memuro, Hokkaido 082-0081, Japan
e-mail: araki-kazuya@hro.or.jp

Keita Arakawa Research Faculty and Graduate School of Agriculture, Hokkaido University, N9-W9, Kita-ku, Sapporo 080-8589, Japan
e-mail: keita-ar@for.agr.hokudai.ac.jp

Junichi Arima Graduate School of Agriculture, Hokkaido University, Kita 9 Nishi 9, Kita-ku Sapporo 060-8589, Japan

Zsófia Bánfalvi Agricultural Biotechnology Center, Gödöllő, Hungary

Annick Bertrand Soils and Crops Research and Development Centre, Agriculture and Agri-Food Canada, 2560 Hochelaga Boulevard, Québec City, QC G1V 2J3, Canada
e-mail: annick.bertrand@agr.gc.ca

Ákos Boldizsár Department of Plant Molecular Biology, Agricultural Institute, Centre for Agricultural Research, Hungarian Academy of Sciences, Brunszvik u. 2., Martonvásár 2462, Hungary

Ferenc Borondics Canadian Light Source, Saskatoon, SK, Canada

Yves Castonguay Soils and Crops Research and Development Centre, Agriculture and Agri-Food Canada, 2560 Hochelaga Boulevard, Québec City, QC G1V 2J3, Canada

Luigi Cattivelli Genomics Research Centre, Consiglio per la ricerca e la sperimentazione in agricoltura (CRA), Fiorenzuola d'Arda (PC), Italy

Alejandra M. Cortes-Barco School of Environmental Sciences, University of Guelph, 50 Stone Road E, Guelph, ON, N1G 2W1, Canada

Cristina Crosatti Genomics Research Centre, Consiglio per la ricerca e la sperimentazione in agricoltura (CRA), Fiorenzuola d'Arda (PC), Italy

Petre Dobrev Laboratory of Hormonal Regulations in Plants, Institute of Experimental Botany, Academy of Sciences of the Czech Republic, Rozvojova, Prague, Czech Republic

Keita Endoh Research Faculty and Graduate School of Agriculture, Hokkaido University, N9-W9, Kita-ku, Sapporo 080-8589, Japan

S. R. Frankel Department of Ecology and Evolutionary Biology, University of Colorado, Boulder, Colorado 80309, USA

Seizo Fujikawa Research Faculty and Graduate School of Agriculture, Hokkaido University, N9-W9, Kita-ku, Sapporo 080-8589, Japan

Yukiharu Fukushi Research Faculty and Graduate School of Agriculture, Hokkaido University, N9-W9, Kita-ku, Sapporo 080-8589, Japan

Gábor Galiba Department of Plant Molecular Biology, Agricultural Institute, Centre for Agricultural Research, Hungarian Academy of Sciences, Brunszvik u. 2., Martonvásár 2462, Hungary
e-mail: galiba.gabor@agrar.mta.hu

Denis Gaudet Agriculture and Agri-Food Canada, Lethbridge Research Centre, 5403 1st Avenue South, Lethbridge, Alberta T1J 4B1, Canada
e-mail: denis.gaudet@agr.gc.ca

Lorenzo Giusti Genomics Research Centre, Consiglio per la ricerca e la sperimentazione in agricoltura (CRA), Fiorenzuola d'Arda (PC), Italy

Paul H. Goodwin School of Environmental Sciences, University of Guelph, 50 Stone Road E, Guelph, ON, N1G 2W1, Canada

Davide Guerra Genomics Research Centre, Consiglio per la ricerca e la sperimentazione in agricoltura (CRA), Fiorenzuola d'Arda (PC), Italy

Bjarni E. Gudleifsson Faculty of Environmental Sciences, Agricultural Univesity of Iceland, Bugardur, Oseyri 2, 601 Akureyri, Iceland
e-mail: beg@lbhi.is

Masaki Hayashi Department of Geoscience, University of Calgary, 2500 University Dr. NW, Calgary, Alberta T2N 1N4, Canada
e-mail: hayashi@ucalgary.ca

Hugh A. L. Henry Department of Biology, University of Western Ontario, London, ON, Canada
e-mail: hhenry4@uwo.ca

Dirk K. Hincha Max-Planck-Institut für Molekulare Pflanzenphysiologie, Am Mühlenberg 1, 14476 Potsdam, Germany
e-mail: hincha@mpimp-golm.mpg.de

Tomoyoshi Hirota Agro-environmental Research Division, NARO Hokkaido Agricultural Research Center, Hitsujigaoka 1, Toyohira-Ku, Sapporo, Hokkaido 062-8555, Japan
e-mail: hirota@affrc.go.jp

Tamotsu Hoshino National Institute of Advanced Industrial Science and Technology (AIST), Higashi-Hiroshima, Japan
e-mail: tamotsu.hoshino@aist.go.jp

Tom Hsiang School of Environmental Sciences, University of Guelph, 50 Stone Road E, Guelph, ON, N1G 2W1, Canada
e-mail: thsiang@uoguelph.ca

Julio C. M. Iehisa Graduate School of Agricultural Science, Kobe University, 1-1 Rokkodai, Nada, Kobe, Hyogo 657-8501, Japan

Ryozo Imai Crop Breeding Research Division, NARO Hokkaido Agricultural Research Center, Sapporo, Hokkaido, Japan
e-mail: rzi@affrc.go.jp

Satoshi Inoue Agro-environmental Research Division, NARO Hokkaido Agricultural Research Center, Hitsujigaoka 1, Toyohira-Ku, Sapporo, Hokkaido 062-8555, Japan
e-mail: ino@affrc.go.jp

Yasushi Ishihama Institute of Advanced Life Sciences, Keio University, Yamagata, Japan

Graduate School of Pharmaceutical Sciences, Kyoto University, Kyoto, Japan

Yukiyoshi Iwata National Institute for Rural Engineering, NARO, Kanondai 2-1-6, Tsukuba, Ibaraki, 305-8609, Japan
e-mail: iwatayuk@affrc.go.jp

Tsutomu Kajiyama Agriculture Research Department, Hokkaido Research Organization Tokachi Agricultural Experiment Station, Shinsei Minami, 9-2, Memuro, Hokkaido 082-0081, Japan
e-mail: kajiyama-tsutomu@hro.or.jp

Hunseung Kang Department of Plant Biotechnology, Chonnam National University, 300 Yongbong-dong, Buk-gu, Gwangju 500-757, South Korea
e-mail: hskang@jnu.ac.kr

Akira Kawakami NARO Kyushu Okinawa Agricultural Research Center, Suya, Koshi, Kumamoto 861-1192, Japan

Yukio Kawamura Cryobiofrontier Research Center, Faculty of Agriculture, Iwate University, Morioka, Japan

Kenichi Kida Graduate School of Life and Environmental Sciences, Osaka Prefecture University, Sakai, Japan

Myung Hee Kim Crop Breeding Research Division, NARO Hokkaido Agricultural Research Center, Sapporo, Hokkaido, Japan

Fuminori Kobayashi Plant Genome Research Unit, National Institute of Agrobiological Sciences, 2-1-2 Kannondai,Tsukuba, Ibaraki 305-8602, Japan

Shion Kobayashi Cryobiofrontier Research Center, Faculty of Agriculture, Iwate University, Morioka, Japan

Gábor Kocsy Department of Plant Molecular Biology, Agricultural Institute, Centre for Agricultural Research, Hungarian Academy of Sciences, Brunszvik u. 2., Martonvásár 2462, Hungary

Juergen Kreyling BayCEER, Biogeography, University of Bayreuth, Universitätsstraße 30, 95448 Bayreuth, Germany
e-mail: juergen.kreyling@uni-bayreuth.de

Chikako Kuwabara Research Faculty and Graduate School of Agriculture, Hokkaido University, N9-W9, Kita-ku, Sapporo 080-8589, Japan

André Laroche Agriculture and Agri-Food Canada, Lethbridge Research Centre, 5403 1st Avenue South, Lethbridge, Alberta T1J 4B1, Canada

Jun Liu Department of Plant Sciences, University of Saskatchewan, Saskatoon, SK, Canada

David P Livingston III Crop Science, USDA ARS and North Carolina State University, 840 Method Rd, Unit 3, Raleigh, NC 27695, USA
e-mail: dpl@ncsu.edu

Ryan C. Lynch Department of Ecology and Evolutionary Biology, University of Colorado, Boulder, Colorado 80309, USA

Kenji Maezuka Tokachi Federation of Agricultural Cooperative Associations, 14, Nishi-3, Minami-7, Obihiro, Hokkaido 080-0013, Japan
e-mail: k_maezuka@nkrtwosv.nokyoren.or.jp

Naoyuki Matsumoto Graduate School of Agriculture, Hokkaido University, Sapporo, Japan
e-mail: nowmat@for.agr.hokudai.ac.jp

Shota Masumoto The Graduate University for Advanced Studies (SOKENDAI), Tachikawa, Japan

Elisabetta Mazzucotelli Genomics Research Centre, Consiglio per la ricerca e la sperimentazione in agricoltura (CRA), Fiorenzuola d'Arda (PC), Italy
e-mail: elisabetta.mazzucotelli@entecra.it

Kenji Miura Faculty of Life and Environmental Sciences, University of Tsukuba, Tsukuba, Japan
e-mail: kmiura@gene.tsukuba.ac.jp

Yoichi Motomura Graduate School of Agricultural Science, Kobe University, 1-1 Rokkodai, Nada, Kobe, Hyogo 657-8501, Japan

Ryo Murakami Graduate School of Life Sciences, Hokkaido University, Sapporo, Japan

Timothy Murray Department of Plant Pathology, Washington State University, Pullman, WA 99164-6430, USA
e-mail: tim.murray@wsu.edu

Brady T. Nash School of Environmental Sciences, University of Guelph, 50 Stone Road E, Guelph, ON, N1G 2W1, Canada

Manabu Nemoto Agro-environmental Research Division, NARO Hokkaido Agricultural Research Center, Hitsujigaoka 1, Toyohira-ku, Sapporo, Hokkaido 062-8555, Japan
e-mail: nemo@affrc.go.jp

Su Jung Park Department of Plant Biotechnology, College of Agriculture and Life Sciences, Chonnam National University, Gwangju, South Korea

Scott C. Peck Department of Biochemistry and Bond Life Sciences Center, University of Missouri, Columbia, MO, USA

Division of Biochemistry, University of Missouri-Columbia, Columbia, MO, USA

Péter Poór Department of Plant Biology, University of Szeged, Szeged, Hungary

Odd Arne Rognli Department of Plant and Environmental Sciences, CIGENE, Norwegian University of Life Sciences, P.O. Box 5003, 1432 Ås, Norway
e-mail: odd-arne.rognli@umb.no

Kentaro Sasaki Crop Breeding Research Division, NARO Hokkaido Agricultural Research Center, Sapporo, Hokkaido, Japan

Shunya Sato Crop Breeding Research Division, NARO Hokkaido Agricultural Research Center, Sapporo, Hokkaido, Japan

Steven K. Schmidt Department of Ecology and Evolutionary Biology, University of Colorado, Boulder, Colorado 80309, USA
e-mail: steve.schmidt@colorado.edu

Kazuo Shinozaki Gene Discovery Research Group, RIKEN Plant Science Center, Ibaraki, Japan

Masaki Shirahata Hokkaido Research Organization Central Agricultural Experiment Station, Higashi-6, Kita-15, Naganuma, Hokkaido 069-1395, Japan
e-mail: sirahata-masaki@hro.or.jp

Yutaka Sonoda Crop Breeding Research Division, NARO Hokkaido Agricultural Research Center, Sapporo, Hokkaido, Japan

Naoyuki Sugiyama Institute of Advanced Life Sciences, Keio University, Yamagata, Japan

Shinji Suzuki Faculty of Regional Environment Science, Tokyo University of Agriculture, 1-1-1, Sakuragaoka, Setagaya, Tokyo, Japan
e-mail: s4suzuki@nodai.ac.jp

Takeshi Suzuki Hokkaido Research Organization Central Agricultural Experiment Station, Higashi-6, Kita-15, Naganuma, Hokkaido 069-1395, Japan
e-mail: suzuki-takeshi@hro.or.jp

Fuminori Takahashi Gene Discovery Research Group, RIKEN Plant Science Center, Ibaraki, Japan

Shigeo Takumi Graduate School of Agricultural Science, Kobe University, 1-1 Rokkodai, Nada, Kobe, Hyogo 657-8501, Japan
e-mail: takumi@kobe-u.ac.jp

Karen Tanino Department of Plant Sciences, University of Saskatchewan, Saskatoon, SK, Canada
e-mail: karen.tanino@usask.ca

Irma Tari Department of Plant Biology, University of Szeged, Szeged, Hungary

Anja Thalhammer Max-Planck-Institut für Molekulare Pflanzenphysiologie, Am Mühlenberg 1, 14476 Potsdam, Germany

Oleg B. Tkachenko Plant Protection Department with Quarantine Service, Federal State Budgetary Institution for Science, Main Botanical Gardens n. a. N. V. Tsitsin of Russian Academy of Sciences (MBG RAS), Botanicheskaya ul. 4, Moscow, Russia 127276
e-mail: otkach@postman.ru

Katsuyuki Tokura Graduate School of Life and Environmental Sciences, Osaka Prefecture University, Sakai, Japan

Motoaki Tojo Graduate School of Life and Environmental Sciences, Osaka Prefecture University, Sakai, Japan
e-mail: tojo@plant.osakafu-u.ac.jp

Anne Marte Tronsmo Department of Plant and Environmental Sciences, Norwegian University of Life Sciences, 1432 Ås, Akershus, Norway
e-mail: anne-marte.tronsmo@umb.no

Jonathan Tung School of Environmental Sciences, University of Guelph, 50 Stone Road E, Guelph, ON, N1G 2W1, Canada

Tan Tuong Crop Science, USDA ARS and North Carolina State University, 840 Method Rd, Unit 3, Raleigh, NC 27695, USA

Matsuo Uemura Cryobiofrontier Research Center, Faculty of Agriculture, Iwate University, Morioka, Japan

Taishi Umezawa Faculty of Agriculture, Tokyo University of Agriculture and Technology, 2-24-16 Nakacho, Koganei, Tokyo 184-8588, Japan
e-mail: taishi@cc.tuat.ac.jp

Kazuei Usuki Upland-farming Research Division, NARO Hokkaido Agricultural Research Center, Shinsei Minami, 9-4, Memuro, Hokkaido 082-0081, Japan
e-mail: kazuei@affrc.go.jp

Attila Vágújfalvi Department of Plant Molecular Biology, Agricultural Institute, Centre for Agricultural Research, Hungarian Academy of Sciences, Brunszvik u. 2., Martonvásár 2462, Hungary

Radomira Vanková Laboratory of Hormonal Regulations in Plants, Institute of Experimental Botany, Academy of Sciences of the Czech Republic, Rozvojova, Prague, Czech Republic

R. L. Wagner Department of Ecology and Evolutionary Biology, University of Colorado, Boulder, Colorado 80309, USA

Donghui Wang Research Faculty and Graduate School of Agriculture, Hokkaido University, N9-W9, Kita-ku, Sapporo 080-8589, Japan

Nan Xiao Kochi Institute for Core Sample Research, Japan Agency Marine-Earth Science and Technology (JAMSTEC), Nankoku, Japan

Yuka Yajima National Institute of Advanced Industrial Science and Technology (AIST), Higashi-Hiroshima, Japan

Yosuke Yanai NARO Institute of Vegetable and Tea Science, 3-1-1 Kanondai, Tsukuba, Ibaraki 305-8609, Japan
e-mail: yosukey@affrc.go.jp

Tomotsugu Yazaki Upland-farming Research Division, NARO Hokkaido Agricultural Research Center, Shinsei Minami, 9-4, Memuro, Hokkaido 082-0081, Japan
e-mail: yazaki@wetlands.jp

Midori Yoshida NARO Hokkaido Agricultural Research Center, Hitsujigaoka, Sapporo 062-8555, Japan
e-mail: midori@affrc.go.jp

Global Change in Winter Climate and Agricultural Sustainability

Timothy Murray and Denis Gaudet

Introduction

There is no question that our climate is changing. Average air temperature of Earth has increased about 0.8 °C since 1880 and Earth area covered by snow and ice has decreased by 2.7 % per decade since 1978 (Hansen et al. 2006; IPCC 2007a). In addition, unusually warm or cool seasons, defined as those where average temperatures are two to three standard deviations greater than the mean, are becoming more common (Hansen et al. 2012b). Current models indicate that climate change will continue at the same rate for the next 15–20 years even with immediate efforts to mitigate greenhouse gas (GHG) production (IPCC 2007a).

All evidence indicates that human activities are a contributing factor to current climate change (IPCC 2007a), especially the production of the GHG: carbon dioxide (CO_2), methane (CH_4), chlorofluorocarbons (CFC), and nitrous oxide (N_2O), which have been increasing even before the industrial revolution (IPCC 2007a). Agriculture contributes about 10 % of global GHG emissions; animal production, manure management, and soil cultivation contribute 52 % of CH_4 and 82 % of N_2O (Eaglesham and Hardy 2009). Changes in land use and deforestation also contribute to increased atmospheric CO_2 concentration (IPCC 2007a). Although there is a strong correlation between industrialization, fossil fuel use, and increased CO_2 emissions, some scientists hypothesize that human agricultural activities began to impact climate as long as 8,000 years ago, long before the industrial revolution (Ruddiman et al. 2011).

T. Murray (✉)
Department of Plant Pathology, Washington State University, Pullman, WA 99164-6430, USA
e-mail: tim.murray@wsu.edu

D. Gaudet
Agriculture and Agri-Food Canada, Lethbridge Research Centre, 5403 1st Avenue South, Lethbridge, AB T1J 4B1, Canada
e-mail: denis.gaudet@agr.gc.ca

R. Imai et al. (eds.), *Plant and Microbe Adaptations to Cold in a Changing World*, DOI 10.1007/978-1-4614-8253-6_1,

There are many questions and great concern about how climate change will impact human activities, including our ability to sustain food production at the rate needed to feed a growing population. As scientists focusing on biotic and abiotic stresses of crop plants grown in the northern hemisphere, we have many questions about how climate change will impact crop productivity in these areas. One of the consequences of global climate change is the earlier appearance of spring and the delayed onset of fall, which effectively lengthens summer and shortens winter (IPCC 2007a). We will review some of the data for past climate change and its causes, and discuss future climate change and its potential impact on agricultural sustainability. Emphasis will be placed on changes in winter climate and the potential to influence plant and microbe interactions at low temperature during winter.

This discussion is not intended as an exhaustive review of the literature; that body of literature is large, growing rapidly, and beyond the scope or intent of this discussion. Interested readers are encouraged to consult the Intergovernmental Panel on Climate Change Fourth Assessment Report (FAR), Working Groups I and II (IPCC 2007a, b), and the other technical references cited herein for more information on climate change. In addition, the Fifth Assessment report will be published beginning in September 2013.

Past Climate Change

Climate and weather represent the same meteorological attributes of a region including temperature, precipitation, relative humidity, atmospheric pressure, wind speed and direction, atmospheric particle count, and others (IPCC 2007a). The difference between climate and weather is the time-scale over which these attributes are measured; climate represents long-term averages usually measured over a 30-year period, whereas weather represents the short-term variation in these same attributes from day-to-day or year-to-year. Consequently, the discussion of climate change focuses on significant, long-term changes in the average distribution of the meteorological variable in question and not the short-term changes that may occur over the course of a single year or season. Many discussions of climate change use the period from 1950 to 1980 as a baseline for comparison (Hansen et al. 2012a, b), but different time periods are used in some studies.

Temperature is the most frequently discussed meteorological attribute in the context of "global warming." However, it is important to remember that temperature is just one attribute and that other important changes are also occurring including the seasonal distribution of precipitation, length of the growing season, amount of snow fall, and others that have the potential to influence agricultural productivity (IPCC 2007a). It is also important to realize that climate change can occur relatively quickly in response to perturbations of the global energy balance and such changes have had significant impacts on human civilization in the past (Zhang et al. 2011). For example, the Little Ice Age in the northern hemisphere extended from about 1350–1850 AD, and was characterized by colder winters in Europe and North

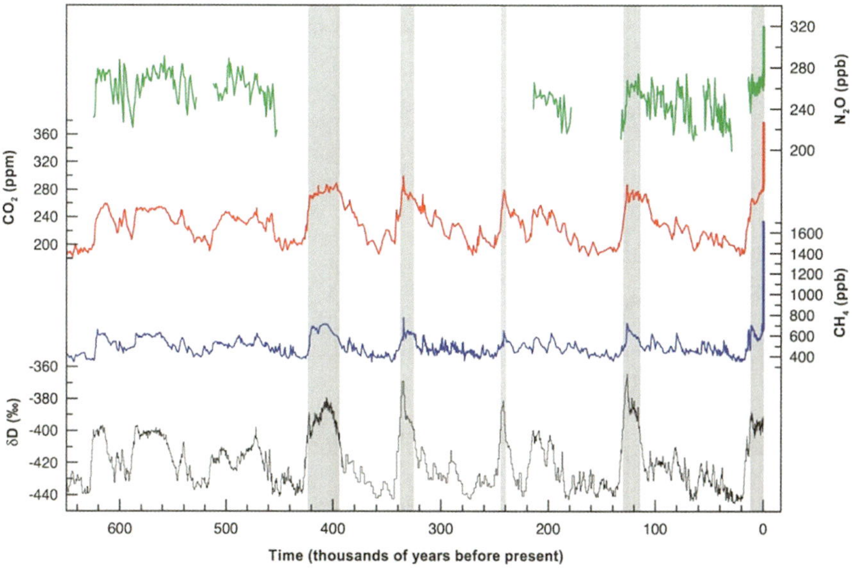

Fig. 1 Variations deuterium (δD; *black*) concentration in Antarctic ice, a proxy for local temperature, and atmospheric concentrations of the greenhouse gases: carbon dioxide (CO_2; *red*), methane (CH_4; *blue*), and nitrous oxide (N_2O; *green*) derived from air trapped in ice cores from Antarctica and recent atmospheric measurements. *Vertical shading* indicates recent interglacial warm periods. (IPCC 2007a, Fig. TS1)

America with locally greater snowfall and worldwide growth of glaciers (Nesje and Dahl 2003). Agricultural productivity and food supply responded to these changes quickly; other human crises including social disturbance, war, migration, famine, epidemics lagged by 5–30 years (Zhang et al. 2011). In the current context, climate change has the potential to threaten food security and destabilize government systems.

Paleological Climate Change Paleoclimate reconstruction is based on temperature proxies, which include air trapped in ice cores, tree rings, coral rings, lake and ocean sediments, and others. Based on these data, Earth's climate has changed many times over the past 650,000 years alternating between relatively warm and cold periods (Fig. 1) (IPCC 2007a). Using deuterium trapped in Antarctic ice cores to reconstruct local temperature change has identified five interglacial periods in the past 450,000 years, each followed by a cold period. The concentration of greenhouse gases has similarly varied over this time interval (Fig. 1), but current concentrations of atmospheric CO_2 (379 ppm) and CH_4 (1,774 ppb) are greater than preindustrial times and at any time in the past 650,000 years (IPCC 2007a).

Holocene Climate Variation About 12,000 years before present, global temperatures increased as Earth emerged from the most recent ice age (Fig. 2) and marked the beginning of the Holocene; the current interglacial period in which we live now.

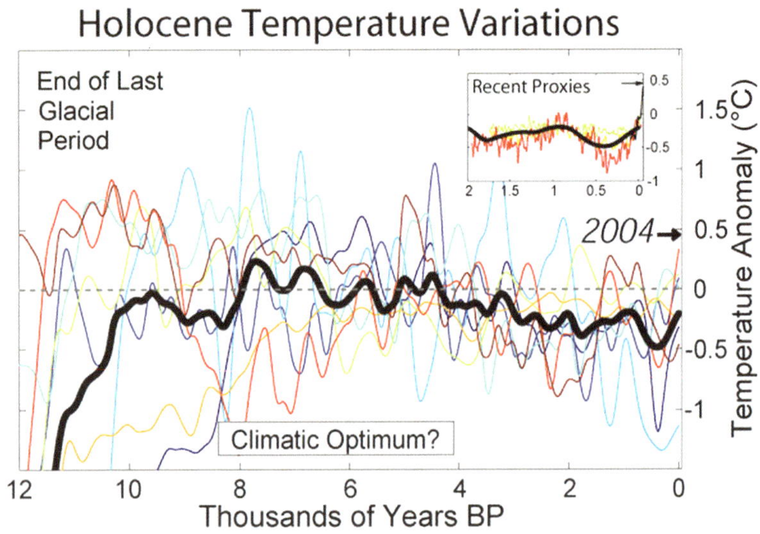

Fig. 2 Eight records (*thin colored lines*) of local temperature variability for the period from 10,000 BC to 2,000 CE (12,000 BP to the present) throughout the Holocene, and the average of these records (*black line*). Data are plotted relative to the mid-twentieth century average temperature. The global average temperature in 2004 is indicated. The inset plot compares the most recent two millennia of the average to other recent reconstructions. (Image created by Robert A. Rohde, Global Warming Art, http://en.wikipedia.org/wiki/File:Holocene_Temperature_Variations.png)

It is difficult to measure temperatures accurately during this period and there is no generally agreed upon method of determining Holocene temperatures; however, it is generally agreed that temperatures since the end of the last glacial period have been relatively stable compared with prior times. The beginning of the Holocene and its stable, warmer temperatures correspond roughly with the emergence of agriculture.

Instrumental Record The instrumental record is the period when humans began recording temperatures directly, first with thermometers and later by remote sensing from satellites. The oldest and longest continuous instrumental temperature record dates from 1659 for the Midlands region of England (Manley 1974), but it encompasses a limited geographic area. Mostly reliable global temperature records began about 1850 and three major databases maintain temperature records from the 1850s onward; the HADCRUT3 dataset at the United Kingdom Met Office Hadley Centre, GISTEMP dataset at the United States National Aeronautics and Space Administration (NASA), and GHCN database at the National Oceanic and Atmospheric Administration (NOAA) (Hansen et al. 2010). All of these databases provide global monthly mean surface temperatures collected from hundreds or thousands of observation sites. Analysis of temperature change during the instrumental period begins about 1880 because that is when the number of locations was sufficient to provide representative global coverage and magnitude of the error rate in measurement became less than temperature change (Hansen et al. 2010). Changes in temperature

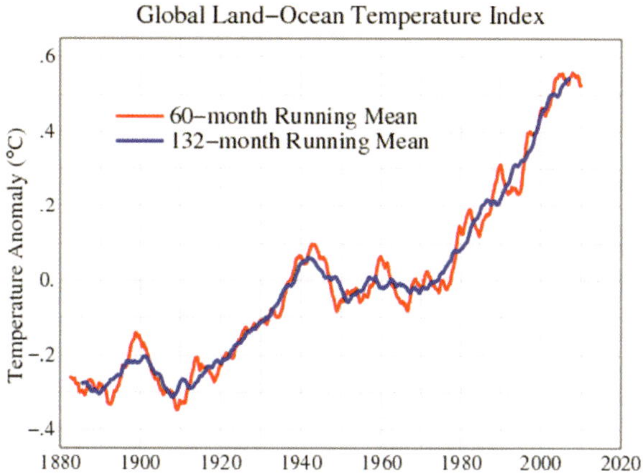

Fig. 3 Five-year (60-month) and 11-year (132-month to minimize the solar cycle effect) running means of the surface temperature deviation from the 1951–1980 mean. (With permission from: Sato and Hansen, http://www.columbia.edu/~mhs119/Temperature/)

during the instrumental record are often compared to the mean from 1950 to 1980 (Hansen et al. 2006, 2012a, b) or from 1961 to 1990 (IPCC 2007a).

Average temperature of Earth has increased about 0.8 °C since 1880 (Fig. 3), and approximately 0.2 °C per decade since 1975, accounting for two-thirds of the warming since 1880 (Hansen et al. 2010). Regression analysis of global temperature change over different time scales shows an increasing rate of warming from 1880 to the present (Fig. 4; IPCC 2007a). Rates of warming during the instrumental record ranged from a low of 0.045 °C per decade for 150 years to a high of 0.177 °C per decade for the most recent 25-year period ending in 2005. Hansen et al. (2010) concluded that despite the perception that global warming declined in the past decade, the rate of warming from 2000 to 2010 is as great as it was in the previous 2 decades. Hansen et al. (2012a) also note that 9 of the 10 warmest years in the instrumental record have occurred since 2000, and the record high global 12-month running mean temperature occurred in 2010. Thus, global warming has not diminished during the early years of the twenty-first century (Hansen et al. 2010) and will very likely continue to increase over the next 15–20 years (IPCC 2007a).

Severe or extreme weather events are usually defined as those two or more standard deviations different than the mean. Heat waves, drought, heavy precipitation, and tropical cyclones are becoming more common with climate change (IPCC 2007a). Other indicators of global climate change especially relevant to agriculture include the number of cold days, cold nights, and frost events, which have all decreased since records have been kept, and the number of hot days and hot nights, which have increased over most land areas (IPCC 2007a).

Winter Climate Change Spatial and temporal distribution of climate change is not uniform around Earth. Warming has been greater over land than sea and in northern latitudes than in low (tropical) or southern latitudes (Fig. 5) (Hansen et al. 1999, 2010; IPCC 2007a). Since 1900, average temperatures have increased about 0.9, 0.5, and 0.4 °C in the northern, low, and southern latitudes, respectively. Since 1950, the 50-year average temperature is very likely higher than at any other time in the

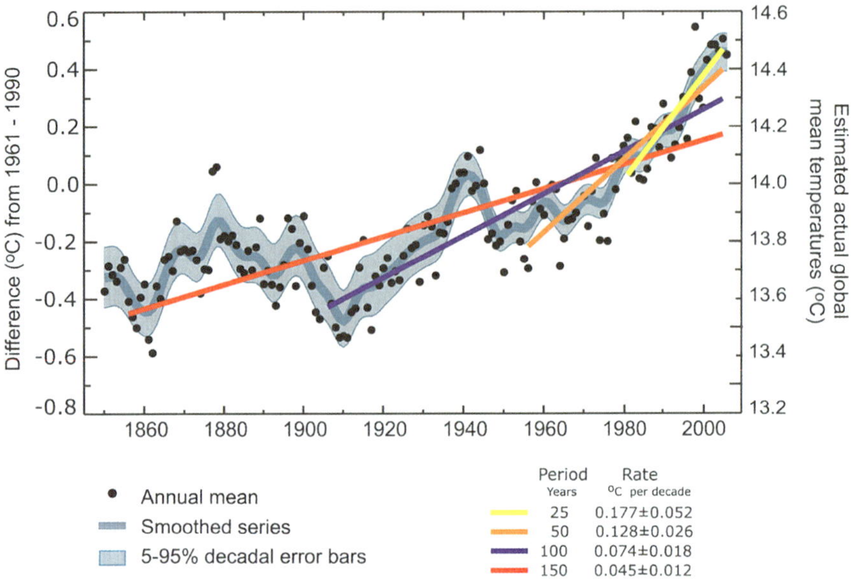

Fig. 4 Annual global mean temperatures (*black dots*) with linear fits to the data. The left axis is temperature anomalies relative to the 1961–1990 average and the right axis is estimated actual temperatures. Linear trends are shown for the last 25 (*yellow*), 50 (*orange*), 100 (*purple*), and 150 years (*red*). The smooth *blue curve* shows decadal variations, with the 90% error range shown as a *pale blue band* about that line. The total temperature increase from the period 1850–1899 to the period 2001–2005 is 0.76±0.19 °C. (IPCC 2007a, Fig. TS6)

past 500 years and the highest in the past 1,300 years (IPCC 2007a). The northern latitudes have greater land mass than mid latitudes or southern latitudes and a greater percentage of global industrialization, which partially explains the greater increase in temperature (Hansen et al. 1999). It is also likely that temperature change in arctic regions is underestimated due to the limited number of observation stations and methods by which temperatures are extrapolated (Hansen et al. 2010).

Increasing temperatures have resulted in decreased snow cover in the northern hemisphere in every month except November and December, with a total decrease in spring snow cover of about 5% since the 1980s (Fig. 6). On average, snow melt in spring was almost 2 weeks earlier in 2000 than in 1972 (IPCC 2007a). Similarly, the average area covered by seasonally frozen soil has decreased about 7% since 1901, and freeze-up and break-up dates for river and lake ice are 5.8 days/century later and 6.5 days/century earlier, respectively (IPCC 2007a). In other words, winter is getting shorter and spring and fall are getting longer.

As a result of increasing temperatures and decreased duration of winter, the intensity of biotic and abiotic stresses that affect agricultural crops in the northern hemisphere will change in response to local climate change, with specific stressors increasing in some areas and decreasing in others.

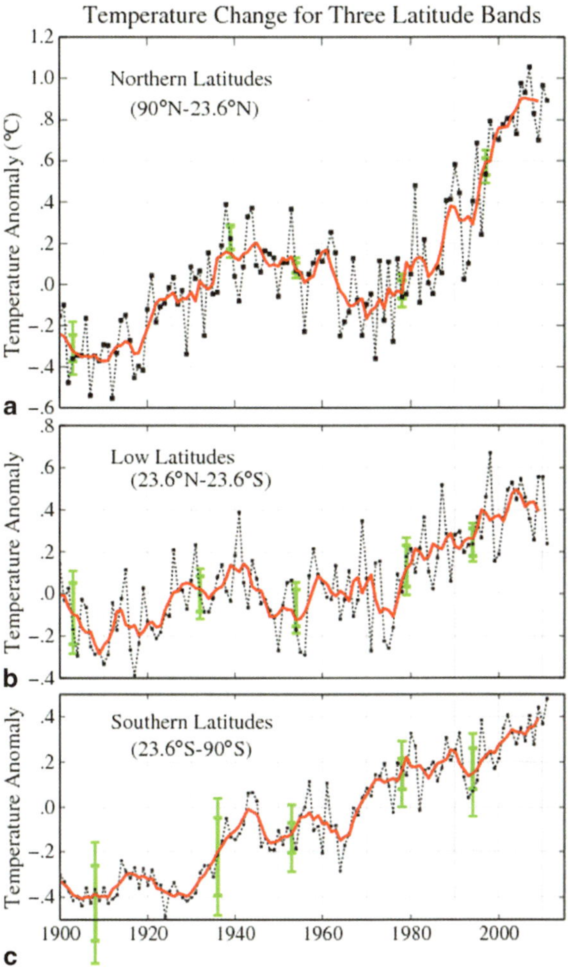

Fig. 5 Annual and 5-year running mean temperature changes relative to the base period 1951–1980 for three latitude bands. *Green bars* are 95 % confidence intervals based on spatial sampling analysis. Mean temperature increase for the northern latitudes is about 0.9 °C compared with 0.5 and 0.4 °C for the mid and southern latitudes, respectively. (With permission from: J. Hansen http://data.giss.nasa.gov/gistemp/graphs_v3/)

Causes of Climate Change

It is clear from the preceding discussion that significant changes have occurred in the global climate. Climate change is the result of changes in Earth's energy balance, which is the net difference by energy absorbed from the sun and energy radiated back into space (IPCC 2007a). Warming occurs when less energy is reflected back into space and conversely, cooling occurs when more energy is reflected back into space than is received. In climate science language, the factors affecting the energy balance are known as drivers that contribute positively or negatively to climate forcing, i.e., changes in the energy balance. Changes in the Paleoclimate were driven by natural processes, but climate change since 1850 also has a human-caused or anthropogenic contribution.

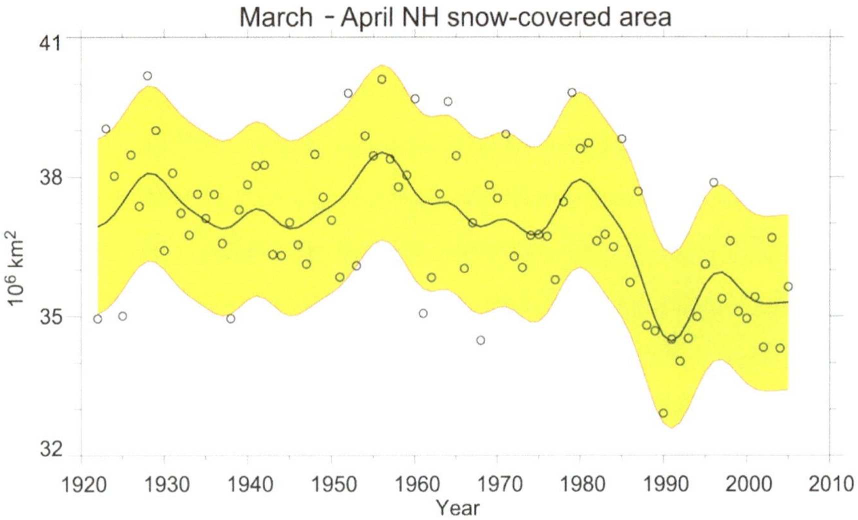

Fig. 6 Northern hemisphere March-April snow-covered area from a station-derived snow cover index prior to 1972 and satellite data from1972 onwards. The total decrease in snow-covered area is about 5 % since the 1980s. *Open circles* are annual means, the *smooth curve* is the 10-year mean, and the *shaded area* shows decadal variations with the 5–95 % data range shaded in *yellow*. (IPCC 2007a, Fig. 4.2)

Natural Processes Past climate change was driven by natural processes only, the largest of which is total solar radiation received by Earth. Total solar radiation is affected by changes in energy output of the sun, which cycles from high to low and back to high over the 11-year solar cycle with a total change in energy output of about 0.1 % (Fröhlich and Lean 2004). Total solar radiation is also affected by Earth's orbital oscillation around the sun, or the Milankovitch cycles, which includes distance to the sun (precession), axis tilt (obliquity), and eccentricity of Earth's orbit (IPCC 2007a). Consequently, solar forcing resulting from orbital oscillation is cyclical with periodicities ranging from 19,000 to 400,000 years.

Volcanic eruptions also impact Earth's energy balance. Data collected since 1980 demonstrate significant changes in atmospheric temperatures following the eruptions of El Chichón in 1982 and Pinatubo in 1991 resulting from the particulate matter and sulfur-containing compounds emitted into the atmosphere that effectively block solar radiation from reaching Earth's surface and at the same time, absorb radiation reflected from Earth's surface, thus preventing it from reaching space; the net effect is surface warming (IPCC 2007a).

The ocean thermohaline circulation pattern, also called meriodonal overturning circulation (MOC) is another natural force that influences Earth's climate (IPCC 2007a). The MOC is a series of deep and shallow currents that traverse Earth's oceans, driven by changes in temperature and density resulting from changes in salinity. One complete circuit around the globe takes about 1,600 years and results in warm surface water being distributed to cooler areas near the poles and cold water

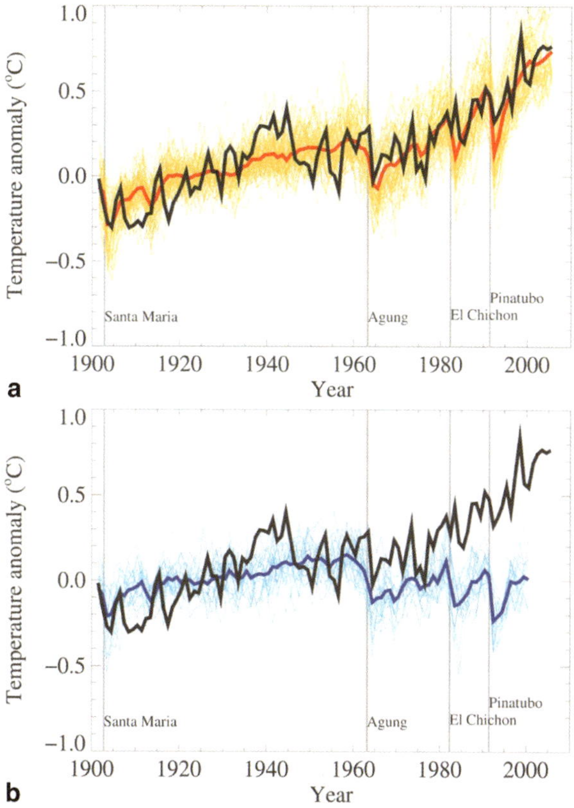

Fig. 7 Comparison between global mean surface temperature anomalies (°C) from observations (*black line*) and Atmosphere–Ocean General Circulation Model simulations using **a** anthropogenic and natural forcings and **b** natural forcings only. Data are global mean temperature anomalies relative from 1901 to 1950. *Vertical gray lines* indicate major volcanic events. Data in **a** are obtained from 58 simulations produced by 14 models with both anthropogenic and natural forcings showing the multimodel mean (*thick red line*) and individual simulations (*thin yellow lines*). Data in **b** are from 19 simulations produced by five models with natural forcings only showing the multimodel mean (*thick blue line*) and individual simulations (*thin blue lines*). (IPCC 2007a, Fig. 9.5)

from the deep oceans to warmer areas of Earth. The MOC effectively regulates Earth's temperature, especially in northern and southern latitudes, by redistributing energy received by the sun around the globe; anything that interferes with the MOC has potential to influence climate (IPCC 2007a).

Anthropogenic Processes The observed changes in average global temperatures since the 1950s cannot be explained by natural process alone (IPCC 2007a). In contrast to the observed global temperatures, Atmosphere–Ocean General Circulation Model simulations of global temperature change predict a slight cooling beyond 1960 (Fig. 7b). It is only when anthropogenic causes are included in these simulations that the results correspond to actual observations (Fig. 7a).

Production of GHG and depletion of stratospheric ozone have had the greatest impact on global temperature change since the mid 1900s, but land-use changes and the production of aerosols also have had significant impacts on global climate change (IPCC 2007a). Among the several GHG produced, CO_2 and CH_4 have had the greatest positive climate forcing effect, with CFC and N_2O also having significant contributions. In 2005, atmospheric concentration of CO_2, CH_4, and N_2O were 379 ppm, 1,774 ppm, and 319 ppb, respectively, which are the highest of any time in the past 20,000 years (IPCC 2007a). Since 1995, atmospheric concentration of CO_2 and N_2O has increased steadily, whereas CH_4 and CFC have decreased.

Fossil fuel use and cement production have likely contributed about three-fourths of the total CO_2 increase, and land-use changes have contributed the remaining one-fourth (IPCC 2007a). Land-use changes, primarily deforestation in the tropics to make room for agriculture, contribute to climate change both from increased CO_2 production as a result of biomass burning and decomposition of soil organic matter and greater reflectance (albedo) of the surface that results in positive radiative forcing.

Agriculture is a significant source of GHG that contributed about 10–12% of total emissions in 2005, but about 52% of CH_4 and 82% of N_2O (Eaglesham and Hardy 2009; IPCC 2007a). Agricultural sources of CH_4 and N_2O mainly include wetlands, rice agriculture, ruminant animals, biomass burning, and nitrogen-based fertilizers (Desjardins 2009; Eaglesham and Hardy 2009; IPCC 2007a). Although agriculture has contributed significantly to CO_2 emissions in the past as a result of deforestation and intensive grassland cultivation, CO_2 fixed by plants now nearly balances the net output from other agricultural sources (Desjardins 2009; Ortiz-Monasterio et al. 2012).

Future Climate Change

There is considerable uncertainty about the rate and magnitude of future climate change. Predicting future change depends on how human society reacts to the causes of climate change, and specifically the issue of GHG emissions. Given that there is no globally agreed upon solution to the problem, it is impossible to predict how the future will unfold relative to climate change. Consequently, the IPCC produced the Special Report on Emissions Scenarios (SRES) that provides four "storylines" with possible outcomes based on the degree to which society adopts environmental versus economic and global versus regional solutions to the problem of GHG emissions (IPCC 2000). Population growth, economy, technology, energy, land-use, and agriculture were the driving forces behind these scenarios. The scenarios were developed as a tool to envision what might happen given certain assumptions and enable an analysis of potential actions to adapt to and mitigate climate change.

SRES Emission Scenarios The four SRES scenarios, labeled A1, A2, B1, and B2, each has a subset of scenarios constituting a "family" that projects potential global temperature increase, sea-level rise, GHG emissions, and other climate parameters out to 2,100. Six "marker scenarios" (A1F1, A1B, A1T, A2, B1, and B2) were selected to represent a wide range of possible responses for use by climate modelers to represent a range of outcomes (Arnell et al. 2004). For example, the A1 storyline family assumes a future with very rapid economic growth; within this family, A1F1 assumes fossil fuel-intensive energy, A1T nonfossil fuel energy, and A1B a balance across energy sources (IPCC 2000). The best estimates of global tempera-

Table 1 Projected global average surface warming at the end of the twenty-first century. (Adapted from IPCC 2007a)

Scenario	Temperature Change (°C) 2090–2099 relative to 1980–1999	
	Best estimate	Likely range
Constant year 2000 GHG concentrations	0.6	0.3–0.9
B1	1.8	1.1–2.9
A1T	2.4	1.4–3.8
B2	2.4	1.4–3.8
A1B	2.8	1.7–4.4
A2	3.4	2.0–5.4
A1F1	4.0	2.4–6.4

GHG greenhouse gas

ture increases by 2090–2099 under these models range from 1.8 °C for B1 to 4.0 °C for A1F1 by the end of the twenty-first century compared with 0.6 °C if year 2000 GHG concentrations are held constant (Table 1). All of the models indicate that climate change will continue at the same rate for the next 15–20 years even with immediate efforts to mitigate GHG production (IPCC 2007a).

Agricultural Sustainability

The concept of sustainable agriculture has evolved over the past 30 years and consequently, means different things to different people. In the United States, sustainable agriculture was defined in the 1990 farm bill as "an integrated system of plant and animal production practices having a site-specific application that will, over the long term: satisfy human food and fiber needs; enhance environmental quality and the natural resource based upon which the agricultural economy depends; make the most efficient use of nonrenewable resources and on-farm resources and integrate, where appropriate, natural biological cycles and controls; sustain the economic viability of farm operations; and enhance the quality of life for farmers and society as a whole." This definition integrates concepts of ecology and environmental health, economic profitability, and social and economic equity to provide stable food production systems. This discussion will focus on the sustainability of satisfying human food and fiber needs in the context of northern hemisphere climate change.

Most climate change predictions are based on global circulation models and consequently, need to be downscaled to predict local or regional change to estimate the impact on agricultural production (Arnell et al. 2004; Jarvis et al. 2012; Parry et al. 2004). Climate change will affect agriculture globally, but not uniformly; the impact of climate change will vary geographically within regions and across latitudes depending on the magnitude of change and predominate crops grown in those regions. The impact of climate change will also vary economically with countries having the greatest Gross Domestic Product devoted to agriculture at the greatest risk for negative impacts (Jarvis et al. 2012). In general, the fertilization effect of increased atmospheric CO_2 concentration, reduced snow cover, longer growing seasons, and increased rainfall in the northern latitudes likely will have positive impacts on crop production in the short term, whereas increased temperature and reduced rainfall in low latitudes will negatively impact crop production. The IPCC FAR (2007a) concluded that global food production would continue to increase with temperature increases up to 3 °C, but decrease thereafter with increasing temperature.

Climate Change and Agriculture The impact of climate change on agricultural production will be mediated through the direct effects of temperature, CO_2 concentration, and other changes in the environment that influence plant growth (e.g., length of growing season), as well as through indirect effects on pests and diseases. In regions where crops are growing at or near their temperature optima, increasing average temperature likely will result in reduced yields and/or transition to other crops; however, for crops growing below their temperature optima, yields may increase (Lobell and Burke 2012). For example, wheat yields in the northern hemisphere likely will increase with increasing average temperature up to 2 °C (Fig. 8) and production will expand into previously unsuitable areas becoming suitable for crop production (IPCC 2007b). However, wheat yields in low latitudes will decrease with any increase in temperature. Similarly, yield of maize will decrease at both mid latitudes to high latitudes and low latitudes (Fig. 8).

Estimates of climate change impacts on agriculture include the fertilization effect of increased CO_2 concentration on crop productivity; however, most of the research on CO_2 and enhanced crop productivity is based on studies conducted under controlled environmental conditions where water and nutrients are not limiting (Jarvis et al. 2012). More studies in open-air chambers that better simulate field conditions are needed to determine whether and to what extent CO_2 may enhance plant productivity, especially under conditions where water and nutrients are not limiting (Long 2012). In addition, studies on the effect of other GHG and specifically elevated ozone concentration on crop productivity have not been considered in these controlled environment studies and may negatively impact crop yield (Long 2012). Another limitation in some estimates of climate change on agricultural productivity is the impact of pests and diseases. Parry et al. (2004) concluded that world agriculture would likely be able to continue feeding the planet for the rest of the twenty-first century. However, their projections assume that all pests and diseases are controlled and that soil conditions resulting from climate change will not be limiting. Lastly, the role and impact of severe weather events on agricultural productivity is unknown and impossible to predict. Given that the frequency and severity of heat waves, drought events, and heavy precipitation are expected to increase, it seems likely that there will be some negative impact on crop productivity.

Climate Change and Plant Diseases The plant disease triangle tells us that a susceptible plant and pathogen must be present in a favorable environment for a

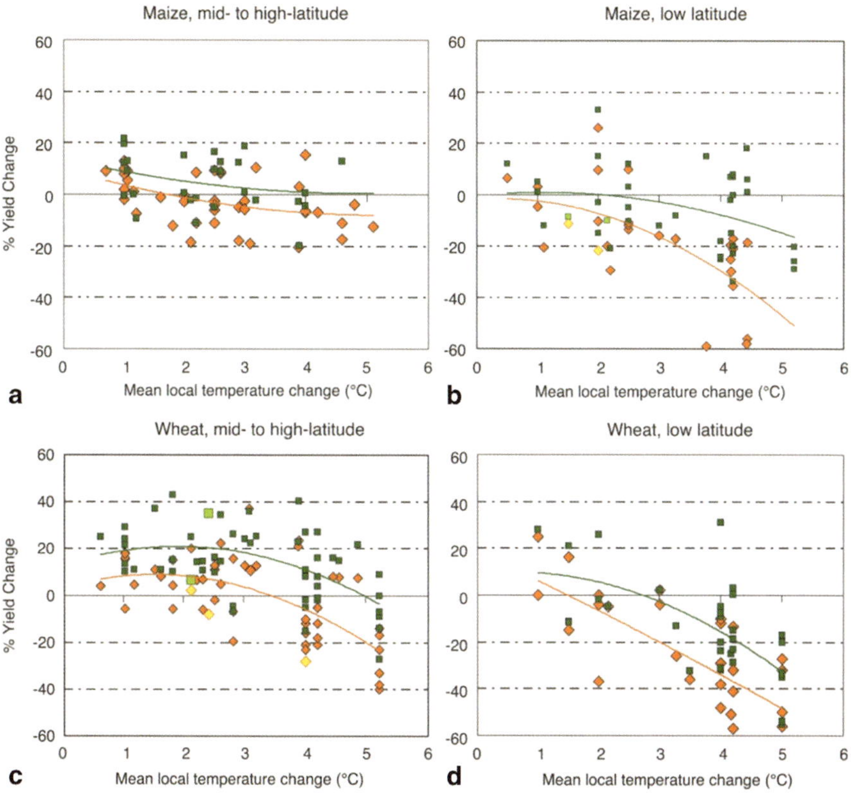

Fig. 8 Response of maize and wheat yield to local temperature change and mid latitude to high latitude and low latitudes. Responses include cases without adaptation (*orange dots*) and with adaptation (*green dots*) across a range of precipitation changes and CO_2 concentrations, and vary in how they represent future changes in climate variability. For example, *yellow-colored dots* in **b** and **c** represent responses of rain-fed crops under climate scenarios with decreased precipitation. (IPCC 2007b, Fig. TS7)

plant disease to occur. In this context, temperature and moisture are the two most important environmental factors influencing the occurrence and severity of plant diseases (Gaudet et al. 2012). Just as crop production will shift as a result of increasing average temperatures, the distribution and importance of pests and diseases endemic to a specific area will shift with increasing temperatures and projected changes in rainfall amount and distribution (Jarvis et al. 2012). Given that average temperatures in the northern hemisphere have increased more than other parts of Earth and are projected to continue on this track, those of us studying cold-weather stresses should expect to see less frequent but perhaps more severe outbreaks of snow mold diseases and winter injury as a result of increased snow cover in some years and unusually cold temperatures with reduced snow cover in others. In the U.S. Pacific Northwest wheat-producing region, which transitions from areas where snow cover does not persist to areas where it persists for > 100 days, the occurrence of severe snow mold epidemics is less frequent now than in the 1940s when it was a limiting factor for winter wheat production. Furthermore, diseases common in areas of Washington State where snow cover does not persist are occurring with greater regularity in the traditional snow mold areas.

Concluding Remarks

There is considerable uncertainty about the magnitude of future climate change and its impact on agricultural sustainability that result from the highly complex interactions among atmospheric components governing climate change, the uncertain direct effects of climate change on plant growth and productivity, and the indirect effects of climate change on pests and diseases. Clearly, shifts in disease and pest spectra within regions are anticipated in response to climate change. It is likely that within a given region, some pests and diseases will become less important as a result of climate change, whereas others will become more important. The specific changes and their impact are not known because they depend on the magnitude of future change.

The impact of climate change on agricultural productivity in the northern hemisphere where winter climates impose unique abiotic and biotic stresses on plants is likely to be significant since the impacts of climate change will likely be greatest in the northern high latitudes. As scientists working in this region, it is important to anticipate changes in these stresses and work proactively to develop appropriate strategies to limit damage. Given that many of the current strategies to limit winter stress damage make use of plant genetic resources and the fact that plants and pathogens have adapted to climate change many times in the past, it seems likely that the genetic variation exists within plants to successfully adapt to anticipated changes.

Acknowledgments The authors wish to thank OECD for providing financial support of the PMAC conference.

References

Arnell NW, Livermore MJL, Kovats S, Levy PE, Nicholls R, Parry ML, Gaffin SR (2004) Climate and socio-economic scenarios for global-scale climate change impacts assessments: characterising the SRES storylines. Global Environmental Change 14:3–20

Desjardins RL (2009) The impact of agriculture on climate change. In: Eaglesham A, Hardy RWF (eds) National Agricultural Biotechnology Council (NABC Report) pp 29–39. National Agricultural Biotechnology Council, Ithaca, NY

Eaglesham A, Hardy RWF (2009) Overview of NABC 21: Adapting Agriculture to Climate Change. In: Eaglesham A, Hardy RWF (eds) National Agricultural Biotechnology Council (NABC Report) pp 3–10. National Agricultural Biotechnology, Ithaca, NY

Fröhlich C, Lean J (2004) Solar radiative output and its variability: evidence and mechanisms. Astron Astrophys Rev 12:273–320

Gaudet DA, Tronsmo AM, Laroche A (2012) Climate change and plant diseases. In: Storey KB, Tanino KK (eds) Temperature adaptation in a changing climate: nature at risk pp 144–159. CABI

Hansen J, Ruedy R, Glascoe J, Sato M (1999) GISS analysis of surface temperature change. J Geophys Res 104:30997–31022

Hansen J, Sato M, Ruedy R, Lo K, Lea DW, Medina-Elizade M (2006) Global temperature change. Proc Nat Acad Sci U S A 103:14288–14293

Hansen J, Ruedy R, Sato M, Lo K (2010) Global surface temperature change. Rev Geophys 48: RG4004

Hansen J, Ruedy R, Sato M, Lo K (2012a) Global Temperature in 2011, Trends and Prospects. http://data.giss.nasa.gov/gistemp/2011/

Hansen J, Sato M, Ruedy R (2012b) Perception of climate change. Proc Nat Acad Sci U S A 109: E2415–2423

Intergovernmental Panel on Climate Change IPCC (2000) Emissions Scenarios: Summary for Policymakers. In: Nakicenovic N, Swart R (eds). Cambridge University Press, Cambridge, p. 20

Intergovernmental Panel on Climate Change IPCC (2007a) Climate change 2007: The physical science basis. In: Solomon S, Qin D, Manning M, Chen Z, Marquis M, Averyt KB, Tignor M, Miller HL (eds) Contribution of Working Group II to the Fourth Assessment Report of the Intergovernmental Panel on climate change. Cambridge University Press, Cambridge, p. 996

Intergovernmental Panel on Climate Change IPCC (2007b) Climate change 2007: Impacts, adaptation and vulnerability. In: Parry ML, Canziani OF, Palutikof JP, van der Linden PJ, Hanson CE (eds) Contribution of Working Group II to the Fourth Assessment Report of the Intergovernmental Panel on climate change. Cambridge University Press, Cambridge, p. 976

Jarvis A, Ramirez J, Anderson B, Leibing C, Aggarwal P (2012) Scenarios of climate change within the context of agriculture. In: Reynolds MP (ed) Climate change and crop production. CABI, Cambridge, pp. 9–37

Lobell D, Burke M (2012) Economic impacts of climate change on agriculture to 2030. In: Reynolds MP (ed) Climate change and crop production. CABI, Cambridge, pp. 38–49

Long SP (2012) Virtual Special Issue on food security— greater than anticipated impacts of near-term global atmospheric change on rice and wheat. Glob Change Biol 18:1489–1490

Manley G (1974) Central England temperatures: monthly means 1659 to 1973. Q J Roy Meteor Soc 100:389–405

Nesje A, Dahl SO (2003) The 'Little Ice Age'—only temperature? The Holocene 13:139–145

Ortiz-Monasterio I, Wassmann R, Govaerts B, Hosen Y, Katayanagi N, Verhulst N (2012) Greenhouse gas mitigation in the main cereal systems: rice, wheat and maize. In: Reynolds MP (ed) Climate change and crop production. CABI, Cambridge, pp 151–176

Parry ML, Rosenzweig C, Iglesias A, Livermore M, Fischer G (2004) Effects of climate change on global food production under SRES emissions and socio-economic scenarios. Glob Environ Change 14:53–67

Ruddiman WF, Kutzbach JE, Vavrus SJ (2011) Can natural or anthropogenic explanations of late-Holocene CO_2 and CH_4 increases be falsified? The Holocene 21:865–8879

Zhang DD, Lee HF, Wang C, Li B, Pei Q, Zhang J, An Y (2011) The causality analysis of climate change and large-scale human crisis. Proc Nat Acad Sci U S A 108:17296–17301

Soil Freezing Dynamics in a Changing Climate: Implications for Agriculture

Hugh A. L. Henry

Introduction

Soil freezing can directly damage the overwintering shoots and root systems of winter cereals, perennial crops and pasture and turfgrasses, increasing mortality, and decreasing yield (Ouellet 1976; Fowler 2008). However, it can also have numerous indirect effects on crop yield via changes in abiotic factors such as soil moisture (Iwata et al. 2010), nutrients (Elliott and Henry 2009), bulk density, and aggregates (Oztas and Fayetorbay 2003) or biotic factors such as weeds, insects, microfauna, or microbes (Fig. 1). These indirect effects can either be positive for crops, as in the cases of decreased soil bulk density or damage to weeds and insect pests, or negative, as in the cases of damage to beneficial microorganisms or losses of soil nutrients. In addition, organisms can benefit from a given intensity of freezing yet suffer from another; for example, although insect pests can be killed by extreme soil freezing, they can also benefit from reduced energy expenditure at mild-to-moderate freezing relative to overwintering at above-freezing temperatures (Irwin and Lee 2003).

Overall, the biological effects of soil freezing on crops are a function not only of freezing intensity, but of soil freezing frequency, timing, rate, and depth (Henry 2007), which depend largely on air temperature and solar radiation. Precipitation can also play a large role by affecting snow cover, soil moisture and cloud cover, and live plant and litter cover can modulate the effects of air temperature and solar radiation on soil freezing (Fig. 2). In soil, the freezing point is an important threshold because it represents a shift in liquid water availability, with plants often becoming inactive and microbial activity decreasing sharply (Schimel and Mikan 2005; Fig. 3a). However, winter cereals and pasture grasses grown in cold regions can typically endure temperatures much lower than 0 °C before succumbing to lethal cold (Malyshev and Henry 2012), and thin water films remain around soil particles

H. A. L. Henry (✉)
Department of Biology, University of Western Ontario, London, ON, Canada
e-mail: hhenry4@uwo.ca

R. Imai et al. (eds.), *Plant and Microbe Adaptations to Cold in a Changing World*,
DOI 10.1007/978-1-4614-8253-6_2, © Springer Science+Business Media New York 2013

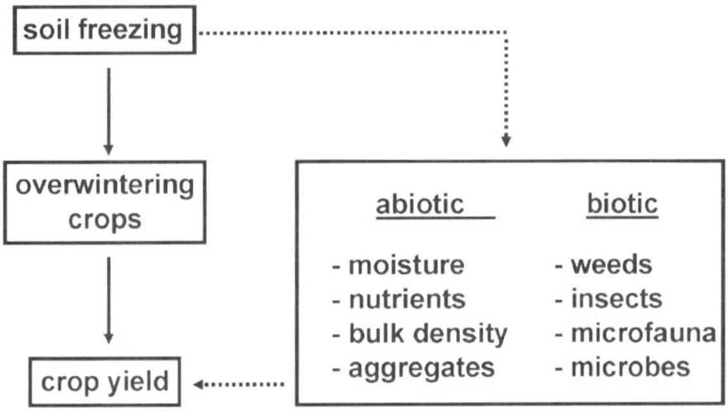

Fig. 1 Summary of direct (*solid lines*) and indirect (*dotted lines*) effects of soil freezing on crop yield

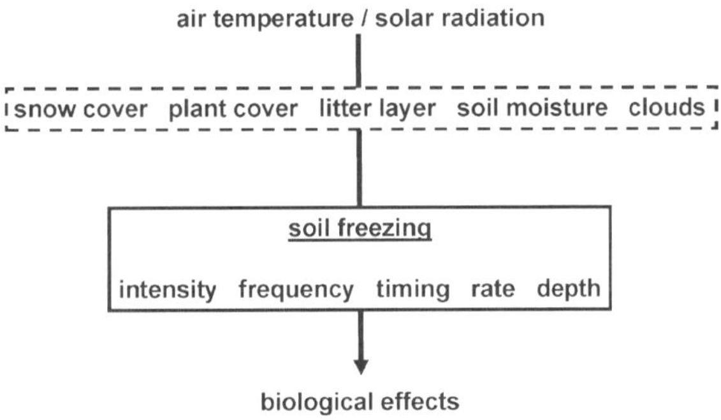

Fig. 2 Summary of factors modulating the effects of air temperature and solar radiation on solar freezing, and the facets of soil freezing that ultimately determine biological effects

below 0 °C, allowing microbial activity to continue until lethal temperatures are reached. Nevertheless, in between 0 °C and the threshold of lethal cold (Fig. 3b) sublethal damage to plants can occur. The dynamics of soil freezing (Fig. 3c) can also be important for determining biological effects. For example, in Fig. 3, panels (a) and (c) exhibit the same mean temperature, yet in (c) soil organisms are exposed to repeated cycles of freezing and thawing, which can be destructive to cells.

The objectives of this synthesis are to (1) summarize the observed and predicted responses of soil freezing dynamics to climate change, (2) assess the potential effects of climate change on freezing severity, and (3) explore options for the management of snow and plant cover to modulate the effects of climate on soil freezing. As a starting point, I briefly discuss the findings of Henry (2008), where soil freezing responses to variation in winter temperature and precipitation were examined for a large range of sites across Canada. I then examine the extent to which the findings of this chapter can be generalized and extrapolated more broadly to other regions. I draw from examples from the forestry and ecology literature in addition to the agriculture literature, based on the expectation that soil freezing across all of these systems is governed by some common principles, despite potential differences in ground cover.

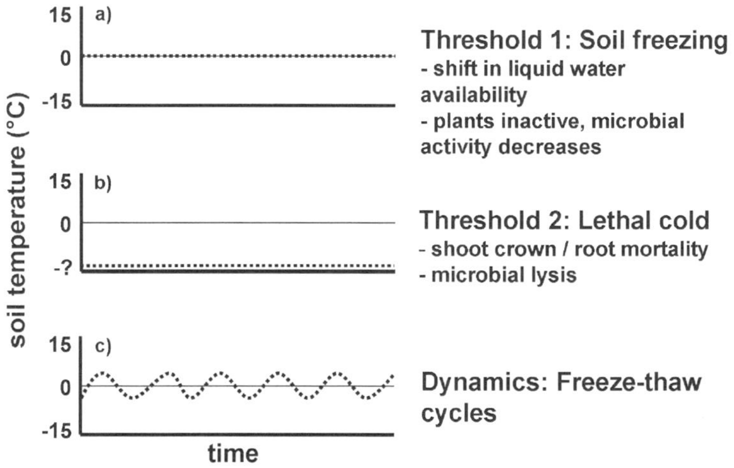

Fig. 3 Important thresholds and dynamics of soil freezing

Freeze-Thaw Cycles (FTC) in Air Versus Soil

In the late fall and early spring, when snow cover is absent or intermittent, soil temperatures often track air temperatures, exhibiting diel FTC; however, soil temperatures can be decoupled substantially from air temperatures over winter when snow cover is present (Sharratt et al. 1992). Therefore, the months of the year that feature the coldest air temperatures often feature relatively mild and stable soil temperatures. Warmer and more variable air temperatures over winter could reduce snow cover and expose soil to cold air overnight or during cold spells, leading to the prediction that in some regions we may observe "colder soils in a warmer world" (Groffman et al. 2001). It is anticipated that increased soil freezing may intensify microbial cell lysis and root mortality, increasing soil hydrological and gaseous fluxes of nitrogen in the spring (Henry 2007). Although there is some ambiguity as to whether it is the freeze or the thaw that is most stressful for soil organisms (Jefferies et al. 2010), or whether increased soluble nitrogen availability over winter results from increased mineralization, FTC can indeed increase soluble nitrogen concentrations in soil (Elliott and Henry 2009).

Frequency of Soil Freeze-Thaw in Canadian Soils

Henry (2008) examined historical weather and soil temperature data for 31 sites in Canada to explore whether there was indeed a relationship between warmer (or drier) winters and increased soil FTC. The sites in the study ranged from arctic to temperate regions, and the weather records spanned up to 40 years. Data were collected from well-kept grass turf on a level surface remote from trees or buildings, and a soil depth of 5 cm was chosen based on the assumption that it best represents the core depth of rooting and soil biological activity.

As predicted, the number of days with snow on the ground decreased in both warm and dry winters, and the effect of warming on snow cover was particularly strong in the warmest sites. The latter trend was explained by the location of the warmest sites along a snow ecotone (i.e., the transition zone between sites that experience much snow over winter and those that experience very little), where the snow pack often melts multiple times in response to warm spells over winter. A decreasing number of days of frozen soil in warm winters was also observed, indicating that soils will not remain frozen as long in a colder world, a result consistent with those of Jylha et al. (2008), based on climate model projections for Europe. However, as predicted, the number of FTC experienced per winter increased in both warm and dry winters. Projected temperature and precipitation increases were then applied to the regression equations derived from the historical data. With respect to days of frozen soil, the northern temperate sites were predicted to experience the greatest changes, despite projected warming being mild at these latitudes relative to arctic regions. Among all sites, more than 80 % were projected to experience an increased frequency of soil FTC with climate warming.

Regional Patterns in the Frequency of Soil FTC

Kreyling and Henry (2011) conducted a follow-up study of 177 soil stations in Germany, and surprisingly, an opposite trend of a decreasing number of FTC with warming was observed for recent decades. However, many of these sites experienced warmer winters than the warmest Canadian sites, such that the German data were an extrapolation of the Canadian data to warmer sites, rather than a contradiction. A number of these German sites are expected to experience vanishing winters, where snow cover and soil freezing will become mostly absent. Beyond these two studies, the question arises as to whether this trend of increased FTC in cold temperate regions (e.g., Canada) and decreased FTC in warmer temperate regions (e.g., Germany) with climate warming can be generalized. The answer appears to be yes, with Decker et al. (2003), Mellander et al. (2007), Campbell et al. (2010), and Fortin (2010), all conducted in relatively cold regions, showing a trend of increased FTC with warming (Table 1). Sinha and Cherkauer (2008) observed a similar trend for the northernmost sites in their study, but similar to Kreyling and Henry (2011), they observed a decreased frequency of FTC with warming in their southernmost sites. With respect to precipitation, Isard and Schaetzl (1998) observed a similar trend as Henry (2008), with an increased number of FTC in dry years.

Decker (2003) observed an increased frequency of FTC in warm years, but commented that many of these FTC occurred under the snow and were of a very small amplitude. This observation raises the question of whether changes in the number

Table 1 Regional summary of changes in freeze-thaw cycle frequency, severity and depth in response to warming (or changes in precipitation, shown in parentheses)

Reference	Location	Latitude and longitude	Sites	Cover	Freeze-thaw cycle		
					Frequency	Severity	Depth
Isard and Schaetzl (1998)	Michigan, USA	43.7, −85.5	10	Forest	↑ (Dry)		↑
Venalainen et al. (2001)	Finland	60.2–68.6, 25.0–27.4	13	Bare soil			↓
Decker et al. (2003)	Vermont, USA	44.5, −72.5	1	Forest	↑	↑	
Hirota et al. (2006)	Hokkaido, Japan	42.8 N, 143.0	11	Agricultural			↓Wet
Isard et al. (2007)	Michigan, USA	43.8–45.3, −83.7 to −86.4	39	Forest		↑	
Mellander et al. (2007)	Northern Sweden	64.2, 19.7	8	Forest	↑		
Henry (2008)	Canada	42.0–74.7, −52.8 to −128.8	31	Grass	↑	–	
Sinha and Cherkauer (2008)	Indiana, USA	38.5–41.4, −86.4 to −86.7	3	Agricultural	↑(N), ↓(S)	↓	
Wang et al. (2009)	Qinghai-Tibet, China	34.7–34.8, 92.8–93.1	2	Alpine meadow			↑
Campbell et al. (2010)	New Hampshire, USA	43.9, −71.8	5	Forest	↑		
Fortin (2010)	Quebec, Canada	46.5–46.8, −71.1 to −71.9	4	Grass	↑		
Sinha et al. (2010)	Midwestern USA	38.5–45.5, −85.5 to −95.0	15	Grass/bare soil			↓
Kreyling and Henry (2011)	Germany	47–55 N, 5–16	177	Bare soil	↓	↓	

↑ increase, ↓ decrease, – no change, *N* northern sites, *S* southern sites, *dry* in dry years, *wet* in wet years of FTC are biologically relevant if the FTC are mild. In some cases multiple FTC have increased losses of soil N (Joseph and Henry 2008), although in other cases the pool of FTC-sensitive materials appears to have become exhausted, with subsequent FTC having less of an effect than the initial FTC (Henry 2007). Large amplitude soil FTC have usually produced stronger effects than small amplitude FTC (Henry 2007), and freezing depth can also be an important factor (Groffman et al. 2001). Therefore, it is clear that other facets of FTC other than frequency must be considered in order to assess possible effects on crops.

Changes in Soil FTC Severity and Depth

In Henry (2008), although the minimum annual soil temperature decreased in dry winters, there was only a marginally significant effect of warm winters on the minimum annual soil temperature ($P=0.07$). The coupling of minimum annual soil surface temperature to mean winter air temperature may be weaker than FTC frequency because extreme freezing can be driven by episodic events (i.e., a cold snap with no snow), whereas FTC frequency is a measure that is cumulative over the whole winter, and thus more representative of mean winter temperature. Nevertheless, increases in the severity of soil freezing with warming have been observed in some northern temperate regions (i.e., Decker et al. 2003; Isard et al. 2007; Table 1). Much like trends in FTC frequency, this trend reverses to one of less severe soil freezing in sites that currently experience mild winters (Sinha and Cherkauer 2008; Kreyling and Henry 2011).

An alternative method to assess soil freezing severity is to measure frost depth. There is a long lag time between changes in air temperature and temperature responses deep in the soil. Therefore, frost depth is a more balanced function of both the severity of air temperatures and the duration of cold over the course of winter. The responses of frost depth to climate change have been highly regional, with decreased frost depth in response to warming in Venalainen et al. (2001) and Sinha et al. (2010), but increased frost depth noted by Isard and Schaetzl (1998) and Wang et al. (2009), and no change noted by Campbell et al. (2010;Table 1).

Deceased snow cover is expected to have a particularly strong influence on frost depth, but with variable effects among continuous permafrost regions, discontinuous/sporadic permafrost regions, and seasonally frozen ground regions (Zhang 2005). Increased snow cover in localized regions can also have strong effect on soil frost depth, as demonstrated by the dramatic decreases in soil freezing depth observed in response to heavy snow brought about by changing air circulation patterns in Hokkaido, Japan, in recent decades (Hirota et al. 2006). Decreased frost depth in this region has increased the overwinter survival of unharvested potatoes that emerge as weeds in the year following planting (discussed further in the following sections, and elsewhere in this book). Similarly, on the Tibetan Plateau, increased snow cover over the last 50 years has retarded frost penetration at depth despite lower than normal air temperatures over this period, although the increased snow cover also delays the date of spring thaw (Yang et al. 2008).

Increased Variability in Air Temperatures and Precipitation

Although the discussion to this point has dealt with changes in mean climate, increased variability in air temperatures and precipitation is also predicted with climate change, which could further stress plants by exposing them to episodic frost events in early fall or late spring when plant cold acclimation is low (Rigby and Porporato 2008; Joseph and Henry 2009). Several studies have noted that the largest changes in the timing and duration of freezing have been occurring in the spring, rather than in the fall (Sharratt 1993; Han et al. 2010). Although these predictions call for fewer extreme frost events and an increase in warm extreme temperature events (Degaetano and Allen 2002; Kharin et al. 2007), early spring warming can promote deacclimation, leaving plants vulnerable if cold early spring conditions return (Gu et al. 2008). Similarly, extreme warming events in the middle of winter can cause plants to deacclimate, leaving them vulnerable to subsequent cold events (Bokhorst et al. 2011). Changes in atmospheric circulation can also counter the general trend of warming locally, resulting in increased extreme cold air temperature events in some regions (Vavrus et al. 2006).

Physical Effects of FTC

In addition to the effects of temperature fluctuations on crops, plants can be sensitive to changes in physical processes in response to climate change (Thorsen et al. 2010). For example, frost heave can damage root systems, and ice encasement can reduce gas exchange, leading to the accumulation of respiration products (Gudleifsson and Larsen 1993). The effects of low temperature on available soil moisture have long been appreciated in agricultural systems (Wilner 1955), and changes in snow depth and soil freezing are expected to interact to modify water infiltration and soil moisture (Hardy et al. 2001). Increased soil frost can reduce water infiltration into soil, increasing runoff and creating a greater potential for soil erosion (Zheng and Flerchinger 2001; Sinha and Cherkhauer 2008; Wang et al. 2009). However, freezing can also increase the quantity of water drawn from deeper layers into the topsoil (Iwata et al. 2010). In addition, episodic winter rain events followed by sudden temperature decreases can impede soil water infiltration substantially, independent of changes in soil frost depth (Iwata et al. 2011). Changes in soil water infiltration over winter are expected to interact with soil nitrogen losses (see elsewhere in this book).

Management of Snow and Plant Cover to Modulate Climate Change Effects on Soil Freezing

Models predicting soil temperatures have been used for decades to assess potential winter cereal mortality (Larsen et al. 1988), and for annual crops, the simplest response to climate change may be simply one of switching to more frost-tolerant cultivars or crops. Alternatively, frost-sensitive crops could still be maintained by modifying snow or litter cover to insulate soils (Fig. 3). For example, winter kill of wheat by frost can be reduced by increasing stubble height, which traps more snow (Malhi et al. 1992), and likewise, the addition of residues or mulch (Gusev and Yasitskiy 1991; Sharratt 2002; Chen et al. 2007) or tillage practices (Hay 1977; Larsen et al. 1987) can alter soil freezing. Commercially available protective covers have also been used for frost control in turf grasses for golf courses (Dionne et al. 1999). The tradeoff of these insulative layers can be one of the increased risk of snow mold infection (see elsewhere in this book). As described above, there are also situations where increased frost depth is desired to control weeds or pests. Increased frost depth can be achieved using snow removal techniques (Hirota et al. 2011), and dust can also be applied to snow to accelerate spring melt (Steltzer et al. 2009).

Conclusions

Overall, the prediction of "colder soils in a warmer world" (Groffman et al. 2001) appears to be likely for many northern temperate systems, although regional changes in the timing and thickness of snow cover will also play an important role in determining soil freezing dynamics independent of changes in air temperature. Similarly, increased climate variability may prove to have a more important influence than mean changes in temperature and precipitation, given that increased variability can result in frost events that occur when plants are not cold acclimated, and warm spells encourage plants to deacclimate prematurely. The trend of decreased soil freezing in southern temperate regions has important implications, both for releasing weeds and pests from frost pressure, and reducing germination of species that have overwinter chilling requirements. The management of soil frost via the modification of snow cover or other insulating materials presents a possible benefit in some cropping systems, but likely requires careful modeling and cost benefit analysis to optimize these benefits.

Acknowledgments I thank NSERC and the OECD for supporting this research.

References

Bokhorst S, Bjerke JW, Street LE, Callaghan TV, Phoenix GK (2011) Impacts of multiple extreme winter warming events on sub-Arctic heathland: phenology, reproduction, growth, and CO_2 flux responses. Glob Change Biol 17:2817–2830

Campbell JL, Ollinger SV, Flerchinger GN, Wicklein H, Hayhoe K, Bailey AS (2010) Past and projected future changes in snowpack and soil frost at the Hubbard Brook Experimental Forest, New Hampshire, USA. Hydrol Process 24:2465–2480

Chen SY, Zhang XY, Peil D, Sun HY, Chen SL (2007) Effects of straw mulching on soil temperature, evaporation and yield of winter wheat: field experiments on the North China Plain. Ann Appl Biol 150:261–268

DeGaetano AT, Allen RJ (2002) Trends in twentieth-century temperature extremes across the United States. J Clim 15:3188–3205

Decker KLM, Wang D, Waite C, Scherbatskoy T (2003) Snow removal and ambient air temperature effects on forest soil temperatures in Northern Vermont. Soil Sci Soc Am J 67:1234–1243

Dionne J, Dube PA, Laganiere M, Desjardins Y (1999) Golf green soil and crown-level temperatures under winter protective covers. Agron J 91:227–233

Elliott AC, Henry HAL (2009) Freeze-thaw cycle amplitude and freezing rate effects on extractable nitrogen in a temperate old field soil. Biol Fertil Soils 45:469–476

Fortin G (2010) Variabilité´et fréquence des cycles de gel-dégel dans la région de Québec, 1977–2006. Can Geogr 54:196–208

Fowler DB (2008) Cold acclimation threshold induction temperatures in cereals. Crop Sci 48:1147–1154

Groffman PM, Driscoll CT, Fahey TJ, Hardy JP, Fitzhugh RD, Tierney GL (2001) Colder soils in a warmer world: a snow manipulation study in a northern hardwood forest ecosystem. Biogeochemistry 56:135–150

Gu L, Hanson PJ, Post WM, Kaiser DP, Yang B, Nemani R, Pallardy SG, Meyers T (2008) The 2007 eastern US spring freeze: increased cold damage in a warming world? BioScience 58:253–262

Gudleifsson B, Larsen A (1993) Advances in plant cold hardiness. CRC Press, Boca Raton

Gusev YM, Yasitshiy SV (1991) Effect of mulch on winter soil temperature conditions. Soviet Soil Sci 23:93–101

Han L, Tsunekawa A, Tsubo M (2010) Monitoring near-surface soil freeze—thaw cycles in northern China and Mongolia from 1998–2007. Int J Appl Earth Obs Geoinfo 12:375–384

Hardy JP, Groffman PM, Fitzhugh RD, Henry KS, Welman AT, Demers JD, Fahey TJ, Driscoll CT, Tierney GL, Nolan S (2001) Snow depth manipulation and its influence on soil frost and water dynamics in a northern hardwood forest. Biogeochemistry 56:151–174

Hay RKM (1977) Effect of tillage and direct drilling on soil temperature in winter. J Soil Sci 28:403–409

Henry HAL (2007) Soil freeze-thaw cycle experiments: trends, methodological weaknesses and suggested improvements. Soil Biol BioChem 39:977–986

Henry HAL (2008) Climate change and soil freezing dynamics: historical trends and projected changes. Clim Change 87:421–434

Hirota T, Iwata Y, Hayashi M, Suzuki S, Hamasaki T, Sameshima R, Takayabu I (2006) Decreasing soil-frost depth and its relation to climate change in Tokachi, Hokkaido, Japan. J Meteorol Soc Jpn 84:821–833

Hirota T, Usuki K, Hayashi M, Nemoto M, Iwata Y, Yanai Y, Yazaki T, Inoue S (2011) Soil frost control: agricultural adaptation to climate variability in a cold region of Japan. Mitig Adapt Strat Glob Change 16:791–802

Irwin JT, Lee RE (2003) Cold winter microenvironments conserve energy and improve overwintering survival and potential fecundity of the goldenrod gall fly, Eurosta *solidaginis*. Oikos 100:71–78

Isard SA, Schaetzl RJ (1998) Effects of winter weather conditions on soil freezing in southern Michigan. Phys Geogr 19:71–94
Isard SA, Schaetzl RJ, Andresen JA (2007) Soils cool as climate warms in the Great Lakes region: 1951–2000. Ann Assoc Am Geogr 97:467–476
Iwata Y, Hayashi M, Suzuki S, Hirota T, Hasegawa S (2010) Effects of snow cover on soil freezing, water movement, and snowmelt infiltration: A paired plot experiment. Water Resour Res 46:W09504
Iwata Y, Nemoto M, Hasegawa S, Yanai Y, Kuwao K, Hirota T (2011) Influence of rain, air temperature, and snow cover on subsequent spring-snowmelt infiltration into thin frozen soil layer in northern Japan. J Hydrol 401:165–176
Jefferies RL, Walker NA, Edwards KA, Dainty J (2010) Is the decline of soil microbial biomass in late winter coupled to changes in the physical state of cold soils? Soil Biol BioChem 42:129–135
Joseph G, Henry HAL (2008) Soil nitrogen leaching losses in response to freeze-thaw cycles and pulsed warming in a temperate old field. Soil Biol BioChem 40:1947–1953
Joseph G, Henry HAL (2009) Retention of surface nitrate additions in a temperate old field: implications for atmospheric nitrogen deposition over winter and plant nitrogen availability. Plant Soil 319:209–218
Jylha K, Fronzek S, Tuomenvirta H, Carter TR, Ruosteenoja K (2008) Changes in frost, snow and Baltic sea ice by the end of the twenty-first century based on climate model projections for Europe. Clim Change 86:441–462
Kharin VV, Zwiers FW, Zhang XB, Hegerl GC (2007) Changes in temperature and precipitation extremes in the IPCC ensemble of global coupled model simulations. J Clim 20:1419–1444
Kreyling J, Henry HAL (2011) Vanishing winters—trends in soil frost dynamics 1 and snow cover in Germany and their ecological implications. Clim Res 46:269–276
Larsen JK, Brun LJ, Enz JW, Cox DJ (1987) Minimizing the risk of producing winter wheat in North Dakota. North Dakota Farm Res 44:9–13
Larsen JK, Brun LJ, Enz JW, Cox DJ (1988) Predicting soil temperatures to indicate winter wheat mortality. Soil Sci Soc Am J 52:776–780
Malhi SS, Bhalla MK, Piening LJ, Caldwell CD (1992) Effect of stubble height and tillage on winter soil temperature in central Alberta. Soil Tillage Res 22:243–251
Malyshev AV, Henry HAL (2012) N uptake and growth responses to sub-lethal freezing in the grass *Poa pratensis* L. Plant Soil 360:175–185. doi:10.1007/s11104-012-1233-4
Mellander PE, Löfvenius MO, Laudon H (2007) Climate change impact on snow and soil temperature in boreal Scots pine stands. Clim Change 85:179–193
Ouellet CE (1976) Winter hardiness and survival of forage crops in Canada. Can J Plant Sci 56:679–689
Oztas T, Fayetorbay F (2003) Effect of freezing and thawing processes on soil aggregate stability. Catena 52:1–8
Rigby JR, Porporato A (2008) Spring frost risk in a changing climate. Geophys Res Lett 35:L12703
Schimel JP, Mikan C (2005) Changing microbial substrate use in Arctic tundra soils through a freeze-thaw cycle. Soil Biol BioChem 37:1411–1418
Sharratt BS (1993) Freeze-thaw and winter temperature of agricultural soils in interior Alaska. Cold Reg Sci Technol 22:105–111
Sharratt BS (2002) Corn stubble height and residue placement in the northern US Corn Belt Part I. Soil physical environment during winter. Soil Tillage Res 64:243–252
Sharratt BS, Baker DG, Wall DB, Skaggs RH, Ruschy DL (1992) Snow depth required for near steady-state soil temperature. Agric For Meteorol 57:243–251
Sinha T, Cherkauer KA (2008) Time series analysis of soil freeze and thaw processes in Indiana. J Hydrometeorol 9:936–950
Sinha T, Cherkauer KA, Mishra V (2010) Impacts of historic climate variability on seasonal soil frost in the Midwestern United States. J Hydrometeorol 11:229–252

Steltzer H, Landry C, Painter TH, Anderson J, Ayres E (2009) Biological consequences of earlier snowmelt from desert dust deposition in alpine landscapes. Proc Natl Acad Sci U S A 106:11629–11634

Thorsen SM, Roer A, Oijen M van (2010) Modelling the dynamics of snow cover, soil frost and surface ice in Norwegian grasslands. Polar Res 29:110–126

Vavrus S, Walsh JE, Chapman WL, Portis D (2006) The behaviour of extreme cold air outbreaks under greenhouse warming. Int J Climatol 26:1133–1147

Venalainen A, Tuomenvirta H, Heikinheimo M, Kellomaki S, Peltola H, Strandman H, Vaisanen H (2001) Impact of climate change on soil frost under snow cover in a forested landscape. Clim Res 17:63–72

Wang JS, Zhang HF, Liu ZG, Huang B (2009) Distribution of winter frozen soil depth in Qilian Mountain and its response to temperature change. 2009 IEEE international geoscience and remote sensing symposium book series, pp 837–840

Wilner J (1955) The effect of low temperatures on available soil moisture during winters on the Canadian Prairies. Agron J 47:411–413

Yang MX, Yao TD, Nelson FE, Shiklomanov NI, Guo DL, Wang CH (2008) Snow cover and depth of freeze-thaw on the Tibetan Plateau: a case study from 1997–1998. Phys Geogr 29:208–221

Zhang TJ (2005) Influence of the seasonal snow cover on the ground thermal regime: an overview. Rev Geophys 43:RG4002

Zheng XQ, Flerchinger GN (2001) Infiltration into freezing and thawing soils under differing field treatments. J Irrig Drain Eng 127:176–182

Winter Climate Change and Ecological Implications in Temperate Systems

Juergen Kreyling

Winter climate is changing. Winters get warmer and wetter in the temperate zone of the northern hemisphere. The loss of snow cover combined with increased air temperature variability complicates projections of growing conditions. Frost events in winter and spring, however, will occur even in a warmer climate.

The interaction between climatic drivers and particularly the presence or absence of snow cover drives ecological response to winter climate change. Yet, vegetation response to winter climate is complex, even in simplified experiments. The reactions depend on interactions between climate drivers in addition to simple frequency and magnitude of the climate manipulations as well as on species interactions, species identities, and within-species differentiation (local adaptations).

Diversification at all levels (within species, in community compositions, and in land use) appears to be a promising adaptation strategy in forestry, agronomy, and nature conservation. Assisted colonization of genotypes is but one aspect of such a strategy.

Winter Climate Change

Burning of fossil fuel and land use change increase radiative forcing, thereby leading to global warming with current atmospheric greenhouse gas concentrations being highest in at least the last 650,000 years (Solomon et al. 2007). The growth rate of greenhouse gas emissions since 2000 is greater than even the most fossil-fuel-intensive scenario of the Intergovernmental Panel on Climate Change emissions scenarios (A1 FI), increasing from 1.1 % per year for 1990–1999 to more than 3 % per year for 2000–2004 (Raupach et al. 2007). A leveling off of climate change can therefore not be expected. Regional to local projections of climate change are

J. Kreyling (✉)
BayCEER, Biogeography, University of Bayreuth, Universitätsstraße 30,
95448 Bayreuth, Germany
e-mail: juergen.kreyling@uni-bayreuth.de

R. Imai et al. (eds.), *Plant and Microbe Adaptations to Cold in a Changing World*,
DOI 10.1007/978-1-4614-8253-6_3,

still a major challenge. Larger-scale geographic patterns of observed and projected warming, however, show the strongest warming signals over land at high northern latitudes (Solomon et al. 2007).

Climate change is further expected to differ between seasons with strongest changes projected during winter (Solomon et al. 2007). Daily temperature variability increased between 1977 and 2000, mainly due to an increase in warm extremes, with almost no change in cold extremes (Tank and Konnen 2003). Yet, fewer air frost days (−30 to −80 %) that are more scattered over time are projected for Europe within the current century (Jylha et al. 2008).

Snow cover is a major aspect in winter as snow works as a temporal water storage and insulates the soil against air temperature extremes (Beniston 2003). The annual extent of snow cover over the northern hemisphere (about 50 million km^2 in February) has already declined by nearly 10 % during the period 1972–2003 (Walsh et al. 2005). This is of special relevance, as 30–40 cm snow depth is sufficient to effectively decouple soil and air temperature (Edwards et al. 2007). In many northern temperate regions, a reduced snow cover in winter, and the consequentially decreased insulation of the soils against freezing, is predicted to increase the frequency of soil freezing and freeze-thaw cycles (FTC) (Groffman et al. 2001). Observations from climate data series confirm these expectations (Henry 2008). Regional climate modeling provides further support, as a decrease in snow cover in southern Finland increases the probability of frozen ground (Venalainen et al. 2001), and an increase in the frequency of soil FTC by more than 30 % in Sweden, also associated with decreased snow cover, is projected (Mellander et al. 2007). Nevertheless, for Germany, a warmer temperate region, significant parts of the country are not expected to exhibit snow cover regularly in the future, and there is evidence that air temperatures are becoming sufficiently warm that both the frequency and intensity of FTC have been decreasing (Kreyling and Henry 2011). However, the variability in winter weather conditions is also expected to increase, such that an increase in minimum temperature over winter is unlikely for many regions (Kodra et al. 2011).

Vegetation Response to Winter Climate Change

Phenology

For middle and high latitude ecosystems, increased productivity is expected due to longer growing seasons in response to global warming. Extensive phenological studies document a progressively earlier spring (2.3–5.2 days per decade over the last 30 years), leading to an elongation of the growing period by about 2 weeks in the second half of the twentieth century (e.g., Root et al. 2003, Parmesan 2007). However, growth forms differ in their responsiveness: annual plants respond stronger than congeneric perennials, insect-pollinated species react stronger than wind-pollinated plants, and woody taxa show less response than herbaceous plants (Fitter and Fitter 2002). This finding emphasizes the importance of species and functional composition in response to climate change. Furthermore, increased variability in minimum temperatures can even delay flowering in warm-temperate zones (von Holle et al. 2010).

Species Composition

Ecosystems are dynamic systems and species compositions can therefore be expected to shift with changing environmental conditions. Taxa in natural communities do not respond in synchrony to external pressures, and therefore existing community compositions may become disaggregated. Yet, short-term genetic adaptations are reported for plants (Bone and Farres 2001, Jump et al. 2008) and ecosystems might tolerate some level of future climate change and will persist with some variations in species compositions, as they have done repeatedly during paleoclimatic changes. With regard to the stability of current species compositions, however, climate change has been identified as one of the major drivers for increasing invasibility of plant communities (Walther et al. 2009).

In general, natural and seminatural ecosystems may be more vulnerable to climate change than managed systems as it commonly takes decades for their establishment. Therefore, slower adaptation is expected (Hitz and Smith 2004). Species distribution modeling implies strong potential threats of climate change to biodiversity (Thomas et al. 2004). Modeling the future potential distribution of 1,350 European plant species under various greenhouse gas emission scenarios indicates that more than half of these species could become vulnerable, endangered, critically endangered, or committed to extinction by 2080 if unable to disperse (Thuiller et al. 2005). Meta-analyses over many animal and plant species confirm that poleward shifts are correlated with warming globally (Parmesan and Yohe 2003). Effective migration rates, however, are expected to lag behind the speed of climate change (Nathan et al. 2011). Besides poleward range shifts, altitudinal range shifts in response to increasing temperatures are documented (e.g., Grabherr et al. 1994). Yet, keeping pace with the speed of warming by altitudinal range shifts is physically limited by migration rates and altitudes of the mountains (Kreyling et al. 2010b).

Extreme Events and Interacting Climatic Parameters

It has been proposed that extreme climatic events have stronger ecological implications than changes in mean temperature or precipitation, mainly because extreme events can break inertia by eliminating biomass and organisms on larger scales (Easterling et al. 2000). Research on the ecological impacts of extreme climatic events has consequently increased within the last years (Jentsch et al. 2007). However, interactions between climatic variables are relevant for the response of ecosystems. For instance, winter warming trends interact with the occurrence of late spring frost events in hardly predictable ways (Kreyling et al. 2012b). Similarly, cold hardiness in response to rising CO_2-levels is reported to increase (Dole et al. 2003) or earlier break of dormancy is reported with increased nitrogen (N)-availability (Saebo et al. 2001). The reliability of projections of plant response into the future is clearly limited by such combined effects of climatic drivers.

Warming over Winter

Experimental winter warming can increase primary productivity (Hutchison and Henry 2010) and net N mineralization over summer (Turner and Henry 2010) to the same extent or even stronger as year-round warming in temperate systems. Another example shall highlight, however, that response to winter climate change is not easily predictable: Temperate species showed strong and opposing responses to winter warming pulses which increased soil freezing and thawing in one winter (Kreyling et al. 2010a). Grassland species such as *Holcus lanatus* profited in their biomass production in the first (+156 %) and second (+42 %) growing season after five additional soil FTC. The dwarf-shrubs, such as *Calluna vulgaris*, showed no effect in their biomass production in the first growing season after the FTC-manipulations, while decreasing in their biomass production by 50 %, accompanied with dieback of adult individuals, by the second growing season after the FTC-manipulation. A potential explanation for the delayed dieback is increased root injury during the manipulations which allowed access for pathogens that happened to kill the plants in the long run (Kreyling et al. 2012a). Future research will certainly reveal further examples of surprises and nonlinear responses of vegetation to climate change.

Frost in a Warmer World

Frost has been identified as major ecological and evolutionary driver (Sakai and Weiser 1973; Inouye 2000). Geographic distributions of congeneric species or subspecies have been shown to depend mainly on frost sensitivity, not drought sensitivity in an altitudinal gradient in an arid ecosystem in the Western US (Lambrecht et al. 2007). Single winter frost events can even prevent successful adaptation by range shifts in response to global warming (Jalili et al. 2010). In temperate regions, frost tolerance and cold acclimation have been studied extensively in the context of forestry (e.g., Bigras and Colombo 2001), winter cereal production (e.g., Fowler 2008), and forage and turf grass systems (e.g., Sandve et al. 2011). Especially in grasses, given the positioning of their meristems at the soil surface, modeling frost exposure is complicated by the need to model changes in ice and snow cover (see previous text). In particular, ice encasement can harm overwintering grasses (Gudleifsson 2010), whereas snow cover can protect overwintering grasses from extremes in air temperature (Tompkins et al. 2004).

The study of frost tolerance has focused extensively on frost kill, which causes major economic losses of forage grasses in temperate regions (e.g., Ouellet 1976). The median lethal temperature (LT50) values of forage and turf grasses during winter (i.e., the temperature at which half of the tillers are killed) typically range from approximately −15 to −30 °C (Dionne et al. 2001; Hulke et al. 2008; Hanslin and Hoglind 2009). Roots, however, are usually less frost tolerant than shoots with typical frost hardiness ranging between −3 and −15 °C in temperate and boreal forest trees (Bigras and Dumais 2005). Generally, protection by snow cover is crucial for the winter survival of tree roots and, consequently, the whole plant (Schaberg et al. 2008b).

Plants remain photosynthetically active in winter (Starr and Oberbauer 2003), during which they also take up nutrients (Grogan et al. 2004; Andresen and Michelsen 2005). Protection against extreme frost by snow cover, however, is a precondition to maintain physiological integrity over winter (Saarinen and Lundell 2010). Increased temperature variability and changes to plant cold acclimation cues may challenge plants over the next decades by exposing them to frost at times when they are not fully cold acclimated. For example, evidence suggests that rising winter temperatures accelerate dehardening and bud break, thereby enhancing the risk of frost damage (Hänninen 1991; Rigby and Porporato 2008). It has been shown that current adaptations will be suboptimal for future conditions, with bud-burst generally occurring earlier than the optimal strategy (Bennie et al. 2010), thereby increasing the risk of late frost damage.

Similarly, investigation of arctic tundra species exposed to multiple midwinter warm spells has shown that plant performance, although improving during the first heat wave, deteriorated during and after the second heat wave, probably due to loss of cold resistance and subsequent frost kill (Bokhorst et al. 2009). On the other hand, vanishing winters in temperate regions accompanied with the lack of a sufficient freezing period over winter can disrupt plant dormancy altogether, paradoxically causing extended plant dormancy in the spring (Yu et al. 2010).

The Importance of Biodiversity in Climate Impact Research

Biodiversity experiments have shown that increasing plant diversity enhances primary biomass production and nutrient retention (see reviews by Balvanera et al. 2006; Cardinale et al. 2006). More important in relation to the resilience of ecosystems against climate change, however, is the insurance hypothesis (Yachi and Loreau 1999), which states that more diverse systems are expected to be more resilient against perturbations. By now, empirical validations of the positive relationship between diversity and stability are reported (Tilman et al. 2006, van Ruijven and Berendse 2010). Experiments combining climate impact research and biodiversity research, however, are scarce, despite the probable importance of this combination.

Potential feedbacks between biodiversity and (climatic) disturbance emphasize the importance of reciprocal relationships (Hughes et al. 2007): Diversity of a given community is affected by abiotic disturbance while diversity can also buffer ecosystem functioning against (negative) effects of disturbance. The net effect is controlled by disturbance severity, initial diversity, competition, and recruitment after the disturbance. The net effect of the interaction between diversity and disturbance can work as a stabilizing feedback, as the counteracting forces of diversity reducing the magnitude of realized disturbance, and low disturbance leading to competitive exclusion might cause communities to converge at intermediate levels of diversity. A major research question therefore is how climate change influences this feedback. In particular, alterations in competition and recruitment need to be tested experimentally. Given that human activities accelerate species loss directly and increase climatic stress, the described feedback suggests synergistic effects with accelerated loss of species.

Within-Species Diversity

Especially in the context of climate impacts, within-species diversity is potentially important. Populations within taxa are known to differ in their adaptations to climate. So-called "provenance trials" in forestry have been conducted for more than a century already (e.g., Krahl-Urban 1958) and results imply that within-species differentiation is clearly exhibited. The phenotypic and genetic differences are commonly expressed in local adaptations to climate and other environmental factors such as soil types (Joshi et al. 2001; Hufford and Mazer 2003; Bennie et al. 2010; Chen et al. 2010; Kreyling et al. 2012b, c). Local adaptation can be defined as the higher fitness of local individuals at their home site compared with that of nonlocal individuals of the same species (Biere and Verhoeven 2008). However, case studies imply that local adaptations to climate are species specific and may even be negligible in some species or concerning some climatic gradients (Macel et al. 2007, Weisshuhn et al. 2011).

Studies of frost tolerance in forage and turf grasses have revealed substantial variation among cultivars in both frost tolerance and resistance to dehardening (Hulke et al. 2008), with northern winter-hardy cultivars dehardening more rapidly than less-hardy cultivars, but exhibiting higher initial frost tolerance (Jorgensen et al. 2010). Differentiation in frost hardiness within species can be explained by differences in soluble carbohydrate concentrations (Hanslin and Hoglind 2009). However, in some cases, ecotypes from northern regions, better insulated from extreme subfreezing temperatures by snow cover, have developed less freezing tolerance than those evolving under milder winter climates (Dionne et al. 2001). Similarly, there can be considerable variation in freezing tolerance of shoots among genotypes collected from the same location (Dionne et al. 2001). Furthermore, differences of more than 4 °C in root cold tolerance are reported among provenances of forest trees (Bigras and Dumais 2005).

Frost over winter (Kreyling et al. 2012d) and in spring (Kreyling et al. 2012b, c) is an important stressor with evolutionary impact as expressed by local adaptations within species. At least for the four grass species investigated in these studies, no such local adaptations were found with respect to summer drought or summer temperature (Beierkuhnlein et al. 2011).

Generally, within-species variation in response to climate is relevant. Recent studies imply that within-species variation at a continental scale can surpass between-species differences at one location in forest trees (Kreyling et al. 2012d) and grass species (Beierkuhnlein et al. 2011). This finding has important implications: Treating species as homogeneous groups to project their response to climate change can be strongly misleading. The results furthermore support the view that climate change may impact species not just at the trailing edges but throughout their range due to genetic adaptation of populations to local environments (Chen et al. 2010).

Within-species Diversity and Evolution

High within-species diversity is furthermore a prerequisite for rapid evolution in response to climate stress. Evolutionary adaptation can emerge within two generations in natural populations based on genetic variation already present within the population (Jump et al. 2008). In forest trees, the interval between recruitment events strongly affects the rate of adaptive response, because selection is most severe during the early stages of forest development (Kramer et al. 2008). Modeling results indicate the high adaptive potential to environmental change in *Fagus sylvatica*, the dominant native tree species of Central Europe, if recruitment intervals are short and many mother trees contribute to the next generation, two preconditions which can be influenced by management strategies (Kramer et al. 2008). Genetic adaptation to climate change in forest trees appears possible, yet slow and is expected to lag behind the moving optimum due to inertia in the system provided by surviving current trees, which slow down the establishment of new trees with more suitable genotypes (Savolainen et al. 2007).

Assisted Colonization

Assisted colonization of adapted genotypes (Kreyling et al. 2011) in combination with other management strategies could potentially enable faster adaptation in seminatural and extensively managed ecosystems. However, I warn against the selection of single, best-adapted ecotypes because of three reasons: First, climate change alters several climatic parameters at once, leaving the question to which single parameter we should select for. Second, we have no good understanding of the importance of interacting climatic parameters in novel climates of the future. And finally, local to regional climate projections are uncertain, especially with regard to precipitation (snow). Rather, genetic diversity should be maximized to enable continued local adaptations. This notion is already acknowledged in forestry (Schaberg et al. 2008a), but should be transferred to nature conservation and agronomy.

Yet, adaptation to climate is just one aspect for species performances. The temperate grass species *Holcus lanatus*, for instance, has been shown to be locally maladapted to one of its important fungal antagonists (Cremieux et al. 2008) while local adaptation to climate still led to improved local fitness (Bischoff et al. 2006). In the temperate herb *Plantago lanceolata*, however, local adaptation to a specific herbivore (Cremieux et al. 2008) contrasted with no local adaptation to climate (Bischoff et al. 2006). Again, trophic interactions require better understanding before generalizations can be drawn.

Conclusions

Winter climate is a major driver of plant and ecosystem performance. Vegetation response to winter climate, however, is already complex in simplified experiments, as the reaction depends on species interactions and species identities besides simple frequency and magnitude of the climate manipulations. Future work will benefit from a multiscale approach by combining small-scale experiments to identify processes, large-scale experimental manipulations in natural settings to determine if these processes scale up, and research on nonlinear processes and thresholds. Furthermore, the importance of diversity, be it genetic diversity, species diversity, or differences between vegetation types, need to be included in climate impact research. A better causal understanding will improve ecosystem modeling and enable more reliable projections into the future. Dealing with uncertainty and preparing for surprises, however, seems unavoidable with respect to future climate change. Diversification at any level, e.g., within species, within communities, and in land use is consequently a promising adaptation strategy.

References

Andresen LC, Michelsen A (2005) Off-season uptake of nitrogen in temperate heath vegetation. Oecologia 144(4):585–597

Balvanera P, Pfisterer AB, Buchmann N, He JS, Nakashizuka T, Raffaelli D, Schmid B (2006) Quantifying the evidence for biodiversity effects on ecosystem functioning and services. Ecol Lett 9(10):1146–1156

Beierkuhnlein C, Thiel D, Jentsch A, Willner E, Kreyling J (2011) Ecotypes of European grass species respond specifically to warming and extreme drought. J Ecol 99(3):703–713

Beniston M (2003) Climatic change in mountain regions: A review of possible impacts. Clim Change 59(1–2):5–31

Bennie J, Kubin E, Wiltshire A, Huntley B, Baxter R (2010) Predicting spatial and temporal patterns of bud-burst and spring frost risk in north-west Europe: the implications of local adaptation to climate. Glob Change Biol 16(5):1503–1514

Biere A, Verhoeven KJF (2008) Local adaptation and the consequences of being dislocated from coevolved enemies. New Phytol 180(2):265–268
Bigras FJ, Colombo SJ (eds) (2001) Conifer cold hardiness. Kluwer, Dordrecht
Bigras FJ, Dumais D (2005) Root-freezing damage in the containerized nursery: impact on plantation sites—a review. New For 30(2–3):167–184
Bischoff A, Cremieux L, Smilauerova M, Lawson CS, Mortimer SR, Dolezal J, Lanta V, Edwards AR et al (2006) Detecting local adaptation in widespread grassland species—the importance of scale and local plant community. J Ecol 94(6):1130–1142
Bokhorst SF, Bjerke JW, Tømmervik H, Callaghan TV, Phoenix GK (2009) Winter warming events damage sub-Arctic vegetation: consistent evidence from an experimental manipulation and a natural event. J Ecol 97:1408–1415
Bone E, Farres A (2001) Trends and rates of microevolution in plants. Genetica 112:165–182
Cardinale BJ, Srivastava DS, Duffy JE, Wright JP, Downing AL, Sankaran M, Jouseau C (2006) Effects of biodiversity on the functioning of trophic groups and ecosystems. Nature 443(7114):989–992
Chen P, Welsh C, Hamann A (2010) Geographic variation in growth response of Douglas-fir to interannual climate variability and projected climate change. Glob Change Biol 16(12):3374–3385
Cremieux L, Bischoff A, Smilauerova M, Lawson CS, Mortimer SR, Dolezal J, Lanta V et al (2008) Potential contribution of natural enemies to patterns of local adaptation in plants. New Phytol 180(2):524–533
Dionne J, Castonguay Y, Nadeau P, Desjardins Y (2001) Freezing tolerance and carbohydrate changes during cold acclimation of green-type annual bluegrass (*Poa annua* L.) ecotypes. Crop Sci 41(2):443–451
Dole KP, Loik ME, Sloan LC (2003) The relative importance of climate change and the physiological effects of CO_2 on freezing tolerance for the future distribution of *Yucca brevifolia*. Glob Planet Change 36(1–2):137–146
Easterling DR, Meehl GA, Parmesan C, Changnon SA, Karl TR, Mearns LO (2000) Climate extremes: observations, modeling, and impacts. Science 289(5487):2068–2074
Edwards AC, Scalenghe R, Freppaz M (2007) Changes in the seasonal snow cover of alpine regions and its effect on soil processes: a review. Quat Int 162:172–181
Fitter AH, Fitter RSR (2002) Rapid changes in flowering time in British plants. Science 296(5573):1689–1691
Fowler DB (2008) Cold acclimation threshold induction temperatures in cereals. Crop Sci 48(3):1147–1154
Grabherr G, Gottfried M, Pauli H (1994) Climate effects on mountain plants. Nature 369(6480):448–448
Groffman PM, Driscoll CT, Fahey TJ, Hardy JP, Fitzhugh RD, Tierney GL (2001) Colder soils in a warmer world: a snow manipulation study in a northern hardwood forest ecosystem. Biogeochemistry 56(2):135–150
Grogan P, Michelsen A, Ambus P, Jonasson S (2004) Freeze-thaw regime effects on carbon and nitrogen dynamics in sub-arctic heath tundra mesocosms. Soil Biol Biochem 36(4):641–654
Gudleifsson BE (2010) Ice tolerance and metabolite accumulation of herbage crops in Iceland and impact of climate change. Icel Agric Sci 23:111–122
Hänninen H (1991) Does climatic warming increase the risk of frost damage in Northern trees? Plant Cell Environ 14(5):449–454
Hanslin HM, Hoglind M (2009) Differences in winter-hardening between phenotypes of *Lolium perenne* with contrasting water-soluble carbohydrate concentrations. Grass Forage Sci 64(2):187–195
Henry HAL (2008) Climate change and soil freezing dynamics: historical trends and projected changes. Clim Change 87:421–434
Hitz S, Smith J (2004) Estimating global impacts from climate change. Global Environ Chang 14(3):201–218

Hufford KM, Mazer SJ (2003) Plant ecotypes: genetic differentiation in the age of ecological restoration. Trends Ecol Evol 18(3):147–155
Hughes AR, Byrnes JE, Kirnbro DL, Stachowicz JJ (2007) Reciprocal relationships and potential feedbacks between biodiversity and disturbance. Ecol Lett 10(9):849–864
Hulke BS, Watkins E, Wyse DL, Ehlke NJ (2008) Freezing tolerance of selected perennial ryegrass (*Lolium perenne* L.) accessions and its association with field winter hardiness and turf traits. Euphytica 163(1):131–141
Hutchison JS, Henry HAL (2010) Additive effects of warming and increased nitrogen deposition in a temperate old field: plant productivity and the importance of winter. Ecosystems 13(5):661–672
Inouye DW (2000) The ecological and evolutionary significance of frost in the context of climate change. Ecol Lett 3(5):457–463
Jalili A, Jamzad Z, Thompson K, Araghi MK, Ashrafi S, Hasaninejad M, Panahi P et al (2010) Climate change, unpredictable cold waves and possible brakes on plant migration. Global Ecol Biogeogr 19:642–648
Jentsch A, Kreyling J, Beierkuhnlein C (2007) A new generation of climate change experiments: events, not trends. Front Ecol Environ 5(7):365–374
Jorgensen M, Ostrem L, Hoglind M (2010) De-hardening in contrasting cultivars of timothy and perennial ryegrass during winter and spring. Grass Forage Sci 65(1):38–48
Joshi J, Schmid B, Caldeira MC, Dimitrakopoulos PG, Good J, Harris R, Hector A et al (2001) Local adaptation enhances performance of common plant species. Ecol Lett 4(6):536–544
Jump AS, Penuelas J, Rico L, Ramallo E, Estiarte M, Martinez-Izquierdo JA, Lloret F (2008) Simulated climate change provokes rapid genetic change in the Mediterranean shrub *Fumana thymifolia*. Glob Change Biol 14:637–643
Jylha K, Fronzek S, Tuomenvirta H, Carter TR, Ruosteenoja K (2008) Changes in frost, snow and Baltic sea ice by the end of the twenty-first century based on climate model projections for Europe. Clim Change 86(3–4):441–462
Kodra E, Steinhaeuser K, Ganguly AR (2011) Persisting cold extremes under 21st-century warming scenarios. Geophys Res Lett 38:L08705.
Krahl-Urban J (1958) Vorläufige Ergebnisse der Buchenprovenienzversuche. Allgemeine Forst- und Jagdzeitschrift 129:242–251
Kramer K, Buiteveld J, Forstreuter M, Geburek T, Leonardi S, Menozzi P, Povillon F et al (2008) Bridging the gap between ecophysiological and genetic knowledge to assess the adaptive potential of European beech. Ecol Model 216(3–4):333–353
Kreyling J, Beierkuhnlein C, Jentsch A (2010a) Effects of soil freeze-thaw cycles differ between experimental plant communities. Basic Appl Ecol 11:65–75
Kreyling J, Wana D, Beierkuhnlein C (2010b) Climate warming and tropical plant species—consequences of a potential upslope shift of isotherms in southern Ethiopia. Divers Distrib 16:593–605
Kreyling J, Bittner T, Jaeschke A, Jentsch A, Steinbauer MJ, Thiel D, Beierkuhnlein C (2011) Assisted colonization: a question of focal units and recipient localities. Restor Ecol 19(4):433–440
Kreyling J, Henry HAL (2011) Vanishing winters in Germany: soil frost dynamics and snow cover trends, and ecological implications. Clim Res 46(3):269–276
Kreyling J, Persoh D, Werner S, Benzenberg M, Wöllecke J (2012a) Short-term impacts of soil freeze-thaw cycles on roots and root-associated fungi of *Holcus lanatus* and *Calluna vulgaris*. Plant Soil 353(1–2):19–31
Kreyling J, Thiel D, Nagy L, Jentsch A, Huber G, Konnert M, Beierkuhnlein C (2012b) Late frost sensitivity of juvenile *Fagus sylvatica* L. differs between southern Germany and Bulgaria and depends on preceding air temperature. Europ J Forest Res 131(3):717–725
Kreyling J, Thiel D, Simmnacher K, Willner E, Jentsch A, Beierkuhnlein C (2012c) Geographic origin and past climatic experience influence the response to late spring frost in four common grass species in central Europe. Ecography 35:268–275
Kreyling J, Wiesenberg GLB, Thiel D, Wohlfart C, Huber G, Walter J, Jentsch A et al (2012d) Cold hardiness of *Pinus nigra* Arnold as influenced by geographic origin, warming, and extreme summer drought. Environ Exp Bot 78:99–108

Lambrecht SC, Shattuck AK, Loik ME (2007) Combined drought and episodic freezing effects on seedlings of low- and high-elevation subspecies of sagebrush (*Artemisia tridentata*). Physiol Plantarum 130(2):207–217

Macel M, Lawson CS, Mortimer SR, Smilauerova M, Bischoff A, Cremieux L, Dolezal J et al (2007) Climate vs. soil factors in local adaptation of two common plant species. Ecology 88(2):424–433

Mellander P, Löfvenius MO, Laudon H (2007) Climate change impact on snow and soil temperature in boreal Scots pine stands. Clim Change 85(1–2):179–193

Nathan R, Horvitz N, He Y, Kuparinen A, Schurr FM, Katul GG (2011) Spread of North American wind-dispersed trees in future environments. Ecol Lett 14:211–219

Ouellet CE (1976) Winter hardiness and survival of forage crops in Canada. Can J Plant Sci 56(3):679–689

Parmesan C (2007) Influences of species, latitudes and methodologies on estimates of phenological response to global warming. Glob Change Biol 13(9):1860–1872

Parmesan C, Yohe G (2003) A globally coherent fingerprint of climate change impacts across natural systems. Nature 421(6918):37–42

Raupach MR, Marland G, Ciais P, Le Quere C, Canadell JG, Klepper G, Field CB (2007) Global and regional drivers of accelerating CO_2 emissions. P Natl Acad Sci U S A 104(24):10288–10293

Rigby JR, Porporato A (2008) Spring frost risk in a changing climate. Geophys Res Lett 35(12):L12703–L

Root TL, Price JT, Hall KR, Schneider SH, Rosenzweig C, Pounds JA (2003) Fingerprints of global warming on wild animals and plants. Nature 421(6918):57–60

Saarinen T, Lundell R (2010) Overwintering of *Vaccinium vitis-idaea* in two sub-Arctic microhabitats: a reciprocal transplantation experiment. Polar Res 29(1):38–45

Saebo A, Haland A, Skre O, Mortensen LM (2001) Influence of nitrogen and winter climate stresses on Calluna vulgaris (L.) Hull. Ann Bot 88(5):823–828

Sakai A, Weiser CJ (1973) Freezing resistance of trees in North America with reference to tree regions. Ecology 54(1):118–126

Sandve SR, Kosmala A, Rudi H, Fjellheim S, Rapacz M, Yamada T, Rognli OA (2011) Molecular mechanisms underlying frost tolerance in perennial grasses adapted to cold climates. Plant Sci 180(1):69–77

Savolainen O, Pyhajarvi T, Knurr T (2007) Gene flow and local adaptation in trees. Ann Rev Ecol Evol Syst 38:595–619

Schaberg PG, DeHayes DH, Hawley GJ, Nijensohn SE (2008a) Anthropogenic alterations of genetic diversity within tree populations: implications for forest ecosystem resilience. Forest Ecol Manag 256(5):855–862

Schaberg PG, Hennon PE, D'Amore DV, Hawley GJ (2008b) Influence of simulated snow cover on the cold tolerance and freezing injury of yellow-cedar seedlings. Glob Change Biol 14(6):1282–1293

Solomon S, Qin D, Manning M, Chen Z, Marquis M (eds) (2007) Climate change 2007: the physical science basis. Contribution of Working Group I to the Fourth Assessment Report of the Intergovernmental Panel on Climate Change. Cambridge University Press, Cambridge

Starr G, Oberbauer SF (2003) Photosynthesis of arctic evergreens under snow: implications for tundra ecosystem carbon balance. Ecology 84(6):1415–1420

Tank A, Konnen GP (2003) Trends in indices of daily temperature and precipitation extremes in Europe, 1946–99. J Climate 16(22):3665–3680

Thomas CD, Cameron A, Green RE, Bakkenes M, Beaumont LJ, Collingham YC, Erasmus BFN et al (2004) Extinction risk from climate change. Nature 427(6970):145–148

Thuiller W, Lavorel S, Araujo MB, Sykes MT, Prentice IC (2005) Climate change threats to plant diversity in Europe. P Natl Acad Sci U S A 102(23):8245–8250

Tilman D, Reich PB, Knops JMH (2006) Biodiversity and ecosystem stability in a decade-long grassland experiment. Nature 441(7093):629–632

Tompkins DK, Ross JB, Moroz DL (2004) Effects of ice cover on annual bluegrass and creeping bentgrass putting greens. Crop Sci 44(6):2175–2179

Turner MM, Henry HAL (2010) Net nitrogen mineralization and leaching in response to warming and nitrogen deposition in a temperate old field: the importance of winter temperature. Oecologia 162(1):227–236

Van Ruijven J, Berendse F (2010) Diversity enhances community recovery, but not resistance, after drought. J Ecol 98(1):81–86

Venalainen A, Tuomenvirta H, Heikinheimo M, Kellomaki S, Peltola H, Strandman H, Vaisanen H (2001) Impact of climate change on soil frost under snow cover in a forested landscape. Clim Res 17(1):63–72

von Holle B, Wei Y, Nickerson D, Moen J (2010) Climatic variability leads to later seasonal flowering of Floridian plants. PLOS ONE 5:e11500

Walsh JE, Anisimov O, Hagen JOM, Jakobsson T, Oerlemans J, Prowse TD, Romanovsky V et al (2005) Cryosphere and hydrology. In: Symon C, Arris L, Heal B (eds) Arctic Climate Impact Assessment. Cambridge University Press, Cambridge, pp 183–242

Walther G, Roques A, Hulme PE, Sykes MT, Pysek P, Kuehn I, Zobel M et al (2009) Alien species in a warmer world: risks and opportunities. Trends Ecol Evol 24(12):686–693

Weisshuhn K, Auge H, Prati D (2011) Geographic variation in the response to drought in nine grassland species. Basic Appl Ecol 12(1):21–28

Yachi S, Loreau M (1999) Biodiversity and ecosystem productivity in a fluctuating environment: the insurance hypothesis. P Natl Acad Sci U S A 96(4):1463–1468

Yu H, Luedeling E, Xu J (2010) Winter and spring warming result in delayed spring phenology on the Tibetan Plateau. P Natl Acad Sci U S A 107(51):22151–22156

Possible Change of Water and Nitrate Cycles Associated with the Frost-Depth Decrease Under Climate Change

Yukiyoshi Iwata, Tomoyoshi Hirota, Masaki Hayashi, Junichi Arima, Shinji Suzuki and Manabu Nemoto

Introduction

Ice on the earth is decreasing because of the climate change (Lemke et al. 2007). Groffman et al. (2001) showed that the decreasing snow reduces heat insulation effects in winter, which might contribute to the increase in soil freezing. In contrast, some researchers have concluded that the soil frost depth, the length of the frozen period, and freeze–thaw cycles had decreased in some locations based on results of field measurements (Cutforth et al. 2004; Frauenfeld et al. 2004; Kreyling and Henry 2011). In the Tokachi region, located on the northernmost large island of the Japanese archipelago, a significant decrease in soil frost depth has also been reported (Hirota et al. 2006). However, this phenomenon was not caused by the increase

Y. Iwata (✉)
National Institute for Rural Engineering, NARO, Kanondai 2-1-6, Tsukuba,
Ibaraki, 305-8609, Japan
e-mail: iwatayuk@affrc.go.jp

T. Hirota · M. Nemoto
Agro-environmental Research Division, NARO Hokkaido Agricultural Research Center,
Hitsujigaoka 1, Toyohira-Ku, Sapporo, Hokkaido 062-8555, Japan
e-mail: hirota@affrc.go.jp

M. Hayashi
Department of Geoscience, University of Calgary, 2500 University Dr. NW, Calgary,
Alberta T2N 1N4, Canada
e-mail: hayashi@ucalgary.ca

J. Arima
Graduate School of Agriculture, Hokkaido University, Kita 9 Nishi 9, Kita-ku Sapporo 060-8589,
Japan

S. Suzuki
Faculty of Regional Environment Science, Tokyo University of Agriculture, 1-1-1, Sakuragaoka,
Setagaya, Tokyo, Japan
e-mail: s4suzuki@nodai.ac.jp

M. Nemoto
e-mail: nemo@affrc.go.jp

R. Imai et al. (eds.), *Plant and Microbe Adaptations to Cold in a Changing World*,
DOI 10.1007/978-1-4614-8253-6_4,

in air temperature, but rather by the early snowfall event, which accumulated large amount of snow on the ground and insulated the soil from the cold air before the soil frost penetrated deeper (Hirota et al. 2006). Inoue et al. (2010) estimated the soil frost depth in the future in this region using a global warming scenario. Based on their results, the annual maximum soil frost depths during 2031–2050 might be shallower than those during 1981–2000 because of the increased snow cover thickness. In contrast, the annual maximum soil frost depths during 2081–2100 might be deeper than those in 2031–2050 because of the remarkable decrease in snow cover thickness caused by the warmer air temperature. These results suggest that the increase or decrease of the soil frost depth was determined not only by the trend of the air temperature but also by the timing and amount of snowfall.

As soil permeability decreases during freezing, the development of soil frost sometimes reduces the amount of snowmelt infiltration and engenders a large amount of runoff (Øygarden 2003). In contrast, a substantial amount of snowmelt water might infiltrate into the ground when the frozen layer is thin (Iwata et al. 2008). Therefore, the decreasing soil frost depth attributable to climate change can be a factor instrumental in increasing snowmelt infiltration and also reducing the runoff during and after the spring snowmelt period. In contrast to the snowmelt period, water movement from deeper soil to the freezing front occurs during the cold winter period because soil frost development results in a very low soil matric potential at the freezing front, which induces a large potential gradient (Miller 1980). This tendency implies that a large potential gradient might be generated in the deep soil layer at the field having thick frozen layer, which implies greater upward soil water flux at the deep soil layer. Therefore, soil water movement during the cold winter might also be influenced by the change of soil frost depth. As dissolved materials in the soil such as nitrate nitrogen move with water, a change of soil water movement also implies a change of soil nutrients.

Despite the possible change of soil water and nutrient movement caused by climate change, few reports describe studies of this phenomenon. In this chapter, we present some results obtained from field studies revealing possible changes in the winter agricultural field, which were possibly induced by climate change.

Study Field

A study was conducted at an experimental field owned by Hokkaido Agricultural Research Center (143°05′E/42°53′W), which is affiliated with the National Agriculture and Food Research Organization. Between 1979 and 2006, the mean annual precipitation was 957 mm. The mean monthly air temperature was −8.7 °C for January and 18.0 °C for July at the Memuro meteorological station, located 2.5 km west of the study site. The study site is covered by volcanic ash soil derived from ash accumulated from 30,000 to 1,000 years ago. Underlying the soil at 1 m is a gravel layer. The water table is generally located approximately 8 m below the ground surface in this region (Oka 2000). The soil is characterized by low bulk density and high permeability. The detailed location and soil properties were described by Iwata et al. (2008, 2010a).

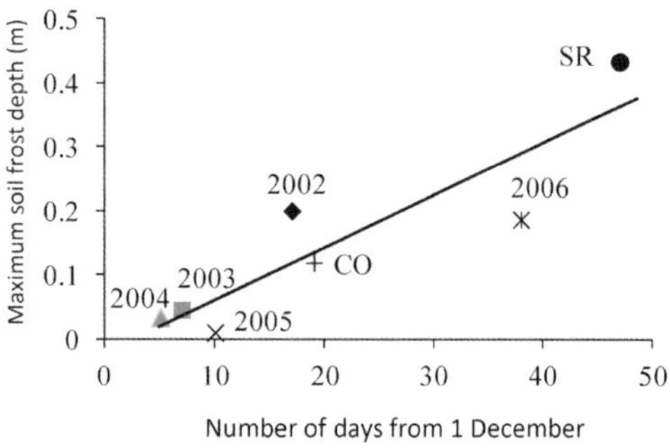

Fig. 1 Relation between annual maximum soil frost depth and number of days from 1 December to the day when the snow cover thickness became greater than 0.2 m

We prepared two experimental sites. One was a long-term observation site, which measured the soil water movement in the natural snow cover condition during July 2002–April 2007. During the observation period, the soil surface was maintained in bare condition using herbicide during summer. The other was a snow-removal experimental site, which measured the soil water and solute movements during November 2005–April 2006. At the snow-removal experimental site, we prepared two study plots and removed snow on one plot to enhance the soil frost depth (SR plot). After the soil frost depth reached the desired value, we replaced the snow to prevent additional penetration of the freezing front. Snow on the other plot was maintained in a natural condition (CO plot). The soil water and nitrate movements in these plots were compared to quantify the influence of the difference of soil frost depth to the water and nutrient cycles during the winter period. Oats were planted before the experiment (i.e., during summer 2005). They were harvested in September. Then soil was plowed with plant residues. The soil surface was left in a bared condition until the end of experiment (i.e., April 2006). A detailed layout of the study plot and methods to quantify the soil water and nitrate movements were described by Iwata et al. (2010a, b, 2013).

Thermal Insulation Effect of Snow Cover

Snow insulates soil from cold air temperatures. Therefore, soil frost depth does not increase after thick snow cover accumulates on the field. Although the threshold of the snow cover thickness preventing additional penetration of the freezing front is generally greater in colder regions, it was reportedly 0.2 m for the Tokachi region (Harada et al. 2009; Hirota et al. 2006). Figure 1 exhibits the relation between the numbers of days from 1 December to the day when snow cover thickness exceeded 0.2 m and the maximum soil frost depths in each winter. Although the air temperatures in early winter were not the same among these winters (e.g., standard deviation of the average air temperature in December during 2002–2006 was ±1.5 °C with the average temperature of −6.5 °C), the relation between them was significant ($P<0.01$; $R=0.90$), which suggests that the timing of the large amount of snowfall is an important factor determining the soil frost depth.

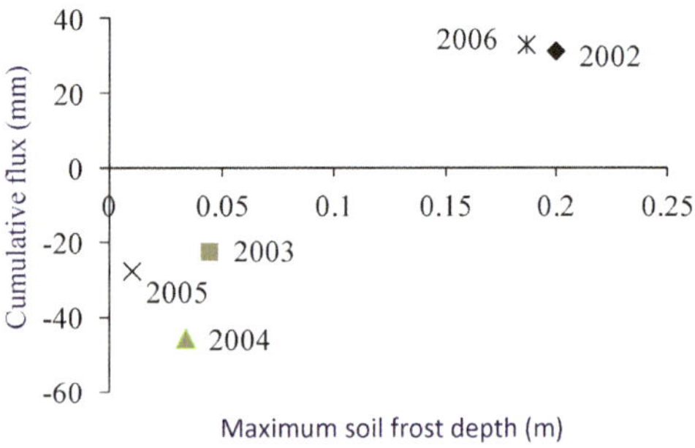

Fig. 2 Cumulative soil water flux at 0.2 m depth during December–February at long-term observation site. The positive and negative values respectively denote upward and downward soil water fluxes. The soil water fluxes during 13–28 February 2003, 21–25 January 2006, 26 December 2006–10 January 2007, and 20–22 January 2007 were not considered because of data gaps

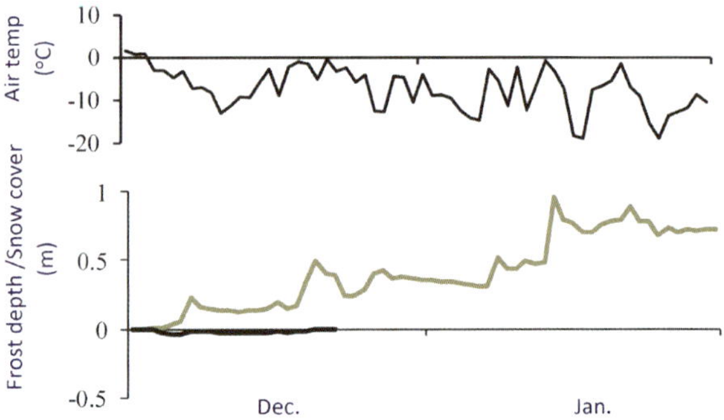

Fig. 3 Time series of air temperature, snow cover thickness (*gray line*), and soil frost depth (*black line*) during December 2003–January 2004

Current Soil Water Movements: Results from the Long-Term Observation Site

At the long-term observation site, the soil frost depth approached 0.2 m in two winters (2002 and 2006 in Fig. 1; hereinafter, "frozen winter"), but it was less than 0.05 m in three winters (2003, 2004, and 2005 in Fig. 1; hereinafter, "unfrozen winter"). In the frozen winters, soil water moved from deep soil layer to the freezing front (Fig. 2) because of the low matric potential at the frozen layer (see the Introduction of this chapter for details). In contrast, soil water moved from the soil surface to the deep soil layer during the cold winter period in unfrozen winters (Fig. 2). In the unfrozen winters, thin frozen layers were formed at the beginning of winter, but they disappeared after the thick snow layer covered the site (Fig. 3) because of the heat supply from the unfrozen soil layer. Reportedly, a substantial amount of snow melted at the base of the snow pack in the thick-snow covered unfrozen fields during the winter including the cold winter period, when the air temperature was lower than 0 °C (e.g., Kojima 1982; Kurashima et al. 1999). This snowmelt water might be a source of the downward soil water fluxes during the unfrozen winter. In both frozen and unfrozen winters, cumulative soil water fluxes during the snowmelt periods were comparable to the amount of snowmelt water (Fig. 4). Despite large amounts of snowmelt infiltration during the spring snowmelt period, Iwata et al. (2010b) showed that the upward cumulative fluxes of the frozen

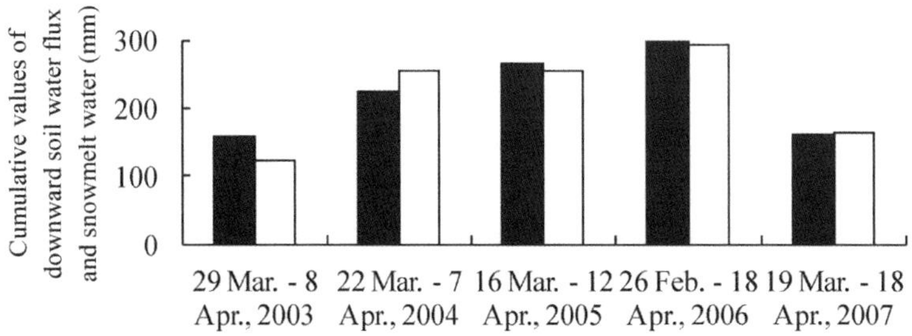

Fig. 4 Cumulative values of downward soil water flux (*solid bars*, positive downward) and snowmelt water (*open bars*) during the spring snowmelt periods at the long-term observation site. Soil water flux at the beginning of the snowmelt period (24–28 March) in 2003 was not considered because of a data gap. For detailed information about the definition of snowmelt period, see Iwata et al. (2010b)

winter and downward cumulative fluxes in the unfrozen winter were 13–25 % and 4–23 %, respectively, of the cumulative downward flux at 0.2 m depth during the snowmelt period, thereby suggesting a substantial amount of soil water movement before the snowmelt.

Soil Water and Nitrate Movement Changes Caused by Climate Change: Results from the Snow-Removal Experimental Site

The maximum soil frost depth of the SR plot became 0.43 m, whereas that of the CO plot was 0.12 m (Fig. 1) because of snow-removal treatment. The average annual maximum soil frost depth during 1986–1996 was 0.38 m based on the dataset from our study field (Hirota et al. 2006). Therefore, the maximum soil frost depth of the SR plot was regarded as the average value of the soil frost depth in the past Tokachi region. Moreover, because of the artificial loading of snow cover, the maximum snow cover thickness of the SR plot was 0.43 m, which was comparable to 0.47 m of the average annual maximum snow cover thickness during 1986–1996 based on the same dataset. We infer that we can reasonably replicate past conditions of soil frost and snow cover at the SR plot.

Soil started freezing on 27 November 2005. We defined the snowmelt period as starting from 26 February 2006, when a 19 mm rain event occurred. During 27 November–25 February, the cumulative upward soil water flux in the SR plot was twice as great as that in the CO plot at 0.4 m depth, which was almost identical to the maximum soil frost depth in the SR plot. The differences were induced mainly during 20 December–13 January, when soil frost depth deepened only at the SR plot (black bars in Fig. 5a). These results imply that a greater amount of soil water moves from the deep soil layer to the freezing front as the freezing front penetrates more deeply. Despite the difference in soil water fluxes, the peak of the nitrate

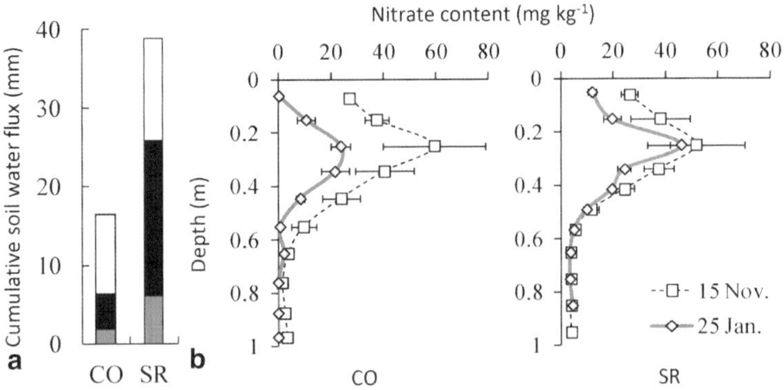

Fig. 5 Soil water flux and vertical distributions of nitrate content during early to mid-winters in the snow-removal experimental site. **a** Cumulative soil water flux at 0.4 m depth during 27 November–19 December, 2005 (*solid gray bars*), when the snow conditions were almost identical between the CO and SR plots, 20 December 2005–13 January 2006 (*solid black bars*), when snow on the SR plot was removed, and 14 January–25 February, 2006 (*open bars*), after the snow was put back on the SR plot. Soil water fluxes are positive upward. **b** Contents of nitrate nitrogen at each soil depth before soil freezing (15 November 2005; *open square*) and in mid-winter (25 January 2006; *solid diamonds*). *Error bars* show the standard error of triplicate soil samples

content was located near 0.25 m depth before and during soil freezing in both the CO and SR plots (Fig. 5b). The amounts of nitrate contents on 25 January were less than those on 15 November, especially in the CO plot. As the amount of rainfall was only 8.5 mm during 15–26 November and because the upward soil water flux occurred after 27 November (Fig. 5a), the nitrate near the soil surface might not be carried to below 1 m. The cold soil temperature in winter also mitigates denitrification in the soil surface layer. Therefore, a smaller amount of nitrate on 25 January was probably induced by the spatial variation of the nitrate content in the study plots. Because of this large spatial variation, it was difficult to discuss fluctuation of the nitrate content from our dataset, especially in the CO plot. However, it might be possible to infer the nitrate movement by the change of the vertical distribution of the nitrate content because the peak depth of the nitrate content might move after a substantial amount of nitrate was carried with soil water. Consequently, the nearly equal depths in both the CO and SR plots on 25 January suggest that remarkable movement of nitrate might not have occurred during 15 November–25 January. These results suggest that much greater cumulative soil water flux might be necessary to induce remarkable movement of soil solutions and that no remarkable difference of nitrate movement might occur during the cold winter period because of climate change in the Tokachi region.

The cumulative soil water fluxes at 0.4 m depth during the snowmelt periods are portrayed in Fig. 6a. The values are negative because downward flux was dominant during the snowmelt periods as a result of the snowmelt infiltration. Because of the artificial loading of snow on the SR plot in mid-January, the snow cover thickness in the SR plot was 0.35 m, which was comparable to that in the CO plot (0.36 m). Results show that snow cover disappeared on 22 March in both plots. Although no

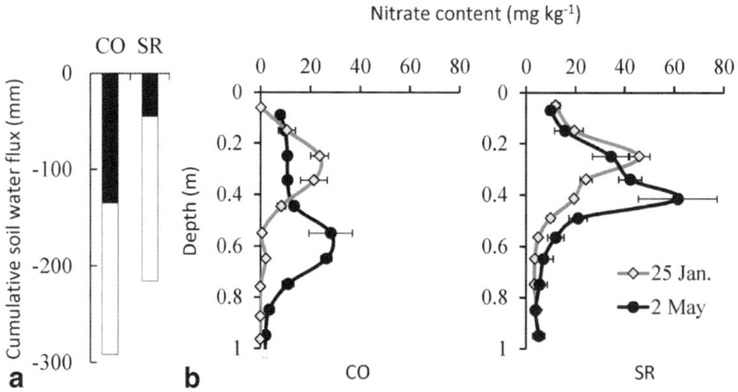

Fig. 6 Soil water flux during spring snowmelt periods and vertical distributions of nitrate content before and after the snowmelt period at the snow-removal experimental site. **a** Cumulative soil water flux at 0.4 m depth during the first snowmelt period (26 February – 22 March, 2006; solid bars) and second snowmelt periods (29 March – 17 April, 2006; open bars). **b** Contents of nitrate nitrogen at each soil depth in mid-winter (25 January 2006; solid gray diamond) and after the snowmelt period (2 May 2006; solid black circle). *Error bars* show the standard error of triplicate soil samples

remarkable snowmelt occurred during 27 February–9 March (Iwata et al. 2010a), we defined the period during 26 February–22 March as the first snowmelt period. The amount of snowmelt water, as calculated using the change of measured snow water equivalent and precipitation during the first snowmelt period, was 136 mm at the CO plot, which was comparable to the amount of downward soil water flux at this plot (solid bar in Fig. 6a). The value is slightly smaller than the 157 mm snowmelt amount during the first snowmelt period at the SR plot. Nevertheless, the amount of downward snowmelt infiltration at the SR plot was only 45 mm, which was less than one-third of the amount of snowmelt water during this period. These results indicate clearly that the thick frozen layer impeded the snowmelt infiltration during the first snowmelt period in the SR plot, whereas most of the snowmelt water infiltrated to deep soil layers during this period in the CO plot.

After the first snowmelt period, a substantial amount of snow fell on 29 March and thawed completely on 17 April in both plots. We defined this period as the second snowmelt period. The amounts of snowmelt water were 142 and 146 mm, respectively, at the CO and SR plots. The cumulative downward flux during the second snowmelt period was 157 mm at the CO plot, which was slightly less than that in the SR plot (open bars in Fig. 6a) probably because of the infiltration by the surface ponding water. However, the total amount of snowmelt water during the first and second snowmelt periods in the SR plot was less than that in the CO plot.

The peak depth of the nitrate content in the SR plot was less than that in the CO plot on 2 May (Fig. 6b). Peak depths of nitrate contents were almost equal between the CO and SR plots on 25 January. Therefore, the difference on 2 May was probably induced by nitrate movement during the first and second snowmelt periods.

Fig. 7 Photographs taken at the past Tokachi region showing surface ponding at the winter wheat field (*left*) and soil erosion caused by runoff (*right*) after the spring snowmelt period

Actually, the amount of the cumulative soil water flux during the first and second snowmelt periods in the CO plot was greater than that in the SR plot, which implies that the larger amounts of snowmelt infiltration induced by thin soil frost depth might carry substantial amounts of nitrate to deeper soil layers in the CO plot. These results imply that remarkable change in nitrate movement might occur during the snowmelt period due to the climate change in the Tokachi region.

Concluding Remarks

The results from the long-term observation site and the CO plot in the snow-removal experimental site indicate that the thin frozen soil layer of less than 0.2 m thickness does not impede snowmelt infiltration. As the soil frost depth did not become greater than 0.2 m in most of the last 15 years in central Tokachi region (e.g., Hirota et al. 2006), a substantial amount of snowmelt water has infiltrated into the deep soil layer in recent years. In contrast, results from the SR plot in the snow-removal experimental site imply that a thick frozen soil layer impeded snowmelt infiltration during the snowmelt period in the past, in the Tokachi region, where the soil frost depth became greater than 0.4 m over the years. The vertical distribution of the nitrate content at this site also suggests that a larger amount of nitrate has infiltrated into the ground in this region in recent years because of the decrease in the soil frost depth resulting from climate change. Some reports have described that a substantial amount of nitrate moved upward with the upward soil water movement generated by the freezing front penetration. We observed a substantial amount of upward soil water movement before the snowmelt period, but no remarkable movement of nitrate nitrogen was observed, even for 0.4 m soil frost depth.

In the past, surface ponding and soil erosion were observed after the snowmelt period for some years in this region (Fig. 7). Our results imply that the drainage of these fields might be improved by the decrease of soil frost depths, whereas the risk of the nitrate leaching to the groundwater might be increased by the climate change in this region.

Because of space limitations, we were not able to present detailed information about the methods, results, and discussion. For more information related to long-term observations, see Iwata et al. (2008, 2010b). Additionally, see Iwata et al. (2010a, 2013) for more information about the snow-removal experimental site.

Acknowledgments We thank Akira Nishimune for photographs. We also appreciate helpful suggestions from Shuichi Hasegawa, Tomotsugu Yazaki, Kazunobu Kuwao, Yosuke Yanai, and Satoshi Inoue. The technical assistance of Masamitsu Fujiwara, Yoshikazu Sato, Norihiro Hirai and other members of the Field Operation Section of National Agricultural Research Center for Hokkaido region is gratefully acknowledged. This study was partly funded by a Global Environment Research Coordination System Grant from the Japanese Ministry of Environment and Global Environment Research Coordination System Grant and the Global Environment Research Fund (A-0807) from the Japanese Ministry of the Environment; by a Research and Development Projects for Application in Promoting New Policy for Agriculture, Forestry and Fisheries (22079) from the Japanese Ministry of Agriculture, Forestry and Fisheries; and by KAKENHI (23380169 and 23221004).

References

Cutforth H, O'Brien EG, Tuchelt J, Rickwood R (2004) Long-term changes in the frost-free season on the Canadian prairies. Can J Plant Sci 84:1085–1091

Frauenfeld OW, Zhang T, Barry RG (2004) Interdecadal changes in seasonal freeze and thaw depths in Russia. J Geophys Res 109(D05101). doi:10.1029/2003JD004245

Groffman PM, Driscoll CT, Fahey TJ, Hardy JP, Fitzhugh RD, Tierney GL (2001) Colder soils in a warmer world: a snow manipulation study in a northern hardwood forest ecosystem. Biogeochemistry 56:135–150

Harada Y, Tsuchiya F, Takeda K, Muneoka T (2009) Characteristics of ground freezing and thawing under snow cover based on long-term observation. Seppyo 71:241–252 (in Japanese with English abstract)

Hirota T, Iwata Y, Hayashi M, Suzuki S, Hamasaki T, Sameshima R, Takayabu I (2006) Decreasing soil frost depth and its relation to climate change in Tokachi, Hokkaido, Japan. JMS J 84:821–833

Inoue S, Hirota T, Iwata Y, Nemoto M (2010) Soil frozen depth prediction in experimental field in Memuro town under global warming. Proceedings of the Cold Region Technology Conference 26:96–100 (in Japanese)

Iwata Y, Hayashi M, Hirota T (2008) Comparison of snowmelt infiltration under different soil-freezing conditions influenced by snow cover. Vadose Zone J 7:79–86

Iwata Y, Hayashi M, Suzuki S, Hirota T, Hasegawa S (2010a) Effects of snow cover on soil freezing, water movement, and snowmelt infiltration: a pared plot experiment. Water Resour Res 46: W09504, doi:10.1029/2009WR008070

Iwata Y, Hirota T, Hayashi M, Suzuki S, Hasegawa S (2010b) Effects of frozen soil and snow cover on cold-season soil water dynamics in Tokachi, Japan. Hydrol Process 24:1755–1765

Iwata Y, Yazaki T, Suzuki S, Hirota T (2013) Water and nitrate movements in an agricultural field having different soil frost depths: field experiments and numerical simulation. A Glaciol 54(63):157–165

Kojima K (1982) Melting at the bottom of a snow cover during the winter in an area with very low temperature and deep snow. Low Temp Sci A41:99–107 (in Japanese with English abstract)

Kreyling J, Henry HAL (2011) Vanishing winters in Germany: soil frost dynamics and snow cover trends, and ecological implication. Clim Res 46:269–276

Kurashima E, Seki M, Kato T, Mukaida Y (1999) Estimation and comparison of bottom snowmelt at an open and a forest site for a snow cover period—a study of bottom snowmelt (2). Trans Jpn Soc Irrigat Drain Reclamat Eng 202:87–94 (in Japanese with English summary)

Lemke P, Ren J, Alley RB, Allison I, Carrasco J, Flato G, Fujii Y, Kaser G, Mote P, Thomas RH, Zhang T (2007) Observations: changes in snow, ice and frozen ground. In: Climate Change 2007: The Physical Science Basis. In: Solomon S, Qin D, Manning M, Chen Z, Marquis M, Averyt KB, Tignor M, Miller HL (eds) Contribution of Working Group I to the Fourth Assessment Report of the Intergovernmental Panel on Climate Change. Cambridge University Press, Cambridge, United Kingdom and New York, New York

Miller RD (1980) Freezing phenomena in soils. In: Hillel D (ed) Applications of soil physics. Academic Press, San Diego, pp 254–299

Oka T (2000) Geological features and its explanation at Middle Tokachi Plain. Geological Survey of Hokkaido, Sapporo, Japan (in Japanese)

Øygarden L (2003) Rill and gully development during an extreme winter runoff event in Norway. Catena 50:217–242

Soil Frost Control: Its Application to Volunteer Potato Management in a Cold Region

Tomoyoshi Hirota, Tomotsugu Yazaki, Kazuei Usuki, Masaki Hayashi, Manabu Nemoto, Yukiyoshi Iwata, Yosuke Yanai, Satoshi Inoue, Takeshi Suzuki, Masaki Shirahata, Tsutomu Kajiyama, Kazuya Araki and Kenji Maezuka

Introduction

Climate change affects agriculture. The development of adaptation strategies is necessary to reduce the loss of agricultural production. In cold regions, the frozen soil depth has been decreasing because of climate warming (IPCC 2007a), presenting important implications for the agricultural environment and practices (IPCC 2007b). Deep soil frost causes (1) a long period of frozen soil with concomitant delay in seed germination, which limits the growing season for plants (e.g., Marcellos and Doyle 1974), (2) a harsh soil environment for overwintering crops (Tsuchiya 1985), (3) reduction of infiltration of snowmelt water, resulting in snowmelt runoff (Bayard et al. 2005; Iwata et al. 2011) and subsequent soil erosion (Nagasawa et al. 1987; Øygarden 2003), and (4) critical triggers for vigorous emissions of nitrous oxide (N_2O) from soils (Yanai et al. 2011). Consequently, deep soil frost has been regarded traditionally as a negative factor affecting agricultural production.

T. Hirota (✉) · M. Nemoto · S. Inoue
Agro-environmental Research Division, NARO Hokkaido Agricultural Research Center, Hitsujigaoka 1, Toyohira-Ku, Sapporo, Hokkaido 062-8555, Japan
e-mail: hirota@affrc.go.jp

T. Yazaki · K. Usuki
Upland-farming Research Division, NARO Hokkaido Agricultural Research Center, Shinsei Minami, 9-4, Memuro, Hokkaido 082-0081, Japan
e-mail: yazaki@wetlands.jp

M. Hayashi
Department of Geoscience, University of Calgary, 2500 University Dr. NW, Calgary, Alberta T2N 1N4, Canada
e-mail: hayashi@ucalgary.ca

Y. Yanai
NARO Institute of Vegetable and Tea Science, 3-1-1 Kanondai, Tsukuba, Ibaraki 305-8609, Japan
e-mail: yosukey@affrc.go.jp

R. Imai et al. (eds.), *Plant and Microbe Adaptations to Cold in a Changing World*,
DOI 10.1007/978-1-4614-8253-6_5,

According to IPCC-AR4, agricultural productivity in cold regions is expected to increase because of global warming. However, climate change often engenders unexpected problems related to agricultural production. In the Tokachi region, Hokkaido, northern Japan, soil frost depth has been decreasing from the late 1980s, facilitating winter survival of small potato (*Solanum tuberosum* L.) tubers that remain unharvested and which emerge as weeds during the subsequent cropping season. Such potatoes are designated as volunteer potatoes. To kill such volunteer potatoes, snow plowing (*yukiwari* in Japanese), which removes snow cover and allows deep soil frost, is spreading as a practical countermeasure in the Tokachi region. However, the technique has been largely empirical, and the level of expertise varies among farmers. The potato tubers survive the winter if the soil frost is too shallow. Contrarily, if the soil frost is too deep, it delays spring agricultural practices and crop growth. To avoid these adverse effects, we developed a soil frost control method by manipulating snowcover thickness using a numerical soil temperature model. Here, we present (1) documentation of the recent reduction in soil frost depths in the Tokachi region associated with the East Asian winter monsoon, (2) its influence on agriculture such as volunteer potatoes, and (3) our developed method of soil frost control and its application to volunteer potato management.

T. Suzuki · M. Shirahata
Hokkaido Research Organization Central Agricultural Experiment Station,
Higashi-6, Kita-15, Naganuma, Hokkaido 069-1395, Japan
e-mail: suzuki-takeshi@hro.or.jp

T. Kajiyama
Agriculture Research Department, Hokkaido Research Organization Tokachi Agricultural Experiment Station, Shinsei Minami, 9-2, Memuro, Hokkaido 082-0081, Japan
e-mail: kajiyama-tsutomu@hro.or.jp

K. Araki
Hokkaido Research Organization Tokachi Agricultural Experiment Station,
Shinsei Minami, 9-2, Memuro, Hokkaido 082-0081, Japan
e-mail: araki-kazuya@hro.or.jp

K. Maezuka
Tokachi Federation of Agricultural Cooperative Associations,
14, Nishi-3, Minami-7, Obihiro, Hokkaido 080-0013, Japan
e-mail: k_maezuka@nkrtwosv.nokyoren.or.jp

M. Nemoto
e-mail: nemo@affrc.go.jp

S. Inoue
e-mail: ino@affrc.go.jp

K. Usuki
e-mail: kazuei@affrc.go.jp

M. Shirahata
e-mail: sirahata-masaki@hro.or.jp

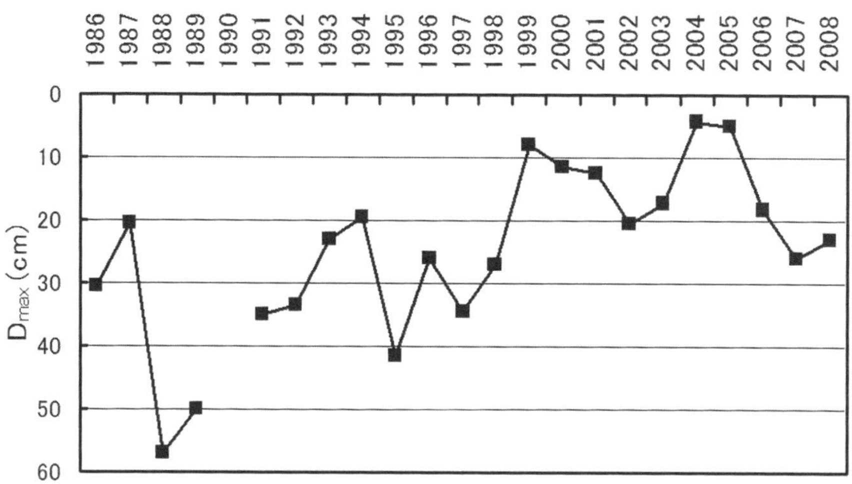

Fig. 1 Annual maximum soil frost depth (D_{max}) at Hokkaido Agricultural Research Center in Memuro, Tokachi region of Hokkaido, Japan

Recent Reduction in Soil Frost Depth in the Tokachi Region Associated with the Eastern Asian Winter Monsoon

Winter climate in Japan changed greatly during the late 1980s. For example, in the Tokachi region, Hokkaido, northern Japan, the soil frost depth decreased remarkably compared with that before the 1980s. Figure 1 shows the annual maximum soil frost depth (D_{max}) at the NARO Hokkaido Agricultural Research Center in Memuro, of the central part of the Tokachi region. The value of D_{max} has decreased drastically during the last 20 years. The D_{max} in 1988 was 0.57 m, but it was only 0.04 m in 2004 and 0.05 m in 2005. From 1998 to the present, D_{max} has remained less than 0.30 m. The soil frost in 2004 and 2005 can be regarded as fundamentally nonexistent from an agricultural perspective because such a shallow frost has a minimal effect on crops. The soil becomes unfrozen by the time of snowmelt. During 1986–1997, the average D_{max} was 0.34 m for the first half period. However, during 1998–2008, the average D_{max} was 0.16 m for the second half. The trend of decrease of D_{max} was significant ($P<0.05$). The decrease in D_{max} resulted from shortened time periods available for freezing, rather than by increasing in air temperature (Hirota et al. 2006). The D_{max} is correlated strongly with soil freezing index F_{20}, which is the cumulative freezing degree-days during which the snowcover thickness was less than 0.20 m. It can be calculated from the daily mean air temperature and snowcover thickness (Fukuda 1982; Tsuchiya 1985). To expand the spatial scale of our analysis using F_{20} as a surrogate of D_{max}, results showed that the decreasing soil frost depth was a regional phenomenon occurring over the Tokachi Plain covering an area of several thousand square kilometers. The long-term trend of the F_{20} for 1955–2005 for Obihiro in Tokachi has been one of decrease since the late 1980s (Fig. 2).

The early onset of snowcover is most likely the result of extratropical cyclones, which tend to develop during weakening of typical winter synoptic patterns characterized by high pressure over Siberia (Siberian high) and low pressure over the Pacific (Aleutian low). The strength of the East Asian winter monsoon (i.e., Siberian high) decreased in the late 1980s. Cyclonic activity in early winter was stronger when the winter monsoon activity was weaker. The strength of the winter monsoon activity is reflected by the amount of snow in the Hokuriku region facing the Sea of Japan, where snowfall is caused primarily by monsoon winds, and by the amount of sea ice in the southern part of the Sea of Okhotsk, which is controlled by the strength of cold monsoon winds coming from Siberia (Tachibana et al. 1996). These two indicators decreased markedly and appeared to shift from high values to low values during the late 1980s (Fig. 2) corresponding to weakening of the winter monsoon activity (Nakamura et al. 2002). The shift in the two indicators coincided with the decrease in D_{max} at Obihiro (Fig. 2). Reduction of soil frost depth started in the late 1980s, coinciding with reduction in snow accumulation depth in the plains area of Hokuriku region and the reduction in drift ice sighting along the Okhotsk Sea coast. Therefore, it is likely that reduction in the soil frost depth since the late 1980s is related to the increased frequency of extratropical cyclones associated with weakening of the East Asian winter monsoon.

Effect of Soil Frost Decreasing on the Volunteer Potatoes Problem

The reduction in soil frost depth means a milder winter condition and warmer soil, which also has some positive effects on agricultural production, for example, positive effects to overwintering crops and extended crop growing season. However, it also has negative effects on agricultural environments. In the Tokachi region, a major potato-producing area in Japan, potato producers have recently noticed a remarkable change in the field. With reduction of the soil frost depth, many volunteer potatoes, which are the numerous small tubers that were not harvested the previous year, abruptly shoot up as weeds in the current year (Fig. 3). The soil used to freeze to a sufficient depth to destroy the left-over tubers, but shallow frost in recent years has allowed some tubers to survive the winter. Volunteer potatoes (1) are weeds that interfere with the growth of current year's rotation crops, (2) can cause pest problems by attracting aphids, transferring viruses, and hosting potato cyst nematode, and (3) cause undesired mixing when two different breeds of potato are planted in consecutive years, which might present a risk in shipping and trading. Therefore, it is necessary to weed out volunteer potatoes. The farm operators have been forced to engage in manual labor of weeding, which requires several tens of hours for each hectare, representing a severe work load for producers in the region, where average farms growing potatoes are larger than several hectares.

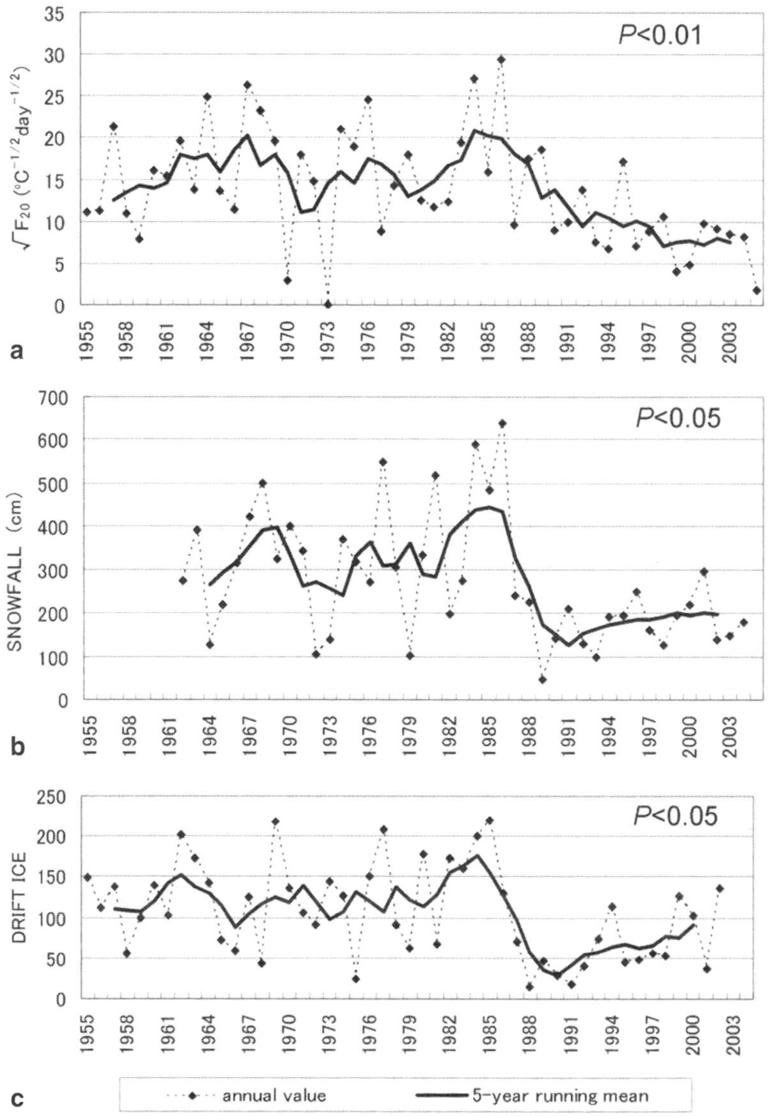

Fig. 2 Long-term changes in **a** modified soil freezing index (F_{20}) and its 5-year running mean at Obihiro, **b** mean total winter accumulated depth of new snowfall (SNOWFALL) and its 5-year running mean in the Hokuriku region facing the Sea of Japan, calculated from the data collected at Fukui, Tsuruga, Wajima, Aikawa, and Takada, **c** an index of drift ice in sight (DRIFT ICE) on the coastal areas facing the Sea of Okhotsk and its 5-year running mean, calculated from the data collected at Esashi, Ohmu, Mombetsu, and Abashiri. The index is the ratio of the observed value to the average over the entire observation period. The average value is set at 100. Recent data were provided courtesy the Office of the Marine Prediction of the JMA. The significance level (P) is based on the Mann–Kendall statistic (Hirota et al. 2006)

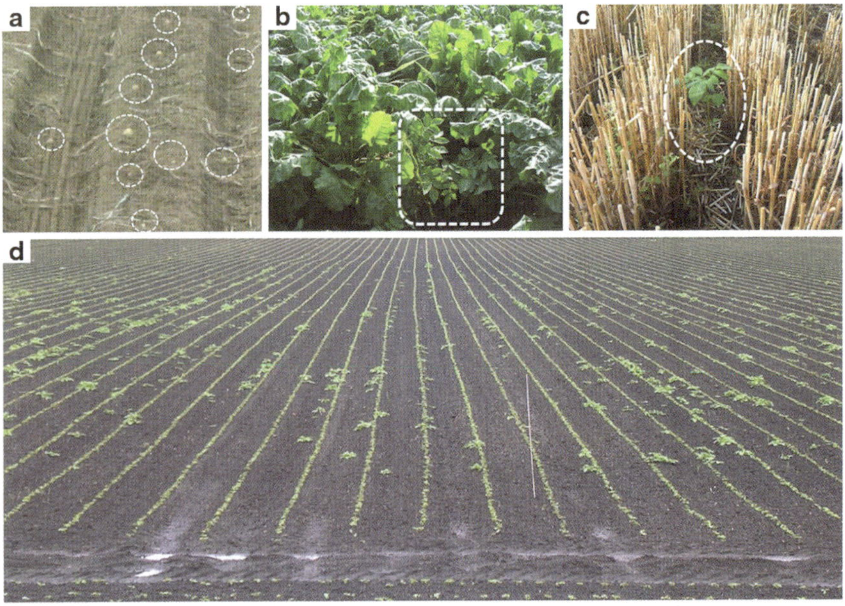

Fig. 3 Photographs showing **a** potatoes left after harvest, **b** volunteer potatoes in a sugar beet field, **c** volunteer potatoes in a wheat field and **d** volunteer potatoes in a bean field (Hirota et al. 2011)

Development of Soil Frost Control Method for Volunteer Potato Management

Some local potato producers in the Tokachi region have devised countermeasures for changing conditions. They have recently started snow plowing (*yukiwari*), which temporarily removes snowcover to allow the frost to penetrate deeper and consequently kill volunteer potato tubers. They can enhance soil frost depth even in a large field by *yukiwari* using machines such as a plow-mounted tractor or a bulldozer. The *yukiwari* consists of the following procedures (Fig. 4):

- Step 1: This step is called "the first-half operation of *yukiwari*." Snow is plowed in a line, leaving snow berms on both sides. Soil frost develops in the exposed part. When the exposed part is snow-covered, the snow is plowed in the line repeatedly.
- Step 2: This step is called "the second-half operation of *yukiwari*." After soil frost reaches the optimal depth at the exposed part in the step 1, the snow berms are subsequently plowed to cover the previously exposed part and expose the soil under the berms. When the exposed part in the Step 2 is snow-covered, the snow is plowed repeatedly.

The *yukiwari* must be conducted at least twice to secure soil frost entirely in the field.

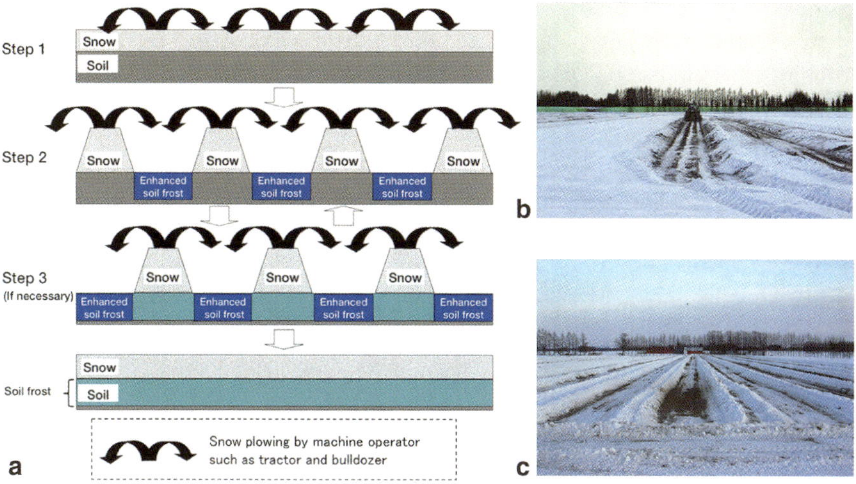

Fig. 4 Schematic diagram of **a** the operation sequence of *yukiwari* (snow plowing) and photographs of **b** a tractor plowing snow and **c** a field after plowing (Hirota et al. 2011)

Although the method appears to be a breakthrough supporting the confrontation of environmental change, this method of weeding is not stably effective because it is based on trial-and-error, relying on the producers' experience and intuition. The potato tubers survive the winter if the soil frost is too shallow. Contrarily, if the soil frost is too deep, it delays spring agricultural practices and crop growth.

To achieve stable outcomes while avoiding adverse effects associated with the delay in spring practices or plant growths, the objective soil frost depth must be determined. Balancing (i.e., optimizing) the two mutually exclusive objectives: (1) allows deep sufficient penetration to kill the potato tubers, and (2) keeps the soil frost depth sufficiently shallow so that it does not cause a delay in seeding or growth of overwintering crops. To achieve soil frost depth optimization, we developed a soil frost control method.

Soil frost depth can be controlled by managing the snowcover thickness, which involves adjustment of the timing and duration of thick snowcover that insulates the soil surface during cold periods (Hirota et al. 2011). *Yukiwari* controls soil frost depth efficiently by enhancing and preventing the penetration of soil frost simultaneously. The timing of *yukiwari* operations between step 1 and step 2 is critical. To achieve optimal soil frost depth, the soil frost (soil temperature) calculation model plays an important role in determining the timing and duration of each step of the *yukiwari*. Our developed soil-temperature model (Hirota et al. 2002) is useful because it can predict the maximum soil frost depth with accuracy of several centimeters using only daily mean air temperature and snowcover thickness as input data.

To determine the optimized soil frost depth, it is crucial to obtain the critical temperature necessary to kill potato tubers. Some reports of previous studies have described that the critical temperature must be below −3 °C to kill potato tubers (Boydston et al. 2006; Hirota et al. 2011; Li et al. 1981). Then, combining the critical temperature and the estimated soil temperature profile, a new indicator of the critical depth to kill the volunteer potatoes was introduced. The critical depth provides information related to the volunteer potato conditions in fields. Based on field observations of tuber survival rates, the optimal soil frost depth of 0.3–0.4 m is proposed as a compromise between elimination of volunteer potatoes and minimum soil frost depth to prevent negative effects on agriculture (Hirota et al. 2011). This agrees with the results of field observation showing that almost all potato tubers were killed when soil frost depth became greater than 0.3 m in the Tokachi region (Yazaki et al. 2012). In winter, soil temperatures decreased at shallower depths. Ideally, potato tubers should remain at a shallow depth of less than several centimeters.

Work Schedule *Yukiwari*

To diffuse the technique of weeding volunteer potatoes by soil frost control, failure in weeding or adverse effects of excessive soil frost formation must be minimized. To attain stable effects using soil frost control, a guideline that facilitates work decision-making is required. Therefore, we obtained the final limit of the work *yukiwari* to attain the target soil frost depth of 0.3 m for the Tokachi region, with interannual fluctuations of winter conditions considered.

We estimated soil frost depths with daily mean air temperature and daily maximum snowcover thickness from November through April, based on the numerical soil-temperature model (Hirota et al. 2002). These input data were obtained from 11 meteorological stations in Tokachi (JMA 2012). The final limit date of the second half-phase of *yukiwari* (Step 2 in Fig. 4) in each year was determined from the final date on which the soil frost depth reaches 0.3 m, assuming that the snowcover thickness after *yukiwari* is 0.02 m until the end of continuous snowcover. For example, in the case of Memuro in 2011 winter, the work of *yukiwari* on February 6 attains 0.3 m soil frost, but the work on February 7 does not. In that case, the final limit date in the 2011 winter is determined as February 6. To obtain the final limit of *yukiwari* even under extremely warm winter conditions, the date in a 30-year return period was calculated by application of the final limit dates for 23 years to a Weibull distribution (Yamada et al. 1998).

In the Tokachi region, the soil frost depth reached 0.30 m by *yukiwari* in most areas except Hiroo on the coast in the southernmost area (Fig. 5). The final limit in a 30-year return period was estimated as around 20 February in mountainous areas in the north and from late January to early February in other areas (Fig. 5). The use of a numerical model for estimating soil temperature facilitates decision-making related to the *yukiwari* work schedule to minimize damage from volunteer potatoes as weeds.

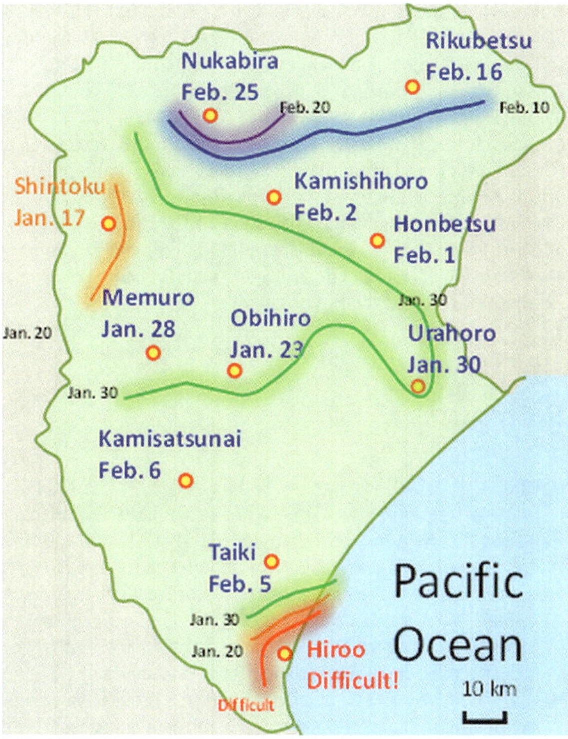

Fig. 5 Final limit of *yukiwari* in a return period of 30 years for the Tokachi region. (Redrawn based on Yazaki et al. 2012)

Concluding Remarks

The motivation for this study was a recent discovery in the northern region of Japan that the soil frost depth had decreased during the past 20 years, thereby allowing unharvested potato tubers to remain unfrozen and survive over winter. These volunteer potatoes pose a severe problem to potato producers in the region. The soil frost depth reduction results from the earlier onset of thick snowcover in recent years, which insulates the ground before the air temperature becomes sufficiently low to freeze the soil.

This study examines the critical soil temperature and soil frost depth that are necessary to kill the potato tubers, and introduces an innovative approach to controlling the soil frost depth by manipulating the snowcover thickness guided by a simple numerical soil–temperature model. We presented the timely development of adaptation technology (and its scientific basis) that is being adopted by agricultural producers in the region. The technology has potential applicability to other aspects of environmental control (e.g., soil water and nutrient transport and greenhouse gas emissions) in addition to potato-tuber removal.

Acknowledgments We thank K. Bitou, T. Taraba, T. Kawai, K. Nishida, N. Miki, T. Michiba, T. Komatsu, K. Kikuchi, and A. Sawazaki for providing information and discussion related to soil frost, volunteer potatoes, and *yukiwari* in the Tokachi region. We also thank T. Hamasaki, R. Sameshima, S. Suzuki, N. Koga, O. Nagata, S. Hasegawa, S. Yokota, and N. Matsumoto for helpful suggestions. We acknowledge M. Fujiwara, K. Shibata, H. Ogawa, S. Takahashi, T. Hirao, M. Kikuchi, S. Souma, E. Takasugi, H. Tokuji, M. Yamaoka, Y. Yamakawa, F. Yagasaki, and M. Hirota for their assistance. The study was funded as projects of Research and Development Project for Application in Promoting New Policy for Agriculture, Forestry and Fisheries (22079) and Evaluation, Adaptation and Mitigation of Global Warming (41230) from the Japanese Ministry of Agriculture, Forestry and Fisheries and of Global Environment Research Coordination System grant and the Environment Research and Technology Development Fund (A-0807) from the Japanese Ministry of the Environment.

References

Bayard D, Stähli M, Parriaux A, Flühler H (2005) The influence of seasonally frozen soil on the snowmelt runoff at two Alpine sites in southern Switzerland. J Hydrol 309:66–84

Boydston RA, Seymour MD, Brown CR, Alva AK (2006) Freezing behavior of Potato (Solanum tuberosum) tubers in soil. Am J Potato Res 83:306–315

Fukuda M (1982) The estimation method of frost penetration depth under snow packs. In: Wakahama G (ed) Research reports of grant-in-aid for special project of Hokkaido University in study on snow distribution and characteristics and its effect on frozen soil depth, vegetation and insect ecology in Hokkaido, pp 55–74 (in Japanese)

Hirota T, Pomeroy JW, Granger RJ, Maule CP (2002). An extension of the force-restore method to estimating soil temperature at depth and evaluation for frozen soils under snow. J Geophys Res 107(D24):4767. doi:10.1029/2001JD001280

Hirota T, Iwata Y, Hayashi M, Suzuki S, Hamasaki T, Sameshima R, Takayabu I (2006) Decreasing soil-frost depth and its relation to climate change in Tokachi, Hokkaido, Japan. J Meteorol Soc Jpn 84:821–833

Hirota T, Usuki K, Hayashi M, Nemoto M, Iwata Y, Yanai Y, Yazaki T, Inoue S (2011) Soil frost control: agricultural adaptation to climate variability in a cold region of Japan. Mitig Adapt Strategies Glob Chang 16:791–802

Intergovernmental Panel on Climate Change (IPCC) (2007a) Climate change 2007: the physical science basis. Contribution of working group I to the fourth assessment. In: Solomon S, Qin D, Manning M et al. (eds) Report of the intergovernmental panel on climate change. Cambridge University Press, Cambridge

IPCC (2007b) Climate Change 2007: impacts, adaptation and vulnerability. Contribution of working group II to the fourth assessment. In: Parry ML, Canziani OF, Palutikof JP et al. (eds) Report of the intergovernmental panel on climate change. Cambridge, Cambridge University Press

Iwata Y, Nemoto M, Hasegawa S, Yanai Y, Kuwao K, Hirota T (2011) Influence of rain, air temperature, and snow cover on subsequent spring-snowmelt infiltration into thin frozen soil layer in northern Japan. J Hydrol 401:165–176

Japanese Meteorological Agency (JMA) (2012) Source of the archived meteorological data http://www.data.jma.go.jp/obd/stats/etrn/index.php. Accessed on 4 January 2012

Li PH, Huner NPA, Toivio-Kinnucan M, Chen HH, Palta JP (1981) Potato freezing injury and survival, and their relationship to other stress. Am J Potato Res 58:15–29

Marcellos H, Doyle AD (1974) Wheat yields—sowing time important on northwest slopes. Agric Gaz N.S.W. 85:14–15

Nagasawa T, Umeda Y, Mizutani T (1987) Soil erosion in the thawing period on agricultural sloping land of snowy, cold region. Mem Res Fac Agric, Hokkaido Univ 15:282–292. (in Japanese with English summary)

Nakamura H, Izumi T, Sampe T (2002) Interannual and decadal modulations recently observed in the Pacific storm track activity and east Asian winter monsoon. J Clim 15:1855–1874
Øygarden L (2003) Rill and gully development during an extreme winter runoff event in Norway. Catena 50:217–242
Tachibana Y, Honda M, Takeuchi K (1996) The abrupt decrease of the sea ice over the southern part of the Sea of Okhotsk in 1989 and its relation to the recent weakening of the Aleutian low. JMSJ 74:579–584
Tsuchiya F (1985) Effect of soil frost and frost penetration on volcanic ash soil in agricultural field in Tokachi, eastern Hokkaido, Japan Dissertation, Hokkaido University, (in Japanese)
Yamada K, Hirota T, Shirooka T (1998) Analyses of continuous snow cover duration and its first and last days using Weibull distribution method. J Agric Meteorol 54:63–69. (in Japanese with English abstract)
Yanai Y, Hirota T, Iwata Y, Nemoto M, Nagata O, Koga N (2011) Accumulation of nitrous oxide and depletion of oxygen in seasonally frozen soils in northern Japan—Snow cover manipulation experiments. Soil Biol Biochem 43:1779–1786
Yazaki T, Hirota T, Suzuki T, Shirahata M, Iwata Y, Inoue S, Usuki K (2012) Work schedule of *yukiwari* (snow plowing) to weed volunteer potatoes by soil-frost control using a numerical model. Clim Bios 12:12–20. (in Japanese with English abstract)

Climatic and Physiological Background of Ice Encasement Damage of Herbage Plants

Bjarni E. Gudleifsson

Introduction

Iceland is situated in the North Atlantic Ocean and was for a long time uninhabited until first settled by Vikings around the year 900 AC. It has been calculated that there were about 40,000 inhabitants at the beginning of the settlement. In the following centuries the number of inhabitants varied between 40,000 and 60,000 depending on the available food supply, including feed for their livestock, which seems to have been closely related to the temperature (Haraldsson and Ólafsdóttir 2007). The Vikings had brought their livestock with them across the ocean, including horses, cattle, and sheep. Due to the harsh winters they had to collect winter feed for the sheep every year. Sheep products have since then dominated the human food consumed in Iceland. The number of inhabitants depended on the number of sheep that survived every year, which on the other hand depended on the available feed, primarily hay from the pastures. Hardship caused by epidemics and volcanic eruptions, coupled with the severe cold climate, decreased the number of inhabitants down to a mere 20,000 around 1400 and 1700 (Finnsson 1970). Thus winter damage of pastures periodically led to a decrease in the population.

Long-Term Temperature Changes

The annual temperatures in Iceland from the time of settlement have been estimated from written annals and by analyzing ice cores sampled from glaciers in Iceland and Greenland (Bergthorsson 1969; Bergþórsson 2003; Gudleifsson 2010; Fig. 1). These estimates indicate that the temperature around the time of the settlement of Iceland, from about 900–1200, was fairly high, but after that the tempera-

B. E. Gudleifsson (✉)
Faculty of Environmental Sciences, Agricultural Univesity of Iceland,
Bugardur, Oseyri 2, 601 Akureyri, Iceland
e-mail: beg@lbhi.is

R. Imai et al. (eds.), *Plant and Microbe Adaptations to Cold in a Changing World*,
DOI 10.1007/978-1-4614-8253-6_6, © Springer Science+Business Media New York 2013

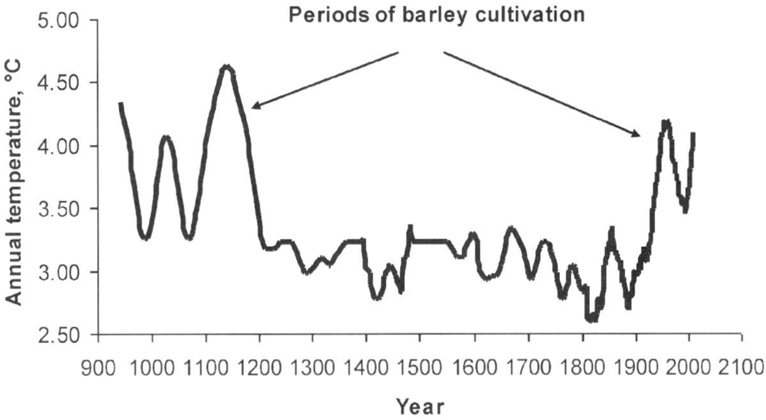

Fig. 1 Estimated annual temperature in Iceland since settlement

ture dropped, though with some fluctuations, until 1900, when it started to increase slowly (Gudleifsson 2010). Today the temperature is similar to the temperature at the time of the settlement, and these are the only two periods in the history of Iceland when barley cultivation has been possible (Fig. 1). During the cold ages, 1200–1900, winter damage to pasture grass was fairly frequent, and according to the annals, 36 % of the years were characterized as having mild or severe winter damage of pastures (Finnsson 1970; Friðriksson 1954). The failure to harvest and store enough feed for the winter could lead to a substantial reduction in the number of livestock and to human famine. As a result of limited grass growth the number of inhabitants never exceeded 60,000 before the year 1850 (Haraldsson and Ólafs-dóttir 2007).

Reliable temperature measurements in Iceland are available for the last century, 1900–2010. In addition, the hay yield was registered on each farm in the same period. Figure 2 shows the mean annual temperature at one specific meteorological station in western Iceland, Stykkishólmur, and the mean annual hay yield for the whole country. It should be stressed that winter damage usually only hit limited areas of the country, whereas the hay yield was collected for the whole country, also from areas without winter damage. Figure 2 clearly demonstrates that a low annual temperature was frequently accompanied by reduced hay yield.

Winter Stress

During winter pasture grasses are subject to many kinds of stresses, some of which may be lethal and result in plant death. The winter stresses may be frost, frost heaving, snow cover, snow molds, desiccation, photoinhibition, flooding, and ice encase-

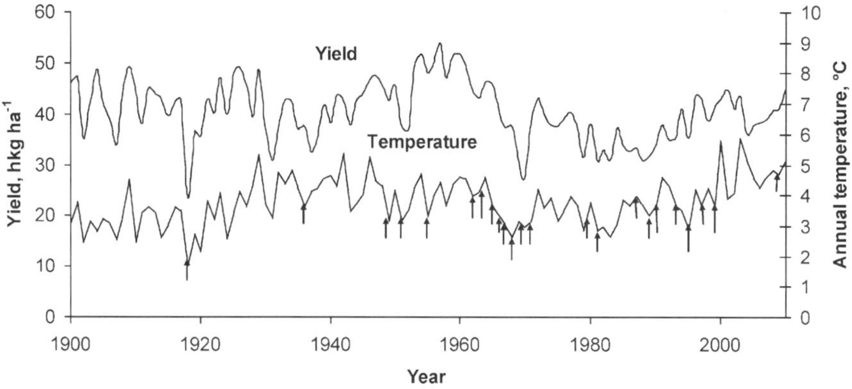

Fig. 2 Annual hay yield in Iceland and mean annual temperature in Stykkishólmur 1900–2010. *Arrows* indicate years with great and extensive winter damage. (Adapted from Gudleifsson 2009)

ment (Griffith et al. 2001). Frequently many of the stresses may appear together, but usually one stress type dominates and is regarded as the main cause of plant death.

Higher temperatures, mainly in the growing season, result in higher yields, as expected. The importance of winter damage on the yield and economy of Icelandic farmers is visualized in the surprising fact that the mean annual hay yield is more closely related to winter temperature than summer temperature (Gudleifsson 1975; Bergthorsson et al. 1987; Gudleifsson 2009). For the period 1900–2010 the correlation coefficients between hay yield and temperature were as follows ($P<0.001$):

Annual temperature (September–August) 0.49***
Summer temperature (June–August) 0.28***
Winter temperature (December–May) 0.57***

Studies on winter killing in Iceland during the last century revealed that direct low temperatures, freezing, were not the main cause of grass death. Ice encasement has been the dominant type of winter damage (Gudleifsson 2009), during which plants are killed under a long-lasting ice sheet. This kind of ice cover is usually formed after a long or short thaw followed by frost. If not only the temperature but also the amount of meltwater and thickness and endurance of the ice cover are taken into consideration, the correlation coefficient between climate parameters and intensity of winter damage increases to 0.70***.

Ice Encasement

Ice encasement damage is most common in northern maritime areas with high winter precipitation and unstable winter temperatures fluctuating around the freezing point. Iceland, the coast of Northern Norway, the east coast of Canada, and

Hokkaido in Japan are examples of areas with that kind of climate, where ice encasement damage has been frequently reported (Andrews et al. 1986; Årsvoll 1973; Friðriksson 1954; Hakamata et al. 1978). The process of ice formation may be as follows (Gudleifsson and Larsen 1993):

Autumn: high precipitation and soil saturation.
Early winter: freezing of soil and accumulation of snow.
Mid-winter: thaw periods with high precipitation followed by freezing of the melt water and ice formation.
Late winter: snow accumulation and possible short thaw periods adding to the ice cover.
Spring: low temperatures prolonging the encasement period followed by standing water or freezing of already damaged plants.

The duration of compact ice cover is critical (Huokuna 1959). Short time encasement may be harmless, but the longer the ice cover lasts the more severe the damage. In Iceland most of the pastures are permanent hayfields composed of a mixture of plant species. The less tolerant plants are killed first and the harder ones survive a longer duration of ice. In Iceland and Finland the rule of thumb is that after 3 months of ice cover most grass plants are killed (Gudleifsson and Larsen 1993; Huokuna 1959).

The topography of the field is crucial. Sloping fields can escape damage as the meltwater runs off of the field before it refreezes, whereas the water stays on flat fields and accumulates especially in depressions. Therefore, the microtopography is very important (Hakamata et al. 1978) and in some cases decisive for the intensity of damage, as illustrated in Fig. 3. When an ice cover melts, the higher parts of the fields escape first from the ice, whereas the ice in the depressions stays longer and this prolonged encasement can become lethal.

Climate Change

The ice tolerance of plants has been measured in field experiments. In the field it is difficult to control the stress because of small microtopographical differences and unpredictable weather. One strong thaw period in the middle of winter will destroy the test. Therefore plants have mainly been tested in the laboratory by encasing them completely in ice for different lengths of time (Gudleifsson et al. 1986). Because of complete ice inclusion plants survive substantially a shorter encasement in the laboratory than in the field (Table 1). Perennial herbage plant species have very variable ice tolerance. Most grass species, such as tufted hairgrass and timothy, are rather tolerant and can tolerate 11–12 weeks of ice in the field. Legumes (white and red clover) can tolerate 3–4 weeks of encasement, whereas *Brassica* species (winter rape and stubble turnip) can tolerate 2–3 weeks, and winter cereals (rye and wheat) tolerate only 1–2 weeks of ice cover (Table 1).

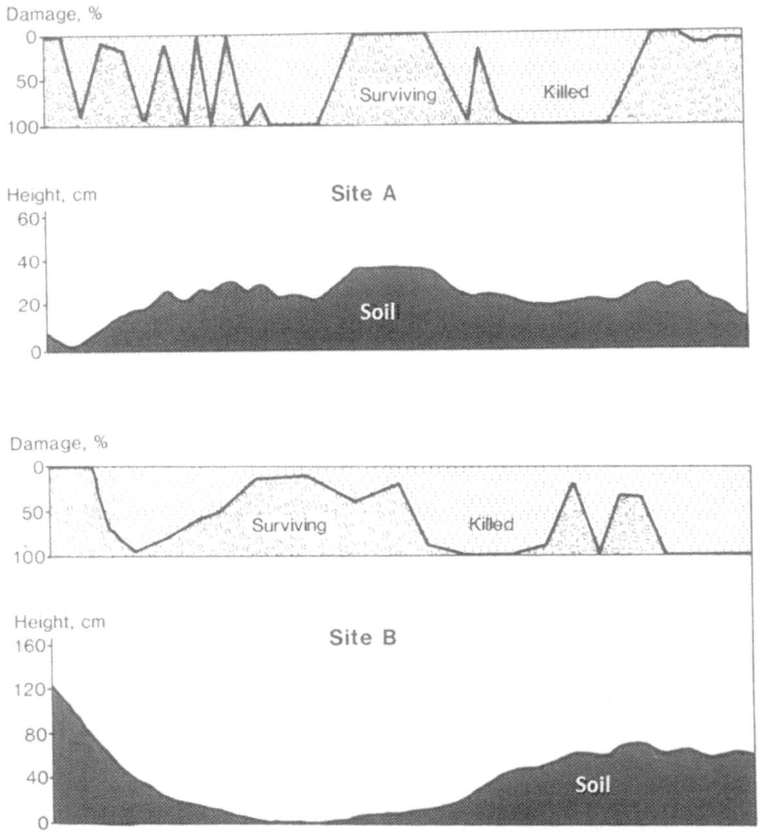

Fig. 3 Impact of microtopography on distribution of ice encasement damage on two hayfields in Iceland. Damage dominates in depressions where ice accumulates, but in the deepest depression (site B), the thick snow layer has not melted and no damage occurs. (Gudleifsson and Larsen 1993)

The climate in Iceland has changed substantially in recent decades. The mean annual temperature has increased (Fig. 2) and the frequency of damage to hayfields caused by ice encasement has decreased (Gudleifsson 2010), most likely the result of the already ongoing global warming. The increase in mean annual temperature in Iceland to the year 2100 is predicted to be 1.9 °C. The temperature rise during summer is predicted to be 1.4 °C and 2.5 °C during winter (Umhverfisráðuneytið 2008). These temperature increases, especially during winter, will probably decrease or even eliminate ice encasement damage on hayfields in the near future. These hayfields are in the lowland agricultural areas. On the other hand, this kind of damage may become a threat to alpine plants at higher elevations (Gudleifsson 2009). The higher winter temperatures may increase the snow melt and the frequency of formations of ice sheets in alpine areas. These alpine plants have, through the ages, been protected during winter by long-lasting snow cover and have not developed ice

Table 1 Frost and ice tolerance in laboratory tests of species representing four plant groups. (Gudleifsson 2009, 2010)

Plant species	Latin name	Frost tolerance, LT_{50} (°C)	Ice tolerance, LD_{50} (days)	Ice tolerance[a] (days)
Alpine arctic chuckweed	*Omalotheca supina* (L.) Cand.	−19	7	?
Winter wheat	*Triticum aestivum* L.	−18	14	7–14
Red clover	*Trifolium pratense* L.	−15	20	21–28
Timothy grass	*Phleum pratense* L.	−13	33	77–85

[a] Ice tolerance experienced in fields in Iceland, *rule of thumb*

encasement tolerance and might therefore in the future become damaged or killed by ice cover. As seen in Table 1 alpine arctic chuckweed, an alpine snow bed plant, has not developed high ice tolerance.

Accumulation of Metabolites

Because ice is a good heat conductor and poor insulator, the soil temperatures under ice cover can decrease close to the lethal point for the plant, but snow cover which often is on the top of the ice will greatly reduce freezing stress. The temperature under ice insulated by snow often fluctuates around 0 °C (Baadshaug 1973; Ouellet and Desjardins 1981). It cannot be excluded that plants under ice cover can finally be killed by frost and it has been demonstrated that the damage caused by ice cover increases with decreasing temperature (Beard and Martin 1970). However, the damage caused by ice cover is not primarily related to freezing or mechanical damage of the plants caused by the ice (Gudleifsson and Larsen 1993). It has been demonstrated that the real cause of death is accumulation of metabolites produced by anaerobic respiration (Pomeroy and Andrews 1986; Andrews and Pomeroy 1989). The main metabolites produced by winter cereals and grasses in anaerobic respiration under ice are CO_2, ethanol, and lactate and some other products such as malate, acetate, propionate, and butyrate (Gudleifsson 1996, 1997). Of these metabolites CO_2 seems to be the most toxic (Andrews and Pomeroy 1979, 1990). On the cellular level the first sign of damage is detected on the cell membranes, appearing in increased microviscosity, increased levels of fatty acids, decreased ion uptake, and electrolytic leakage (Hetherington et al. 1987, 1988). According to Andrews and Pomeroy (1989) damage to the ion transport system is the earliest manifestation of membrane injury due to ice encasement.

During ice encasement the carbohydrates of timothy decrease slowly as a result of respiration, which in this case is mainly anaerobic respiration (Gudleifsson 2010). The plant is unable to release these metabolites into the air therefore they

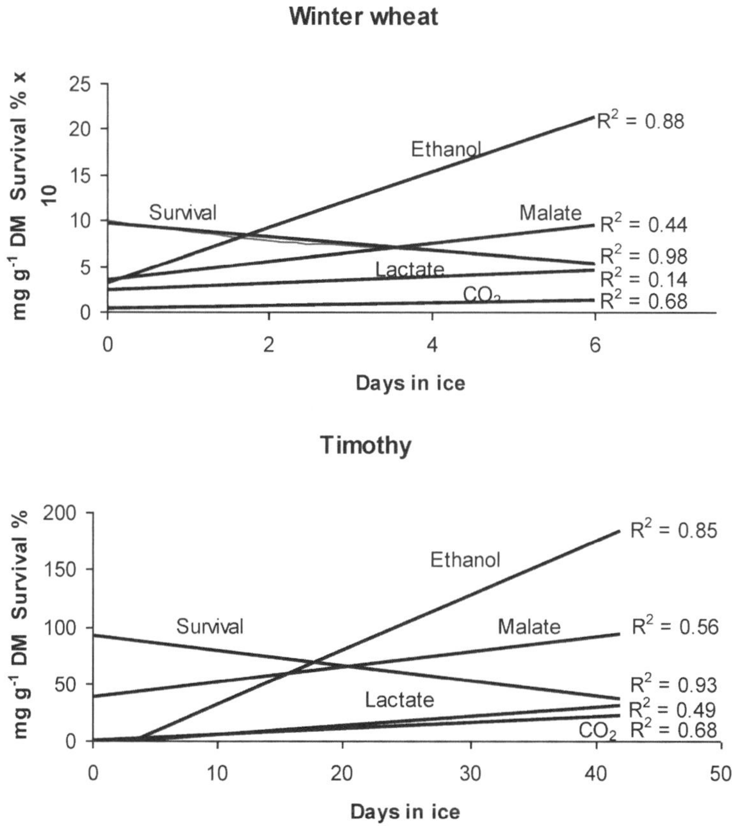

Fig. 4 Survival of winter wheat and timothy grass and accumulation of metabolites during ice encasement. (Adapted from Gudleifsson 2010)

accumulate in the plant. The accumulation of metabolites under ice has been measured in different kinds of herbage crops, mainly winter cereals and meadow grasses (Gudleifsson 2010). Figure 4 presents the accumulation of the main metabolites accumulated in timothy and winter wheat in compact ice at −2 °C, ethanol, malate, lactate, and CO_2. The figure also illustrates the great differences in metabolite concentrations and in ice tolerance between these two crop species, timothy and winter wheat. In addition to the metabolites mentioned, citrate, propionate, pyruvate, fumrate, and shikimate were also detected in these experiments. When long-lasting ice cover melts from the hayfields in Iceland in spring, a strong odor has been detected from the meltwater (Gudleifsson 2009). This type of meltwater from hayfields has been analyzed and the above-mentioned metabolites were identified along with, for example, butyrate, acetate, and malonate (Gudleifsson 1997, 2009).

Reactive Oxygen Species

It has been proposed that plants killed by ice encasement are not killed until after the disappearance of the ice. This conclusion is supported by the observations that plants escaping from lethal ice cover look green and healthy at first but wilt and die within a couple of days (Tanino and McKersie 1985; Gudleifsson and Larsen 1993). The transition from anoxia to air is a great shock for the plant, and reactive oxygen species (ROS) might develop and kill the cells (McKersie and Lesheim 1994). If this is the case antioxidants would help by scavenging the ROS and then reduce the damage. This was the case in an experiment with ice encased timothy. Antioxidant, 5 mM ascorbate, added to the water before icing, increased plant survival from 50–81 % (Gudleifsson 2009). The ROS are supposed to damage cell membranes, as do the accumulated metabolites. Thus toxicity of metabolites and formation of ROS during ice melting might both participate in cell membrane damage in ice encasement.

Phytotoxicity

After occurrence of winter injuries, it is economically important for farmers to renovate hayfields as fast as possible to ensure forage production. When permanent hayfields are totally killed by winter stress, the usual remedy is to renovate the plant cover by plowing and reseeding. Then the top soil is mixed with lower soil layers. When a field is only partly killed, direct drilling of the killed spots has been practiced. Then no plowing is needed and the soil layers are not mixed. This method has some beneficial aspects and is a faster and cheaper method to renovate the hayfield. However, successful establishment of grasses without plowing has been difficult to achieve, both in scientific studies and under practical conditions (Nesheim and Renolen 1992; Gudleifsson 1999). This has been related to some kind of phytotoxins in the top layer of the soil.

In an experiment using bioassays of soil–water extracts after three types of winter damage, the impact on plant growth was studied (Brandsæter et al. 2005). The winter stresses tested were ice encasement, frost, and snow molds. Only ice encasement caused phytotoxicity in the soil. The phytotoxic substances, such as ethanol, lactate, acetate, and butyrate, leak from meltwater into the topsoil in spring and may be a cause of poor establishment of direct-drilled grass. Some of these substances may also originate from the breakdown of killed grass debris and thus be of microbial origin. Many types of microorganisms are active under ice, snow, and water in spring under field conditions (Gudleifsson 1997). The toxicity of the topsoil may decrease later in the summer as the substances are washed out with rain water (Brandsæter et al. 2005; Gudleifsson 1999).

Concluding Remarks

Studies of winter stresses and winter damage on crops have mainly been focused on frost injury, and much is known about that kind of plant damage. Ice encasement may sometimes locally cause more serious problems. The yield of perennial plants may fail and cause economic losses and difficulties in agriculture. This kind of damage is not as extensively studied as frost damage and at present we do not know the exact cause of cell death. Here two causes are pointed out: membrane damage from toxic effects of accumulated metabolites in anaerobic respiration, and the injurious impact of ROS when plants return to the air.

References

Andrews CJ, Pomeroy MK (1979) Toxicity of anaerobic metabolites accumulated in winter wheat seedlings during ice encasement. Plant Physiol 64:120–125

Andrews CJ, Pomeroy MK (1989) Metabolic acclimation to hypoxia in winter cereals. Low temperature flooding increases adenylates and survival in ice encasement. Plant Physiol 91:1063–1068

Andrews CJ, Pomeroy MK (1990) Low temperature anaerobiosis in ice encasement damage to winter cereals. In: Jackson MB, Davies DD, Lambers H (eds) Plant life under oxygen deprivation. SPA Academic Publishing, The Hague, pp 85–99

Andrews CJ, Pomeroy MK, Seaman WL (1986) The response of fall-sown cereals to winter stress in eastern Ontario. Can J Plant Sci 66:25–37

Årsvoll K (1973) Winter damage in Norwegian grasslands, 1968–1971. Meldinger fra Norges Landbrukshögskole 52(3):1–21

Baadshaug OH (1973) Effects of soil type and soil compaction on the wintering of three grass species under different wintering conditions. Acta Agriculturae Scandinavica 23:77–86

Beard JB, Martin DP (1970) Influence of water temperature on submersion tolerance of four grasses. Agron J 62:257–259

Bergthorsson P (1969) An estimate of drift ice and temperature in Iceland in 100 years. Jökull 19:94–101

Bergþórsson P (2003) Lögmál byrst – í tölum talið [Hunger and starving – in figures]. In: Stefánsdóttir S, Kristjánsson K, Hjaltason J (eds) Afmæliskveðja til Háskóla Íslands. Hólar, pp 45–72. [In Icelandic]

Bergthórsson P, Bjornsson H, Dyrmundsson Ó, Gudmundsson B, Helgadottir Á, Jónmundsson J (1987) The effect of climatic variations in agriculture in Iceland. In: Parry ML, Carter TR, Konjin NT (eds) The impact of climatic impacts on agriculture, vol I. Assessment in Cool Temperate and Cold regions. Reidel, Dordrecht, pp 383–509

Brandsæter LO, Haugland E, Helgheim M, Gudleifsson BE, Tronsmo AM (2005) Identification of phytotoxic substances in soils following winter injury of grasses as estimated by a bioassay. Can J Plant Sci 85:115–123

Finnsson H (1970) Mannfækkun af hallærum [Population decrease as result of famine]. In: Eyþórsson J, Nordal J (eds) Almenna Bókafélagið. [In Icelandic], 209 p

Friðriksson S (1954) Rannsóknir á kali túna árin 1951 og 1952 [Studies on winter damage on hayfields 1951 and 1952]. Department of Agriculture, Reports, Series. B no 7. [In Icelandic]

Griffith M, Gudleifsson BE, Fukuta N (2001) Abiotic stresses in overwintering crops. In: Iriki N, Gaudet DA, Tronsmo AM, Matsumoto N, Yoshida M, Nishimune A (eds) Low temperature plant microbe interactions under snow. Hokkaido National Agricultural Station, pp 101–114

Gudleifsson BE (1975) Um kal og kalskemmdir IV. Samband veðurfars og kalskemmda [Winter damage IV. The relationship between winter damage and climate]. Ársrit Ræktunarfélags Norðurlands 72:45–64. [In Icelandic]

Gudleifsson BE (1996) Accumulation of metabolites during ice encasement of herbage plants. Nordisk jordbruksforskning 78(2):75

Gudleifsson BE (1997) Microbes active under snow and ice covers in hayfields in Iceland. Proceedings of International Workshop, Plant-Microbe Interaction at Low Temperature under Snow. Sapporo, Japan

Gudleifsson BE (1999) Ísáning – sáð í gróinn túnsvörð [Direct drilling into hayfields]. Ráðunautafundur 1999, 90–99. [In Icelandic]

Gudleifsson BE (2009) Ice encasement damage on grass crops and alpine plants in Iceland—impact of climate change. In: Gusta L, Wisniewski W, Tanino K (eds) Plant cold hardiness: from the laboratory to the field. CAB International, pp 163–172

Gudleifsson BE (2010) Ice tolerance and metabolite accumulation of herbage crops in Iceland and impact of climate change. Icel Agric Sci 23:111–122

Gudleifsson BE, Larsen A (1993) Ice encasement as a component of winter kill in herbage plants. In: Li PH, Christersson L (eds) Advances in plant cold hardiness. CRC Press, pp 229–249

Gudleifsson BE, Andrews CJ, Bjornsson H (1986) Cold hardiness and ice tolerance of pasture grasses grown and tested in controlled environments. Can J Plant Sci 66:601–608

Hakamata T, Noshiro M, Hirashima T, Nose I (1978) Investigation of actual condition on the winter killing of pasture species in the Nemuro Kushiro district—exploration of factors by the quantification No. 1. J Jpn Soc Grassl Sci 23:280–288

Haraldsson HV, Ólafsdóttir R (2007) Sögulegur fólksfjöldi á Íslandi – Ný nálgun með tilliti til burðargetu lands [Historical population in Iceland simulated from dynamic biological production]. Landabréfið 23(1):57–66. [In Icelandic]

Hetherington PR, McKersie BD, Borochov A (1987) Ice encasement injury to microsomal membranes from winter wheat crowns. I. Comparison of membrane properties after lethal ice encasement and during a post-thaw period. Plant Physiol 85:1068–1072

Hetherington PR, Broughton HL, McKersie BD (1988) Ice encasement injury to microsomal membranes isolated from winter wheat crowns. II. Changes in membrane lipids during ice encasement. Plant Physiol 87:740–743

Huokuna E (1959) The losses caused by ice cover on the Vik pasture experiment in 1956/57. Agricu Res 11:305–311

McKersie BD, Leshem YY (1994) Stress and stress coping in cultivated plants. Kluwer Academic Publishers, Dordrecht

Nesheim L, Renolen H (1992) Fornying av grasmark utan plöying [Grassland renovation without plowing]. FAGINFO 24:30. [In Norwegian]

Ouellet CE, Desjardins RL (1981) Interpretation des relations entre le climate et la survie à l´hiever de la luzerne par l´analyse des correlations. Can J Plant Sci 61:945–954. [In French]

Pomeroy MK, Andrews CJ (1986) Changes in adenine nucleotides and energy charge in isolated winter wheat cells during low temperature stress. Plant Physiol 86:361–366

Tanino KK, McKersie BD (1985) Injury within the crown of winter wheat seedlings after freezing and ice stress. Can J Botany 64:432–436

Umhverfisráðuneytið 2008. Hnattrænar loftslagsbreytingar og áhrif þeirra á Íslandi [Global climate change in Iceland]. Skýrsla vàsindanefndar um loftslagsbreytingar. [In Icelandic], 118 p

Hormones, NO, Antioxidants and Metabolites as Key Players in Plant Cold Acclimation

Gábor Galiba, Radomira Vanková, Irma Tari, Zsófia Bánfalvi, Péter Poór, Petre Dobrev, Ákos Boldizsár, Attila Vágújfalvi and Gábor Kocsy

Introduction

Over-wintering plants must be capable of surviving freezing temperatures in temperate-climate regions. The freezing tolerance (FT), however, is not a constant attribute; it changes over time, in response to changing environmental conditions. Among these, the primary factor is low temperature (Thomashow 1999). Under natural conditions, the gradual decrease in the temperature during autumn serves as a hardening agent to increase FT. The cold-inducible process that results in the manifestation of FT is called cold acclimation. For example, temperate winter cereals, such as wheat or rye, which are grown under room temperatures, can be killed upon freezing at about −5 °C, but after a cold acclimation period wheat can survive −16 °C and rye even −30 °C, respectively. During cold acclimation, the total reconfiguration of the metabolism is still not completely understood. Many of the biochemical and metabolic changes that occur during cold acclimation involve alterations in gene expression. The most known cold regulatory system with a key role in frost tolerance is the CBF/DREB (C-repeat binding factor or dehydration responsive element binding factor) pathway (Jaglo-Ottosen et al. 1998; Liu et al. 1998) which has been well studied both in monocot and dicotyledonous plant

G. Galiba (✉) · Á. Boldizsár · A. Vágújfalvi · G. Kocsy
Department of Plant Molecular Biology, Agricultural Institute, Centre for Agricultural Research, Hungarian Academy of Sciences, Brunszvik u. 2., Martonvásár 2462, Hungary
e-mail: galiba.gabor@agrar.mta.hu

R. Vanková · P. Dobrev
Laboratory of Hormonal Regulations in Plants, Institute of Experimental Botany, Academy of Sciences of the Czech Republic, Rozvojova, Prague, Czech Republic

I. Tari · P. Poór
Department of Plant Biology, University of Szeged, Szeged, Hungary

Z. Bánfalvi
Agricultural Biotechnology Center, Gödöllő, Hungary

R. Imai et al. (eds.), *Plant and Microbe Adaptations to Cold in a Changing World*,
DOI 10.1007/978-1-4614-8253-6_7, © Springer Science+Business Media New York 2013

species (Galiba et al. 2009). This pathway will be not described here, although its interaction—either with hormone metabolism or antioxidant systems—will be highlighted. Considering the molecular changes during the cold acclimation, crucial processes include changes in membrane cryobehaviour, the production of cryoprotective proteins and the accumulation of low molecular weight cryoprotectants such as carbohydrates and proline (Thomashow 1999; Guy et al. 2008). This review covers predominantly the low molecular weight metabolites.

Plant hormones have a central role in the control of cold acclimation. Abscisic acid (ABA) and salicylic acid (SA) have the greatest influence on FT, while the role of gibberellins, especially through their effect on plant growth rate under cold condition, is also not negligible (Kosová et al. 2012). The effect of hormones on growth and defence during acclimation may be mediated by reactive oxygen and nitrogen species (ROS and RNS), but they can control these processes independently from ROS and RNS too.

ROS and RNS, together with antioxidants, are involved in the regulation of FT through their effect on the cellular redox environment as well as signalling processes (Foyer and Noctor 2009). These substances control growth and defence due to their influence on gene and protein activity. The expression of many genes proved to be redox-sensitive. The proteins are affected by the redox changes through the formation or removal of disulphide bridges, and their posttranslational modification by (de)glutathionylation (Szalai et al. 2009).

During cold acclimation, the metabolite composition is reprogrammed. The changes in the amount of the compounds involved in the defence processes are of paramount importance. Thus, water-soluble carbohydrate and certain amino acids are involved in the osmotic adjustment, while the positively charged polyamines protect the negatively charged nucleic acids (Kovács et al. 2010; Vágújfalvi et al. 1999).

The dynamics of changes in FT is given by a complex regulatory network in which the hormone and redox signalling is interconnected and the metabolite composition may be reprogrammed in order to adjust plants to the changing environmental temperatures (Fig. 1).

Plant Hormone Interactions During Different Phases of Cold Stress Response

Plant hormones play important roles in cold responses, which substantially differ in the individual stress phases. The alarm phase (cold shock) is associated with decrease of hydraulic conductivity of roots, which results in decrease of water potential in leaves. Water status needs to be stabilized by stomata closure, which is governed by ABA. Apart from regulation of stomata aperture, ABA stimulates the expression of a number of stress-related genes, including transcription factors (Yamaguchi-Shinozaki and Shinozaki 2006). ABA exhibits other protective functions in cold

Fig. 1 Simplified model of regulatory network of plant cold acclimation. Cold signal-induced interaction among gene expression, metabolism, plant hormones and reactive oxygen/nitrogen species (ROS/RNS) ensures the fine adaptive response

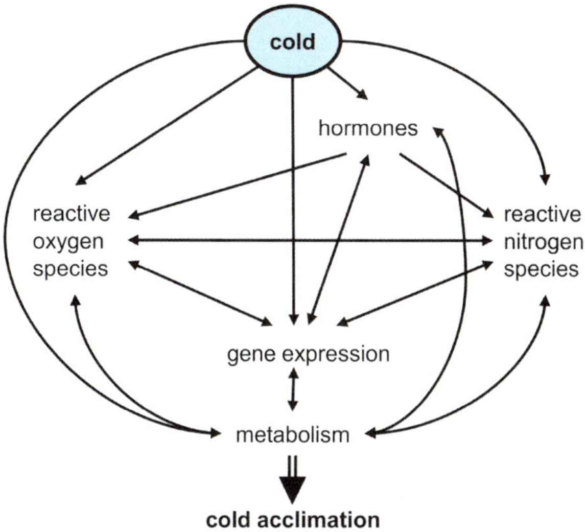

stress defence, especially stabilization of membranes and protection against oxidative stress. The ABA content was reported to increase transiently in the early stage of the cold stress response (Galiba et al. 1993; Janowiak et al. 2002; Pociecha et al. 2009), which is more profound in more tolerant winter cultivars (Veisz et al. 1996).

Increased level of ABA was found to coincide with down-regulation of other stress hormones, SA and jasmonic acid (JA) during an early phase of wheat response to cold stress (Kosová et al. 2012), which indicates antagonistic character between ABA and the other stress hormones. SA is a stress hormone predominantly associated with biotroph infection, while JA is associated with the response to wounding and necrotroph attack. There is a relationship among these stress hormones, however, it is not simple. SA and JA, have been reported to induce the expression of several dehydrins in a wide variety of plant species, including a wild relative of the chickpea *Cicer pinnatifidum*, and barley (Bhattarai and Fettig 2005; Sun et al. 2009). In the case of JA and SA, the resulting effect strongly depends on the actual concentration of the hormone. Treatment with low concentrations of exogenous SA (up to 0.25 mM, 'priming') stimulates barley *Dhn5* mRNA expression as well as protein levels, while higher concentrations (more than 0.25 mM) that mimic severe biotic stress lead to decrease in both *Dhn5* mRNA and protein levels (Sun et al. 2009). Ethylene level, usually estimated via the content of its precursor 1-aminocyclopropane-1-carboxylic acid (ACC), was found to increase rapidly during the alarm phase in the wheat leaves and crowns (Kosová et al. 2012).

Apart from the stimulation of defence mechanisms, an important component of the stress response is modulation of plant growth and development. The cold-inducible CBF transcription factor was found to down-regulate levels of active gibberellins (GAs), hormones positively affecting germination, growth and reproductive development, by stimulation of the expression of genes for the GA-inactivating enzymes (GA 2-oxidases). CBFs also stabilize DELLA proteins, which are the repressors of GA signalling pathway (Achard et al. 2008).

Transient peak of *CBF1* was found to be accompanied by stimulation of the expression of *ARR5*, *ARR6* and *ARR7*, negative regulators of the signalling pathway of cytokinins, plant hormones indispensable for cell division and growth (Jeon et al. 2010). Over-expression of *GhDREB1* in *Arabidopsis* was reported to suppress positive type-B response regulators (Huang et al. 2009). Cytokinin analysis in wheat showed that levels of active cytokinins are down-regulated during the alarm phase as well (Kosová et al. 2012). The rapid decrease of active cytokinins in leaves may also contribute to the regulation of stomata aperture, as cytokinins are known to induce stomata opening (Acharya and Assmann 2009).

The other growth-promoting hormone, auxin (indole-3-acetic acid), showed a decrease in the wheat leaves in alarm phase as well (Kosová et al. 2012). A recent report by Shibasaki et al. (2009) indicated that cold stress (12 h) might affect auxin transport by inhibiting the intracellular trafficking of the auxin efflux carriers PIN2 and PIN3. These authors did not find any change in auxin signalling.

The characteristic features of acclimation phase are an increase in acquired frost tolerance, accumulation of protective proteins, especially dehydrins, down-regulation of ABA levels and an elevation of positive regulators of cell division and growth (i.e. cytokinins, gibberellins and auxin). These changes indicate plant adaptation to low temperature and a readjustment of metabolic activity to less favourable conditions. As cytokinins exert a positive effect on plant photosynthesis, their elevation during the adaptation phase might contribute to energy formation. The application of exogenous cytokinin (N^6-benzyladenine) was found to improve cold tolerance in the sugar beet (Dix et al. 1994). Xia et al. (2009) demonstrated a positive effect of cytokinins on growth at both normal and low temperatures. Constitutive expression of the cytokinin biosynthetic gene (*ipt*) was reported to improve the maintenance of chlorophyll levels as well as enhance cold tolerance in callus of *Festuca arundinacea* (Hu et al. 2005). Similar results have been obtained by Guo et al. (2010), using *ipt*-expressing *Arabidopsis* plants.

The necessity of active cytokinin signalling during prolonged cold stress is also indicated by the fact that the constitutive over-expression of *ARR7* (a negative type-A cytokinin response regulator, the expression of which is stimulated during the initial phase of the cold response) results in hypersensitivity to freezing temperatures for a longer-term (Jeon et al. 2010). An *arr7* knockout mutant was found to exhibit increased FT.

Similarly, elevation of active gibberellins (GA_1, GA_4) was found to occur during the adaptation phase of the stress response in wheat (Kosová et al. 2012). Apart from its positive effect on growth, GA_1 was found to be involved in repression of the promoter activity of *GhDREB1* (Shan et al. 2007).

The decrease in ABA during the acclimation phase coincides with the elevation of other stress hormones during this period. The ACC content was found to exhibit a transient minimum at day 3 and a second maximum at day 7 in both leaves and crowns of winter and spring wheat (Kosová et al. 2012). The second ACC maximum coincided with an elevation of SA. Increase of free and bound SA as well as

O-hydroxycinnamic acid was found in winter wheat (Janda et al. 2007; Scott et al. 2004). Elevation of SA might be related to SA functioning in the regulation of ROS evolution, which is an important part of plant stress defences.

JA was reported to increase in acclimation phase and even more during longer-term exposure to cold. Gaudet et al. (2011) and Goulas et al. (2006) reported up-regulation of the expression of the JA synthetic gene, which encodes allene oxide synthase, between days 21 and 49 of acclimation or after 10 days of cold exposure, respectively. During the same period, these authors found increased resistance of winter wheat to snow mould, powdery mildew and stripe rust pathogens.

Redox Regulation of Freezing Tolerance

Antioxidants and ROS participates in the adjustment of plant growth and development to the changing environmental conditions and are involved in the defence processes due to their effect on the activity of many genes and proteins (Foyer and Noctor 2009). H_2O_2 affected gene expression pattern as it was demonstrated by its exogenous application and in a catalase-deficient *Arabidopsis* mutant (Desikan et al. 2001). In addition, glutathione (GSH) degrading of H_2O_2 influenced the transcription of several stress-responsive genes, too (Ball et al. 2004). The disulphide form of GSH influences the activity of proteins through the oxidation or glutathionylation of their sulphhydryl groups (Rouhier et al. 2008).

In the improvement of tolerance to low temperature stress, H_2O_2 and other ROS are involved since it was shown that in maize seedlings H_2O_2 is accumulated during cold acclimation, and treatment with H_2O_2 improves chilling tolerance (Prasad et al. 1994). In addition, the accumulation of ROS was observed in wheat subjected to cold (Okuda et al. 1991; Soltész et al. 2011). Parallel to the accumulation of ROS, the activation of the antioxidant system, regulating the ROS level, was indicated by changes in the activity of antioxidant enzymes and in the quantity of non-enzymatic antioxidants in cold-treated wheat (Kocsy et al. 2001; Soltész et al. 2011). The importance of GSH in low temperature tolerance was demonstrated in wheat and maize by comparing genotypes with different levels of stress tolerance and by chemically manipulating the GSH concentration (Kocsy et al. 2000; Kocsy et al. 2001). In the course of abiotic stress, the redox state of GSH is influenced by various plant regulators (Szalai et al. 2009), as demonstrated by treating maize with SA and ABA (Kellős et al. 2008). Besides GSH, the amount of ascorbate, which is the other non-enzymatic component of the ascorbate-glutathione cycle, is also increased during hardening, as described in rye (Streb et al. 2002). Cold-induced changes in the activities of the enzymes of this cycle (Scebba et al. 1998), may be related to the increased expression of the corresponding genes, as found in cold-acclimated near-isogenic wheat lines (Baek and Skinner 2003). The observed changes in the antioxidant levels may be involved in cold acclimation indirectly through their participation in redox signalling or directly due to the control of ROS levels (Foyer and Noctor 2009; Ogawa 2005).

ROS and antioxidants are also involved in the regulation of plant development, since the vernalization requirement, necessary for the flowering of certain plant species, could be replaced by the addition of GSH or its precursor, cysteine, while the inductive effect of vernalization on flowering was nullified by the inhibition of GSH synthesis in *Arabidopsis*. The regulatory effect of GSH on flowering was also demonstrated in *Arabidopsis* mutants (Ogawa et al. 2001). Similar to glutathione, ascorbate also affected flowering, since spraying *Arabidopsis* with the precursor of ascorbate (L-galactono-γ-lactone) delayed flowering (Attolico and de Tullio 2006) and the ascorbate-deficient *Arabidopsis* mutants indeed flowered earlier than the wild-type plants (Kotchoni et al. 2009). Treatment with GSH and ascorbate modified the speed of the vegetative/generative transition as it was monitored by the appearance of the double ridges in wheat (Kocsy, unpublished results) indicating that the development of reproductive organs is affected by redox changes in wheat too. The fine regulation of this transition by temperature-dependent redox changes is very important, because the flower primordia are extremely sensitive to the freezing temperatures (Galiba et al. 2009).

As part of the defence processes, the programmed cell death may also occur in the severely damaged tissues and it can be controlled by ROS and antioxidants. The involvement of antioxidants in death process was shown in wheat, where a treatment with GSH and ascorbate modified the frost-induced membrane injury (Kocsy, unpublished results). If the amount of non-enzymatic antioxidants and the activity of antioxidant enzymes, including catalase and ascorbate peroxidases, are not sufficient to detoxify the excess of H_2O_2, it accumulates to toxic levels (Gechev and Hille 2005). Regardless of the way of sensing, it is still unclear how this small molecule is able to trigger such different responses as stress acclimation or programmed cell death and to initiate distinct developmental programs. The answer to H_2O_2 multi-functionality and the complexity of the responses can be found in the interaction of H_2O_2 with antioxidants, amino acids, polyamines, calcium, nitric oxide (NO), lipid and plant hormone signalling pathways. For example, interactions between H_2O_2 and NO are essential for programmed cell death during the hypersensitive response and also for the defence against pathogens, where a fine balance between the two signals modulates cell death (Gechev and Hille, 2005).

The redox state-dependent response to abiotic stresses may also be transmitted by various plant hormones. SA increased the chilling tolerance of maize by inhibiting catalase, thus increasing the H_2O_2 concentration (Horváth et al. 2007). SA regulated cell death- and defence-related processes (Greenberg et al. 2000) and the vegetative/generative transition in *Arabidopsis* (Martinez et al. 2004), the effects of which may be mediated by ROS also. Another stress hormone, ABA, induced changes in ROS concentration in *Arabidopsis*, activating the Ca^{2+} channels of the cell membranes and increasing the Ca^{2+} concentration in the cytosol (Murata et al. 2001).

In summary, redox changes are important in the regulation of development and stress adaptation. These alterations were mainly investigated in extracts made from plant organs. Since compartment-specific redox control of development and stress response could be very important in the effective adaptation to the various environmental conditions, *in situ* investigation of ROS and antioxidants would be very important in the future.

The Role of Nitric Oxide in Cold Acclimation

NO has clearly been established as a key player in the plant response to cold stress. Nitrate reductase (NR) and nitric oxide synthase (NOS)-like enzymes can catalyse the synthesis of NO but evidence has also been accumulated for other NO sources in plant tissues (Spadaro et al. 2010).

RNS, NO, peroxynitrite and S-nitrosoglutathione (GSNO) can react with a number of molecules, which leads to chemical modification of proteins, lipids and nucleic acids. NO and GSNO can induce S-nitrosylation by the addition of NO to reactive Cys thiol groups. These changes may precisely regulate the structure and function of proteins, enzymes or transcription factors and confer specificity to NO-based signalling (Spadaro et al. 2010). Peroxynitrite can also modify proteins by irreversible nitration of tyrozines or tryptophan residues and can also react with DNA leading to formation of 8-nitroguanine (Niles et al. 2006).

NO production may play a role in the cold-induced changes in various stress phases. Fast and transient increase in NO production could be detected during the alarm phase of chilling stress in wheat roots (Majláth et al. 2012). In cucumber fruits NO levels increased until the sixth day of cold storage suggesting its role in the acclimation phase (Dong et al. 2012). NO contents were permanently higher in the apical meristem under cold stress than under control temperatures during the vegetative/generative transition of wheat and this delayed the appearance of "double ridge" (Kolbert and Poór, personal communication).

Cold stress up-regulated the expression of an NR gene, *NIA1*, and stimulated the enzyme activity in *Arabidopsis* seedlings. The NR defective, double mutant *nia1/nia2* seedlings produced less NO and were less tolerant to freezing than wild-type plants (Zhao et al. 2009). NOS-like enzymes can also be sources of NO in pea in response to low temperatures (Corpas et al. 2009).

Apart from the general downstream effects of NO there are other signalling intermediates activated by NO under cold stress. Cantrel et al. (2010) demonstrated that NO could effectively modulate the cold response by inhibiting the phosphorylation of sphingolipids. Some of the cold responsive genes have C-repeat/dehydration-responsive (CRT/DRE) *cis*-elements in the promoter and their transcription is activated by *CRT/DRE*-binding protein factors (CBFs). In tomato fruit it was found that NO participated in the enhanced expression of *LeCBF-1*, and improved cold tolerance of tomato fruit during cold storage (Zhao et al. 2011).

Low temperature stress induced nitrosylation of several proteins involved in photosynthesis, glycolysis and signalling processes. The carboxylase activity of Rubisco was inhibited by S-nitrosylation which correlated well with the inhibition of CO_2 assimilation (Abat and Deswal 2009).

Cold stress affects the expression level of cryoprotective proteins, enzymes involved in the biosynthesis of osmoprotectants (proline, sugars), non-enzymatic antioxidants and ROS-scavenging enzymes (Yamaguchi-Shinozaki and Shinozaki 2006). NO targets include several metal or thiol containing proteins, such as catalase (CAT), peroxidase and superoxide dismutase (SOD) (Lindermayer et al. 2005). The exposure of pepper leaves to chilling temperature was accompanied by increased lipid peroxidation and tyrosine nitration. The oxidative stress and nitrosative stress were eliminated by the accumulation of non-enzymatic antioxidants, GSH and ascorbate and by the increased activity of NADPH-generating dehydrogenases (Airaki et al. 2012). Exogenous application of yeast saccharide to cucumber fruits before cold storage triggered endogenous NO accumulation and reduced oxidative stress by decreasing lipid peroxidation (Dong et al. 2012). This suggests that a well-controlled balance is necessary between various forms of ROS and RNS.

Cold stress up-regulated Δ(1)-pyrroline-5-carboxylate synthetase1 (*P5CS1*) and resulted in enhanced accumulation of proline (Pro) in *Arabidopsis* plants. The stimulation of Pro accumulation can be prevented by NR inhibitors or NO scavengers (Zhao et al. 2009). Exogenous application of sodium nitroprusside, an NO generator, increased the expression of enzymes participating in fructan biosynthesis, the polymerization degree of fructans and total carbohydrate content in wheat cultivars (Li et al. 2013) demonstrating the role of NO in osmotic adaptation during cold treatment.

Many hormone-mediated events such as brassinosteroid (BR)-induced responses have NO implicated in the downstream signalling. Scavenging or inhibition of NO production reduced BR-induced tolerance to cold stress, and partially blocked the BR-induced expression of the genes coding for several antioxidant enzymes, such as ascorbate peroxidase, *CAT*, *SOD* and glutathione reductase suggesting that NO participates in BR-induced antioxidant response during cold acclimation (Cui et al. 2011). These findings imply that cold tolerance of crop plants or vegetables can be enhanced by inducing endogenous NO accumulation.

Metabolite Accumulation upon Cold Acclimation

Most molecular studies on plant responses to cold stress are focused on the mechanism of cold acclimation and found that reprogramming of gene expression results in the accumulation not only of protective proteins but also of hundreds or more metabolites, some of which are known to have protective effects (Zhu et al. 2007).

Majority of cold acclimation studies were carried out on the model plant, *Arabidopsis thaliana*. Exposure to 4 °C is not a lethal stress for *Arabidopsis*. Plants continue to grow, although at a much lower rate, and eventually, flower, set seeds, and successfully complete their lifecycle under these conditions. Large-scale profiling of metabolites by gas chromatography-mass spectrometry (GC-MS) has revealed alterations in steady-state pool sizes of more than 300 metabolites or mass spectral tags in response to cold shock in *Arabidopsis* (Kaplan et al. 2004;

Cook et al. 2004). These and other studies performed by Zuther et al. (2004), Guy et al. (2008) and Usadel et al. (2008) showed extensive modifications supporting cellular changes, increased antioxidant production and the accumulation of metabolites with protective functions such as sucrose, trehalose, fructan, maltose, galactinol, raffinose, proline and glycinebetaine. Cook et al. (2004) showed that there is around 80% overlap between metabolic responses of constitutive overexpression of *CBF3* and cold acclimation, indicating that the changes in alteration of many metabolites are regulated by *CBF* genes (see Introduction) during cold acclimation.

In plants, there is a large overlap between cold-regulated and circadian-regulated genes. Among the circadian-regulated genes are the *CBFs* and several other cold-regulated (*COR*) genes. In *Arabidopsis* it was shown that cold affects the expression of clock oscillator genes (Fowler et al. 2005; Bieniawska et al. 2008). Espinoza et al. (2010) found that about 80% of metabolites that showed diurnal cycles maintained these oscillations during cold treatment.

Although metabolite profiling has indicated a major reprogramming of plant metabolism in the cold, it has remained unclear which of these changes are related to FT. Korn et al. (2010) discovered combinations of metabolites that predict FT and its heterosis in *A. thaliana*. In particular, the pathway leading to raffinose was determined as crucial statistical predictor for FT and its heterosis. In addition, several unidentified metabolites strongly contributed to the prediction of both FT and its heterosis. One of these was identified as a hexose conjugate.

A. thaliana is a geographically widespread species consisting of local accessions differing both genetically and phenotypically. Studying 54 *Arabidopsis* accessions with diverse geographical origin. Zuther et al. (2012) demonstrated that leaf glucose, fructose, sucrose and raffinose contents are correlated with FT in the cold acclimated state, while proline contents are not. Moreover, expression of the *CBF1*, *2* and *3* genes is also not correlated with FT, while that of some CBF-regulated *COR* genes is.

The genus *Thellungiella* is related to *Arabidopsis*, however, this genus is characterized by high tolerance of abiotic stresses. In *Thellungiella*, acclimated FT was correlated with sucrose levels but not with raffinose accumulation. Furthermore, in contrast to that found by Zuther et al. (2012) in *Arabidopsis*, the proline content was in correlation with FT in *Thellungiella* (Lee et al. 2012).

Nonetheless, one might expect differing cold acclimation and FT strategies, when studying different plant organs, and comparing annual and perennial species, metabolome of no other crops than strawberry (*Fragaria vesca*) was investigated so far during cold acclimation. Metabolic changes in *F. vesca* showed a strong modulation of central metabolism, and induction of sugars (fructose, glucose, galactose and raffinose), amino acids (aspartic acid), and amines (putrescine), while cold-induction of proline was absent. Furthermore, changes in metabolite pools of cold-acclimated *F. vesca* were clearly influenced by genotype (Rohloff et al. 2012).

More specific studies than metabolome analysis, have been carried out with other crops as well, including cereals, i.e. wheat, oat and rye (Vágújfalvi et al. 1999; Livingstone et al. 2006). These studies clearly demonstrated transient increases of soluble sugars triggered by low temperature, which was strongly genotype-dependent. Besides sugars, amino acid contents of different crops were also investigated during cold acclimation. It was shown, for example, that proline and arginine levels differed following hardening in the various genotypes of white clover (Svenning et al. 1997). In potato and winter wheat, cold acclimation increased the proline content and the change was greater in tolerant genotypes than in sensitive ones (van Swaaj et al. 1985; Dörrfling et al. 1990; Machackova et al. 2006). Cold treatment resulted in increased free amino acid levels in barley and wheat (Mazzucotelli et al. 2006). The ratio of members of the glutamate family increased during cold acclimation, while the ratio of amino acids belonging to the aspartate family decreased (Kovács et al. 2011). The cold-induced accumulation of glutamine may influence the synthesis of glutathione and polyamine, the increase in amounts of which at low temperature was described in wheat (Kocsy et al. 2000; Kovács et al. 2010).

The protective role of proline against adverse environmental conditions was proved in transgenic plants (Valliyodan and Nguyen 2006) including transgenic wheat plants (Sawahel and Hassan 2002; Vendruscolo et al. 2007). In contrast, despite the cold-induced accumulation of raffinose in several species, varying raffinose concentrations in *Arabidopsis* by overexpression of galactinol synthase or knockout of raffinose synthase did not affect FT (Zuther et al. 2004). This negative result highlights that direct correlation of a multigenic trait such as FT with the concentration of only one or a few metabolites may not be what one would expect. Multigenic traits are regulated by a network of gene-interaction, and to investigate such networks a special approach is needed that incorporates metabolite interconversions and regulation strategies as well. Such a possibility is provided by systems biology techniques, which have rapidly become integrated into metabolic research recently. Using the tools of systems biology Nagele et al. (2011) made a comprehensive overview of interconversion rates of accession-specific regulation strategies in *Arabidopsis* during exposure to low temperature and provided evidence for a pivotal role of sucrose–hexose interconversion in increasing the cold acclimation output. One should use this tool also in crops to see whether it is a general strategy of cold acclimation or specific to *Arabidopsis*.

Conclusions

ROS and RNS, in interaction with plant hormones participate in the control of developmental phase-specific changes in FT due to their effect on reprogramming of transcriptome, proteome and metabolome. The role of plant hormones substantially differs in the individual stress phases, i.e. during alarm phase or acclimation phase, so study and elucidation of their action are meaningful only in a time-dependent manner. Deeper insight into these changes can be gained by system biology approach that allows the determination of regulatory and metabolic networks in full complexity.

Acknowledgements This work was supported by the Hungarian Scientific Research Fund (OTKA CNK80781, K83642, K75528), National Developmental Agency (TÁMOP-4.2.2/B-10/1-2010-0025) to University of Pannonia, Czech Science Foundation (project no. 522/09/2058).

References

Abat JK, Deswal R (2009) Differential modulation of S-nitrosoproteome of *Brassica juncea* by low temperature: change in S-nitrosylation of Rubisco is responsible for the inactivation of its carboxylase activity. Proteomics 9:4368–4380

Achard P, Gong F, Cheminant S, Alioua M, Hedden P, Genschik P (2008) The cold-inducible CBF1 factor-dependent signaling pathway modulates the accumulation of the growth-repressing DELLA proteins via its effect on gibberellin metabolism. Plant Cell 20:2117–2129

Acharya B, Assmann S (2009) Hormone interactions in stomatal function. Plant Mol Biol 69:451–462

Airaki M, Leterrier M, Mateos RM, Valderrama R, Chaki M, Barroso JB, del Rào LA, Palma JM, Corpas FJ (2012) Metabolism of reactive oxygen species and reactive nitrogen species in pepper (*Capsicum annuum* L.) plant under low temperature stress. Plant Cell Environ 35:281–295

Attolico AD, De Tullio MC (2006) Increased ascorbate content delays flowering in long-day grown *Arabidopsis thaliana* (L.) Heynh. Plant Physiol Biochem 44:462–466

Baek K-H, Skinner DZ (2003) Alteration of antioxidant enzyme gene expression during cold acclimation of near-isogenic wheat lines. Plant Sci 165:1221–1227

Ball L, Accotto G-P, Bechtold U, Creissen G, Funck D, Jimenez A, Kular B, Leyland N, Mejia-Carranza J, Reynolds H, Karpinski S, Mullineaux PM (2004) Evidence for a direct link between glutathione synthesis and stress defence gene expression in *Arabidopsis*. Plant Cell 16:2448–2462

Bhattarai T, Fettig S (2005) Isolation and characterization of a dehydrin gene from *Cicer pinnatifidum*, a drought-resistant wild relative of chickpea. Physiol Plantarum 123:452–458

Bieniawska Z, Espinoza C, Schlereth A, Sulpice R, Hincha DK, Hincha DK, Hannah MA (2008) Disruption of the *Arabidopsis* circadian clock is responsible for extensive variation in the cold-responsive transcriptome. Plant Physiol 147:263–279

Cantrel C, Vazquez T, Puyaubert J, Rezé N, Lesch M, Kaizer WM, Dutilleul C, Guillas I, Zachowski A, Bauduin E (2010) Nitric oxide participates in cold-responsive phosphoshingolipid formation and gene expression in *Arabidopsis thaliana*. New Phytol 189:415–427

Cook D, Fowler S, Fiehn O, Thomashow MF (2004) A prominent role for the CBF cold response pathway in configuring the low temperature metabolome of Arabidopsis. Proc Natl Acad Sci U S A 101:15243–15248

Corpas FJ, Palma JM, del Rào LA, Barroso JB (2009) Evidence supporting the existence of L-arginine-dependent nitric oxide synthase activity in plants. New Phytol 184:9–14

Cui JX, Zhou YH, Ding JG, Xia XJ, Shi K, Chen SC, Asami T, Chen Z, Yu JQ (2011) Role of nitric oxide in hydrogen peroxide-dependent induction of abiotic stress tolerance by brassinosteroids in cucumber. Plant Cell Environ 34:347–358

Desikan R, Mackerness SA-H, Hancock JT, Neill SJ (2001) Regulation of the Arabidopsis transcriptome by oxidative stress. Plant Physiol 127:159–172

Dix PJ, Finch I, Burke JI (1994) Genotypic differences in cold tolerance are masked by high sucrose and cytokinin in shoot cultures of sugar-beet. Plant Cell Tiss Org 36:285–290

Dong J, Yu Q, Lu L, Xu M (2012) Effect of yeast saccharide treatment on nitric oxide accumulation and chilling injury in cucumber fruit during cold storage. Postharvest Biol Technol 68:1–7

Dörffling K, Schulenburg S, Lesselich G, Dörffling H (1990) Abscisic acid and proline levels in cold-hardened winter wheat leaves in relation to variety-specific differences in freezing resistance. J Agron Crop Sci 165:230–239

Espinoza C, Degenkolbe T, Caldana C, Zuther E, Leisse A, Willmitzer L, Hincha DK, Hannah MA (2010) Interaction with diurnal and circadian regulation results in dynamic metabolic and transcriptional changes during cold acclimation in Arabidopsis. PLoS One 5:e14101

Fowler SG, Cook D, Thomashow MF (2005) Low temperature induction of Arabidopsis *CBF1, 2,* and *3* is gated by the circadian clock. Plant Physiol 137:961–968

Foyer CH, Noctor G (2009) Redox regulation in photosynthetic organisms: signalling, acclimation, and practical implications. Antioxid Redox Sign 11:861–905

Galiba G, Tuberosa R, Kocsy G, Sutka J (1993) Involvement of chromosome 5A and 5D in cold-induced abscisic acid accumulation in and frost tolerance of wheat calli. Plant Breeding 110:237–242

Galiba G, Vágújfalvi A, Li C, Soltész A, Dubcovsky J (2009) Regulatory genes involved in the determination of frost tolerance in temperate cereals. Plant Sci 176:12–19

Gaudet DA, Wang Y, Frick M, Puchalski B, Penniket C, Quellet T, Robert L, Singh J, Laroche A (2011) Low temperature induced defence gene expression in winter wheat in relation to resistance to snow moulds and other wheat disease. Plant Sci 180:99–110

Gechev TS, Hille J (2005) Hydrogen peroxide as a signal controlling plant programmed cell death. J Cell Biol 168:17–20

Goulas E, Schubert M, Kieselbach T, Kleczkowski LA, Gardestrom P, Schrode W, Hurry V (2006) The chloroplast lumen and stromal proteomes of *Arabidopsis thaliana* show differential sensitivity to short- and long-term exposure to low temperature. Plant J 47:720–734

Greenberg JT, Silverman FP, Liang H (2000) Uncoupling salicylic acid-dependent cell death and defense-related responses from disease resistance in the Arabidopsis mutant *acd5*. Genetics 156:341–350

Guo JC, Duan RJ, Hu XW, Li KM, Fu SP (2010) Isopentenyl transferase gene (ipt) downstream transcriptionally fused with gene expression improves the growth of transgenic plants. Transgenic Res 19:197–209

Guy C, Kaplan F, Kopka J, Selbig J, Hincha DK (2008) Metabolomics of temperature stress. Physiol Plant 132:220–235

Horváth E, Szalai G, Janda T (2007) Induction of abiotic stress tolerance by salicylic acid signalling. J Plant Growth Regul 26:290–300

Hu YL, Jia WL, Wang JD, Zhang YQ, Yang LL, Lin ZP (2005) Transgenic tall fescue containing the *Agrobacterium tumefaciens ipt* gene shows enhanced cold tolerance. Plant Cell Rep 23:705–709

Huang JG, Yang M, Liu P, Yang GD, Wu CA, Zheng CC (2009) *GhDREB1* enhances abiotic stress tolerance, delays GA-mediated development and repress cytokinin signalling in transgenic Arabidopsis. Plant Cell Environ 32:1132–1145

Jaglo-Ottosen KR, Gilmour SJ, Zarka DG, Schabenberger O, Thomashow MF (1998) Arabidopsis *CBF1* overexpression induces *COR* genes and enhances freezing tolerance. Science 280:104–106

Janda T, Szalai G, Leskó K, Yordanova R, Apostol S, Popova LP (2007) Factors contributing to enhanced freezing tolerance in wheat during frost hardening in the light. Phytochemistry 68:1674–1682

Janowiak F, Maas B, Dőrffling K (2002) Importance of abscisic acid for chilling tolerance of maize seedlings. J Plant Physiol 159:635–643

Jeon J, Kim NY, Kim S, Kang NY, Novak O, Ku SJ, Cho C, Lee DJ, Lee EJ, Strnad M, Kim J (2010) A subset of cytokinin two-component signaling system plays a role in cold temperature stress response in Arabidopsis. J Biol Chem 285:23369–23384

Kaplan F, Kopka J, Haskell DW, Zhao W, Schiller KC, Gatzke N, Sung DY, Guy CL (2004) Exploring the temperature-stress metabolome of Arabidopsis. Plant Physiol 136:4159–4168

Kellős T, Tàmár I, Szilágyi V, Szalai G, Galiba G, Kocsy G (2008) Stress hormones and abiotic stresses have different effects on antioxidants in maize lines with different sensitivity. Plant Biol 10:563–572

Kocsy G, Galiba G, Brunold C (2001) Role of glutathione in adaptation and signalling during chilling and cold acclimation in plants. Physiol Plantarum 113:158–164

Kocsy G, Szalai G, Vágújfalvi A, Stéhli L, Orosz G, Galiba G (2000) Genetic study of glutathione accumulation during cold hardening in wheat. Planta 210:295–301

Korn M, Gärtner T, Erban A, Kopka J, Selbig J, Hincha DK (2010) Predicting *Arabidopsis* Freezing Tolerance and Heterosis in Freezing Tolerance from Metabolite Composition. Molecular Plant 3:224–235

Kosová K, Prasil IT, Vitamvas P, Dobrev P, Motyka V, Flokova K, Novak O, Turecková V, Rolcik J, Pesek B, Travnickova A, Gaudinova A, Galiba G, Janda T, Vlasakova E, Prasilova P, Vankova R (2012) Complex phytohormone responses during the cold acclimation of two wheat cultivars differing in cold tolerance, winter Samanta and spring Sandra. J Plant Physiol 169(5):567–576

Kotchoni SO, Larrimore KE, Mukherjee M, Kempinski CF, Barth C (2009) Alterations in the endogenous ascorbic acid content affect flowering time in Arabidopsis. Plant Physiol 149:803–815

Kovács Z, Simon-Sarkadi L, Szücs A, Kocsy G (2010) Differential effects of cold, osmotic stress and abscisic acid on polyamine accumulation in wheat. Amino Acids 38:623–631

Kovács Z, Simon-Sarkadi L, Sovány C, Kirsch K, Galiba G, Kocsy G (2011) Differential effects of cold acclimation and abscisic acid on free aminoacid composition in wheat. Plant Sci 180:61–68

Lee YP, Babakov A, Boer B de, Zuther E, Hincha DK (2012) Comparison of freezing tolerance, compatible solutes and polyamines in geographically diverse collections of *Thellungiella* sp. and *Arabidopsis thaliana* accessions. BMC Plant Biol 12:131

Li C, Li T, Zhang D, Jiang L, Shao Y (2013) Exogenous nitric oxide effect on fructan accumulation and FBEs expression in chilling-sensitive and chilling-resistant wheat. Environ Exp Bot 86:2–8.

Lindermayer C, Saalbach G, Durner J (2005) Proteomic identification of S-nitrosylated proteins in Arabidopsis. Plant Physiol 137:921–930

Liu Q, Kasuga M, Sakuma Y, Abe H, Miura S, Yamaguchi-Shinozaki K, Shinozaki K (1998) Two transcription factors, DREB1 and DREB2, with an EREBP/AP2 DNA binding domain separate two cellular signal transduction pathways in drought- and low-temperature-responsive gene expression, respectively, in Arabidopsis. Plant Cell 10:1391–1406

Livingstone DP, Premakumar R, Tallury SP (2006) Carbohydrate partitioning between upper and lower regions of the crown in oat and rye during cold acclimation and freezing. Cryobiol 52:200–208

Machackova I, Hanisova A, Krekule J (2006) Levels of ethylene, ACC, MACC, ABA and proline as indicators of cold hardening and frost resistance in winter wheat. Physiol Plantarum 76:603–607

Majláth I, Szalai G, Soós V, Sebestyén E, Balázs E, Vankova R, Dobrev PI, Tari I, Tandori J, Janda T (2012) Effect of light on the gene expression and hormonal status of winter and spring wheat plants during cold hardening. Physiol Plantarum 145:296–314

Martànez C, Pons E, Prats G, León J (2004) Salicylic acid regulates flowering time and links defence responses and reproductive development. Plant J 37:209–217

Mazzucotelli E, Tartari A, Cattivelli L, Forlani G (2006) Metabolism of gamma-aminobutyric acid during cold acclimation and freezing and its relationship to frost tolerance in barley and wheat. J Exp Bot 57:3755–3766

Murata Y, Pei Z-M, Mori IC, Schroeder J (2001) Abscisic acid activation of plasma membrane Ca^{2+} channels in guard cells requires cytosolic NAD(P)H and is differentially disrupted upstream and downstream of reactive oxygen species production in *abi1–1* and *abi2–1* protein phosphatase 2C mutants. Plant Cell 13:2513–2523

Nagele T, Kandel BA, Frana S, Meissner M, Heyer AG (2011) A systems biology approach for the analysis of carbohydrate dynamics during acclimation to low temperature in Arabidopsis thaliana. FEBS J 278:506–518

Niles JC, Wishnok JS, Tannenbau SR (2006) Peroxynitrite-induced and nitration products of guanine, 8-oxoguanine: structures and mechanisms of product formation. Nitric Oxide 14:109–121

Ogawa K (2005) Glutathione-associated regulation of plant growth and stress responses. Antioxid Redox Signal 7:973–981

Ogawa K, Tasaka Y, Mino M, Tanaka Y, Iwabuchi M (2001) Association of glutathione with flowering in *Arabidopsis thaliana.* Plant Cell Physiol 42:524–530
Okuda T, Matsuda Y, Yamanaka A, Sagisaka S (1991) Abrupt increase in the level of hydrogen peroxide in leaves of winter wheat is caused by cold treatment. Plant Physiol 97:1265–1267
Pociecha E, Plazek A, Janowiak F, Waligorski P, Zwierzykowski Z (2009) Changes in abscisic acid, salicylic acid and phenylpropanoid concentrations during cold acclimation of androgenic forms of Festulolium (*Festuca pratensis x Lolium multiflorum*) in relation to resistance to pink snow mould (*Microdochium nivale*). Plant Breed 128:397–403
Prasad TK, Anderson MD, Martin BA, Stewart CR (1994) Evidence for chilling-induced oxidative stress in maize seedlings and a regulatory role for hydrogene peroxide. Plant Cell 6:65–74
Rohloff J, Kopka J, Erban A, Winge P, Wilson RC, Bones AM, Davik J, Randall SK, Alsheikh MK (2012) Metabolite profiling reveals novel multi-level cold responses in the diploid model Fragaria vesca (woodland strawberry). Phytochemistry 77:99–109
Rouhier N, Lemaire SD, Jacquot J-P (2008) The role of glutathione in photosynthetic organisms: emerging functions for Glutaredoxins and Glutathionylation. Annu Rev Plant Biol 59:143–166
Sawahel WA, Hassan AH (2002) Generation of transgenic wheat plants producing high levels of the osmoprotectant proline. Biotechnol Lett 24:721–725
Scebba F, Sebastiani L, Vitagliano C (1998) Changes in activity of antioxidative enzymes in wheat (Triticum aestivum) seedlings under cold acclimation. Physiol Plantarum 104:747–752
Scott IM, Clarke SM, Wood JE, Mur LAJ (2004) Salicylate accumulation inhibits growth at chilling temperature in Arabidopsis. Plant Physiol 135:1040–1049
Shan DP, Huang JG, Yang YT, Guo YH, Wu CA, Yang GD, Gao Z, Zheng CC (2007) Cotton GhDREB1 increases plant tolerance to low temperature and is negatively regulated by gibberellic acid. New Phytol 176:70–81
Shibasaki K, Uemura M, Tsurumi S, Rahman A (2009) Auxin response in Arabidopsis under cold stress: underlying molecular mechanisms. Plant Cell 21:3823–3838
Soltész A, Tàmár I, Vashegyi I, Kellős T, Szalai G, Szücs A, Vágújfalvi A, Kocsy G, Galiba G (2011) Redox changes during cold acclimation affect freezing tolerance but not the vegetative/reproductive transition of the shoot apex in wheat. Plant Biol 13:757–766
Spadaro D, Yun B-W, Spoel SH, Chu C, Wang Y-Q, Loake GJ (2010) The redox switch: dynamic regulation of protein function by cysteine modification. Physiol Plantarum 138:360–371
Streb P, Shang W, Feierabend J (2002) Resistance of cold-hardened winter rye leaves (*Secale cereale* L.) to photo-oxidative stress. Plant Cell Environ 22:1211–1223
Sun X, Xi DH, Feng H, Du JB, Lei T, Liang HG, Lin HH (2009) The dual effects of salicylic acid on dehydrin accumulation in water-stressed barley seedlings. Russ J Plant Physiol 56:348–354
Svenning MM, Rosnes K, Junttila O (1997) Frost tolerance and biochemical changes during hardening and dehardening in contrasting white clover populations. Physiol Plantarum 101:31–37
Swaaj AC van, Jacobsen E, Feenstra WJ (1985) Effect of cold hardening, wilting and exogenously applied proline on leaf proline content and frost tolerance of several genotypes of *Solanum.* Physiol Plantarum 64:230–236
Szalai G, Kellős T, Galiba G, Kocsy G (2009) Glutathione as an antioxidant and regulatory molecule in plants under abiotic stress conditions. J Plant Growth Regul 28:66–80
Thomashow MF (1999) Plant cold acclimation: freezing tolerance genes and regulatory mechanisms. Annu Rev Plant Physiol Plant Mol Biol 50:571–599
Usadel B, Blasing OE, Gibon Y, Poree F, Höhne M, Günter M, Trethewey R, Kamlage B, Poorter H, Stitt M (2008) Multilevel genomic analysis of the response of transcripts, enzyme activities and metabolites in Arabidopsis rosettes to a progressive decrease of temperature in the non-freezing range. Plant Cell Environ 31:518–547
Vágújfalvi A, Kerepesi I, Galiba G, Tischner T, Sutka J (1999) Frost hardiness depending on carbohydrate changes during cold acclimation in wheat. Plant Sci 144:85–92
Valliyodan B, Nguyen HT (2006) Understanding regulatory networks and engineering for enhanced drought tolerance in plants. Curr Opin Plant Biol 9:189–195
Veisz O, Galiba G, Sutka J (1996) Effect of abscisic acid on the cold hardiness of wheat seedlings. J Plant Physiol 149:439–443

Vendruscolo EC, Schuster I, Pileggi M, Scapim CA, Molinari HB, Marur CJ, Vieira LG (2007) Stress-induced synthesis of proline confers tolerance to water deficit in transgenic wheat. J Plant Physiol 164:1367–1376

Yamaguchi-Shinozaki K, Shinozaki K (2006) Transcriptional regulatory networks in cellular responses and tolerance to dehydration and cold stresses. Annu Rev Plant Biol 57:781–803

Zhao MG, Chen L, Zhang LL, Zhang WH (2009) Nitric reductase-dependent nitric oxide production is involved in cold acclimation and freezing tolerance in Arabidopsis. Plant Physiol 151:755–767

Zhao R, Sheng J, Lu S, Zheng Y, Zhang J, Yu M, Shen L (2011) Nitric oxide participates in the regulation of LeCBF1 gene expression and improves cold tolerance in harvested tomato fruit. Postharvest Biol Technol 62:121–126

Zhu J, Dong CH, Zhu JK (2007) Interplay between cold-responsive gene regulation, metabolism and RNA processing during plant cold acclimation. Curr Opin Plant Biol 10:290–295

Zuther E, Buchel K, Hundertmark M, Stitt M, Hincha DK, Heyer AG (2004) The role of raffinose in the cold acclimation response of *Arabidopsis thaliana.* FEBS Lett 576:169–173

Zuther E, Schulz E, Childs LH, Hincha DK (2012) Clinal variation in the nonacclimated and cold acclimated freezing tolerance of *Arabidopsis thaliana* accessions. Plant Cell Environ 35:1860–1878

The Function and Evolution of Closely Related COR/LEA (Cold-Regulated/Late Embryogenesis Abundant) Proteins in *Arabidopsis thaliana*

Anja Thalhammer and Dirk K. Hincha

Introduction

Cold has a major influence on plant growth and survival, limiting the geographical distribution of wild species and the yield and growth season of agricultural crops. Considerable effort has, therefore, been directed towards understanding how plants adapt to low temperature. Freezing damage is, however, not a consequence of low temperatures, but the result of cellular dehydration brought about by extracellular ice crystallization (see Steponkus 1984; Thomashow 1999; Xin and Browse 2000 for reviews). Cellular membranes are generally considered the major sites of freezing injury in plants and there is extensive evidence for damage during a freeze–thaw cycle to the plasma membrane (Steponkus 1984) and to chloroplast thylakoid and envelope membranes (Ehlert and Hincha 2008; Hincha et al. 1987). However, the inactivation of enzymes cannot be excluded, although unequivocal evidence for the latter is still missing.

The freezing tolerance of a plant depends strongly on its thermal history. Nonacclimated tolerance describes the ability to survive freezing without prior adaptation. Acclimated tolerance is usually higher in herbaceous plants by approximately 2 to 25 °C (Fennell et al. 1990; Lee et al. 2012; Rumich-Bayer and Krause 1986; Rütten and Santarius 1988; Steponkus et al. 1993; Yelenosky and Guy 1989; Zuther et al. 2012) depending on the species. This reflects the ability of many plant species from temperate climates to increase their freezing tolerance in response to low but nonfreezing temperatures, usually in the range between 10 and 0 °C over several days to weeks.

Genetic approaches have defined some of the key regulatory components of cold acclimation in *Arabidopsis thaliana*. A central role is played by the *C-repeat binding factors (CBF)1*, *2*, and *3* (Gilmour et al. 1998), also known as *dehydration*

D. K. Hincha (✉) · A. Thalhammer
Max-Planck-Institut für Molekulare Pflanzenphysiologie,
Am Mühlenberg 1, 14476 Potsdam, Germany
e-mail: hincha@mpimp-golm.mpg.de

A. Thalhammer
e-mail: thalhammer@mpimp-golm.mpg.de

R. Imai et al. (eds.), *Plant and Microbe Adaptations to Cold in a Changing World*,
DOI 10.1007/978-1-4614-8253-6_8,

responsive element binding1 (DREB1)b, *c*, and *a* (At4g25490, At4g25470, At4g25480), respectively (Liu et al. 1998). These transcription factors are rapidly induced in response to cold (Gilmour et al. 1998; Liu et al. 1998; McKhann et al. 2008; Stockinger et al. 1997) and in turn activate the expression of a set of target genes including a number of COR/LEA protein-encoding genes.

In addition, there are complex physiological and biochemical changes associated with cold acclimation in plants, including changes in growth and water balance, the accumulation of compatible solutes, changes in membrane and cell wall composition, increased antioxidant production, and changes in gene expression and protein levels (see Guy et al. 2008; Steponkus 1984; Thomashow 1999; Xin and Browse 2000 for reviews).

Characteristics of LEA and COR Proteins

General Characteristics of LEA Proteins

LEA (late embryogenesis abundant) proteins were first described more than 30 years ago to accumulate late in cotton seed development, when the embryo becomes desiccation tolerant (Dure et al. 1981). Related proteins were later found in seeds of other plant species and in vegetative plant organs, mostly under abiotic stress conditions such as low temperature, high salt concentrations, or drying, resulting in cellular dehydration (see Tunnacliffe et al. 2010; Tunnacliffe and Wise 2007 for reviews). However, LEA proteins are not plant specific. Closely related proteins have also been found in bacteria, cyanobacteria, and some invertebrates (Hand et al. 2011; Tunnacliffe et al. 2010).

Plant LEA proteins have been separated into different groups sharing various sequence motifs, but the groupings and the nomenclature of the groups are not consistent in the literature (see Hincha and Thalhammer 2012; Tunnacliffe et al. 2010 for recent reviews). In the model plant *Arabidopsis thaliana*, 51 LEA proteins were identified that were separated into nine different groups (Bies-Etheve et al. 2008; Hundertmark and Hincha 2008). Predominantly, members of the Pfam LEA_4 family have also been found in nonplant organisms.

Most LEA proteins are predicted to be IDPs (intrinsically disordered proteins) in the fully hydrated state. However, they may acquire structure and (partially) fold to form α-helices during drying (Hand et al. 2011; Hincha and Thalhammer 2012; Tunnacliffe et al. 2010). Due to their hydrophilic, unstructured nature, most LEA proteins do not aggregate during drying, freezing, or even boiling. Secondary structure determinations by circular dichroism (CD) spectroscopy of recombinant *Arabidopsis* LEA proteins from the Pfam families LEA_4 (Popova et al. 2011; Thalhammer et al. 2010), LEA_5, and LEA_6 (Hundertmark et al. 2012) showed that all eight proteins are 60–80% unstructured in the hydrated state, while the degree of unstructuredness is strongly reduced in the dry state. For several LEA proteins, computational analysis indicated that they may form amphipathic class A α-helices (e.g., Popova et al. 2011; Thalhammer et al. 2010; Tolleter et al. 2007), a motif found in membrane-interacting apolipoproteins (Segrest et al. 1992).

Although the physiological role of most LEA proteins is still obscure, several possible activities of LEA proteins, such as binding to RNA or DNA, water or ion binding, antioxidative activity, or sugar glass stabilization in the dry state have been suggested based on in vitro experiments (Tunnacliffe et al. 2010; Tunnacliffe and Wise 2007). There is, however, good evidence that some LEA proteins can stabilize membranes and/or enzymes during freezing and desiccation.

Some COR Proteins Are Also LEA Proteins

COR (Cold-Regulated) genes were originally identified by differential cDNA screening and DNA sequencing in *Arabidopsis*. Based on sequence comparisons, several COR proteins were simultaneously classified as LEA proteins (Thomashow 1999). It was also realized that many *COR* genes form parts of tandem repeats in the *Arabidopsis* genome. In a genome-wide survey of all *Arabidopsis* genes encoding LEA proteins, this has been shown for the gene pairs *COR47/LTI45* (dehydrin), *COR15A/COR15B* (LEA_4), and *XERO1/XERO2* (*LTI30*, dehydrin; Hundertmark and Hincha 2008). In addition, some cold-induced genes form homeologous pairs in the *Arabidopsis* genome, i.e., they are derived from ancient genome duplications (see section Evolution and Expression Networks of COR15A and COR15B and Their Closest Homologues in *Arabidopsis*). In this case, however, sometimes only one gene in a pair encodes a COR protein. Such homeologous pairs are formed by: *LEA1* (COR) and *LEA27* (LEA_2), *LEA2* and *LEA38* (COR, LEA_3), *LTI45* (COR) and *ERD14* (dehydrin), and *LTI30* and *RAB18* (both COR, dehydrin). These complex genomic relationships pose interesting questions about the evolutionary, regulatory, and functional relationships between different COR and LEA proteins that are explored in more detail for COR15A and COR15B and their closest homologues in section Evolution and Expression Networks of COR15A and COR15B and Their Closest Homologues in *Arabidopsis*.

Functions of COR/LEA Proteins

Protection of Enzymes During Drying and Freezing

Various LEA proteins have been shown to preserve the activity of isolated enzymes during desiccation or freezing (see Battaglia et al. 2008; Tunnacliffe et al. 2010 for reviews) where they may form aggregates and lose their catalytic activity. Some LEA proteins prevent aggregation and inactivation (see, e.g., Chakrabortee et al. 2007; Goyal et al. 2005; Kovacs et al. 2008; Pouchkina-Stantcheva et al. 2007) both in combination with single, isolated enzymes and also in complex protein mixtures, such as the water-soluble proteomes of human and nematode cells (Chakrabortee et al. 2007). The *Medicago truncatula* protein MtPM25 (LEA_SMP family; Hundertmark and Hincha 2008) was, in addition, able to dissolve protein aggregates formed during drying or freezing (Boucher et al. 2010), a property that was not observed in other LEA proteins. Most of the investigated LEA proteins showed no classical chaperone function, i.e., they were not able to prevent heat aggregation and inactivation of heat-sensitive enzymes (Tunnacliffe et al. 2010). However, the two *Arabidopsis* dehydrins ERD (Early Response to Dehydration) 10 (LTI (Low Temperature Inducible) 45) and ERD14 also prevented protein aggregation during heat stress (Kovacs et al. 2008). These data indicate that LEA proteins possess different functional properties in relation to enzyme stabilization that have not been linked to any structural characteristics of LEA proteins or their target proteins yet.

A "molecular shield" mechanism has been proposed to explain the antiaggregation activity of LEA proteins (Goyal et al. 2005). In this model, the unstructured LEA proteins exert an excluded volume effect which would sterically interfere with interactions between partially denatured proteins, thereby reducing aggregation (Liu et al. 2011). However, more recent studies indicate that additional mechanisms may be required to not only prevent aggregation but also loss of activity in enzymes during drying (Chakrabortee et al. 2012; Hughes and Graether 2011).

Stabilization of Membranes During Drying and Freezing

Membrane integrity is an essential factor for cell survival under stress conditions. Although LEA proteins are, in general, highly hydrophilic, some may form amphipathic α-helices during drying, as described earlier. These structures may enable them to interact peripherally with membranes under conditions of partial or complete dehydration. This has been shown by Fourier transform infrared (FTIR) spectroscopy for both rotifer and plant proteins (Popova et al. 2011; Pouchkina-Stantcheva et al. 2007; Rahman et al. 2010; Thalhammer et al. 2010; Tolleter et al. 2010). In all cases, such interactions resulted in shifts of lipid phase transition temperatures that were also influenced by membrane lipid composition. In addition, the protein LEAM from pea seed mitochondria stabilized liposomes during freezing and drying, indicated by increased retention of a fluorescent dye after thawing or rehydration (Tolleter et al. 2010). Significantly, protection was highest for membranes containing the mitochondrial lipid cardiolipin, indicating a certain degree of membrane specificity of this LEA protein.

For some LEA proteins, especially dehydrins, membrane binding in solution has been shown. All dehydrins contain at least one K-segment, a 15-mer Lys-rich sequence with a propensity to fold into a potentially membrane-interacting amphipathic α-helix (Mouillon et al. 2006; Rahman et al. 2010). However, binding to artificial membrane vesicles has been found to depend critically on the presence of negatively charged lipid headgroups on the membrane surface (Eriksson et al. 2011;

Koag et al. 2003; Koag et al. 2009; Kovacs et al. 2008). Indeed, it has been shown for the *Arabidopsis* dehydrin LTI30 that such electrostatic interactions are mediated by the histidine groups in the protein (Eriksson et al. 2011). In this case, binding leads to aggregation of the lipid vesicles. A similar effect has also been shown for the basic *Arabidopsis* protein LEA18, which, in addition, induces leakage of soluble contents from the vesicles (Hundertmark et al. 2011). Whether such interactions contribute to the dehydration tolerance of cells remains to be resolved.

Structure, Potential Functions, and Evolution of the *Arabidopsis thaliana* COR15 Proteins and Their Closest Homologues

Functional Characteristics of COR15A and COR15B

The *Arabidopsis* LEA proteins COR15A and COR15B are 70 % identical in their amino acid sequences (Thalhammer et al. 2010) and are encoded by a tandem repeat pair of genes (Wilhelm and Thomashow 1993). Both genes are cold induced (Hundertmark and Hincha 2008) and the relative expression of *COR15A* is positively correlated with the cold-acclimated freezing tolerance of a collection of 54 *Arabidopsis* accessions (Zuther et al. 2012). A membrane-stabilizing function, in vivo, has only been shown for COR15A (Artus et al. 1996). Overexpression of this protein in transgenic *Arabidopsis* plants leads to increased leaf-freezing tolerance without cold acclimation. In addition, it was shown that COR15A is localized in the chloroplast stroma (Lin and Thomashow 1992b) and the same localization was predicted for COR15B from the full-length amino acid sequence (Hundertmark and Hincha 2008).

FTIR spectroscopy provided evidence for interactions of both COR15A and COR15B with the chloroplast-specific glycolipid MGDG in the dry state (Thalhammer et al. 2010). COR15A was also able to prevent the formation of the interbilayer hexagonal II phase in membrane dispersions containing MGDG (Steponkus et al. 1998). However, membrane binding and protein folding in the (partially) hydrated state still remain to be shown for both proteins. In addition, it has been shown that COR15A is able to stabilize the enzyme lactate dehydrogenase (LDH) from rabbit muscle against inactivation during repeated freeze–thaw cycles between liquid nitrogen and room temperature (Lin and Thomashow 1992a; Nakayama et al. 2007; Nakayama et al. 2008). Whether COR15A and/or COR15B have an enzyme-stabilizing function in plants under milder, more realistic freezing conditions, however, remains to be experimentally elucidated.

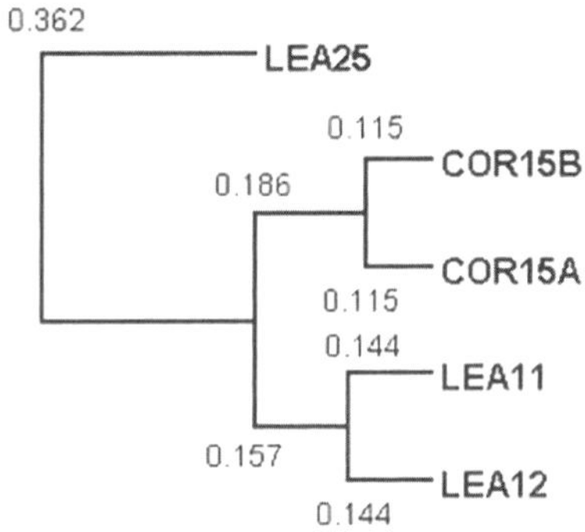

Fig. 1 An unrooted dendrogram of the *COR15A*, *COR15B*, *LEA11*, *LEA12*, and *LEA25* nucleotide sequences was created using SplitsTree4 (Huson and Bryant 2006) with the UPGMA consensus tree algorithm. Evolutionary distances expressed as nucleic acid exchanges per site were calculated with the Uncorrected P algorithm. ClustalW (Goujon et al. 2010; Larkin et al. 2007) was used to calculate pairwise identity scores between all investigated LEA genes. Branch lengths are drawn proportional to evolutionary distances which are additionally indicated numerically

Evolution and Expression Networks of COR15A and COR15B and Their Closest Homologues in Arabidopsis

The nucleotide sequences of the *COR15A* and *COR15B* genes are closely related to three other genes that are also located on chromosome 2 of *Arabidopsis*. The genetic distances of these five LEA_ 4 genes, *COR15A* (At2g42540), *COR15B* (At2g42530), *LEA11* (At2g03740), *LEA12* (At2g03850), and *LEA25* (At2g42560) are visualized in an unrooted dendrogram (Fig. 1). *COR15A* and *COR15B*, as well as *LEA11* and *LEA12,* each form a tandem repeat gene pair and the two pairs are more closely related to each other than to *LEA25*. Both genes in each tandem repeat pair share a nucleotide sequence identity of more than 80 %, while all five genes show an identity of more than 65 %.

Consequently, a multiple sequence alignment of the five encoded protein sequences (Fig. 2) shows several sequence domains that are common to all proteins. The N-termini of COR15A, COR15B, LEA11, and LEA12 were predicted to be chloroplast transit peptides (Hundertmark and Hincha 2008) that direct the import of the nuclear-encoded proteins into the chloroplast stroma. Chloroplast localization has only been shown experimentally for COR15A (Lin and Thomashow 1992a; Nakayama et al. 2007), but according to the N-terminal amino acid sequences, this seems to be a common feature of all four proteins.

Both *COR15* genes are highly (more than 80-fold) induced in leaves after cold treatment and to a lesser extent in response to other abiotic stresses (Hundertmark and Hincha 2008). The other three genes (*LEA11*, *LEA12*, *LEA25*) are not stress-regulated, but rather show tissue-specific expression. This is visualized in Fig. 3a where we used the GeneCAT expression-profiling tool linked to a database that contains developmental expression data from large-scale microarray experiments (Mutwil et al. 2008; Schmid et al. 2005). These data show that *LEA11* and *LEA12* are exclusively expressed in flower buds, albeit at rather low levels. Interestingly, the two stress-related genes *COR15A* and *COR15B* also show prominent expression in flower tissues, where the expression of *COR15A* is up to fourfold higher than the expression of the other three *LEA* genes. Also, *COR15A* is expressed during all

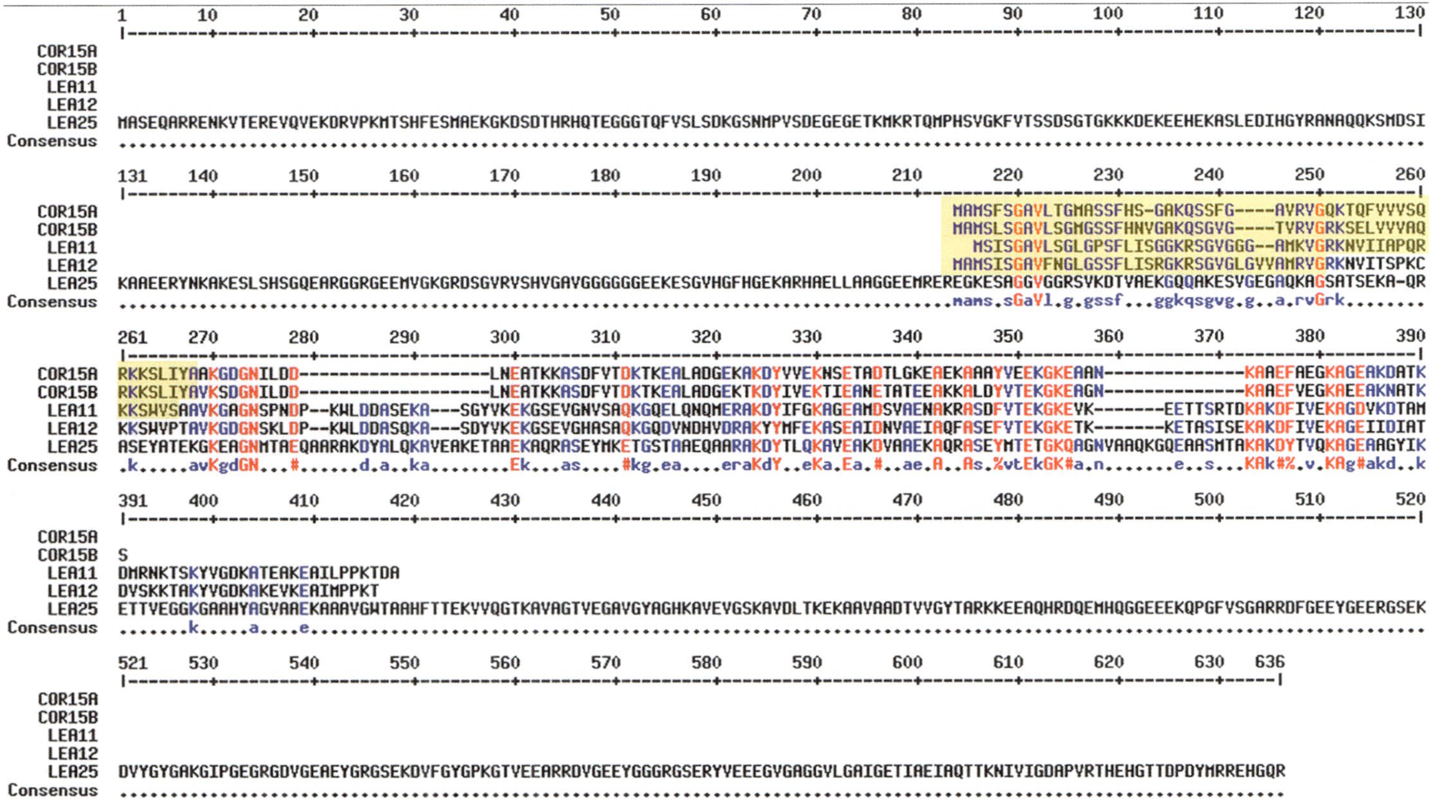

Fig. 2 Multiple amino acid sequence alignment of the proteins COR15A, COR15B, LEA11, LEA12, and LEA25. The alignment was performed using the web tool MultAlin (Corpet 1988) with the default settings. Chloroplast transit peptides were predicted with TargetP (Emanuelsson et al. 2007) and are highlighted in *yellow*. *Blue* and *red* letters indicate high (>90 %) and low (>50 %) consensus levels, respectively. In the consensus sequence, *dots* indicate positions without conserved residues, while *upper-* and *lowercase letters* refer to high and low consensus levels indicated above

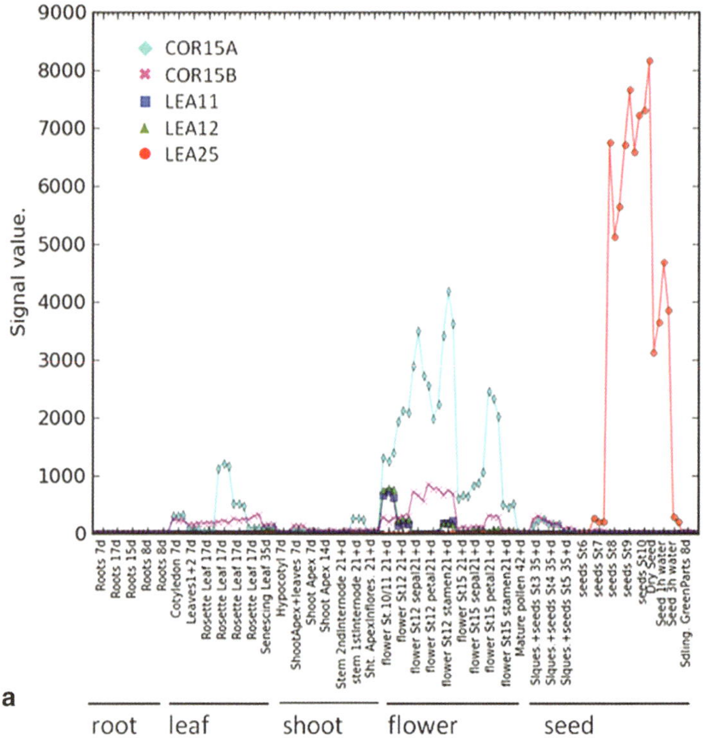

	COR15A	COR15B	LEA11	LEA12	LEA25
COR15A	1				
COR15B	0,93017	1			
LEA11	0,24646	0,23587	1		
LEA12	0,14623	0,10038	0,69703	1	
LEA25	-0,1831	-0,2106	-0,0948	-0,0588	1

1,0-0,75
0,75-0,5
0,5-0,25
0,25-0
0--0,25
-0,25--0,5
-0,5--0,75
0,75--1,0

b

Fig. 3 **a** Expression profiles of the five LEA genes of interest were generated using the GeneCAT software (Mutwil et al. 2008) on gene expression data covering *Arabidopsis* development using the Affymetrix ATH1 microarray (Schmid et al. 2005). **b** Pearson correlation coefficients (*r*-values) indicating the degree of coexpression in pairwise comparisons between all five LEA genes

stages of flower development, whereas *LEA11*, *LEA12*, and *COR15B* expression seems to be limited to the early stages. However, the role of all four proteins in flowers is unclear. *LEA25*, on the other hand, is exclusively expressed in seeds and while this may point to a functional role in seed desiccation tolerance or seed longevity, experimental evidence for such a role is also still missing.

Since the database linked to GeneCAT contains no stress-related expression data (Mutwil et al. 2008; Schmid et al. 2005), the well-known *COR15* gene response to

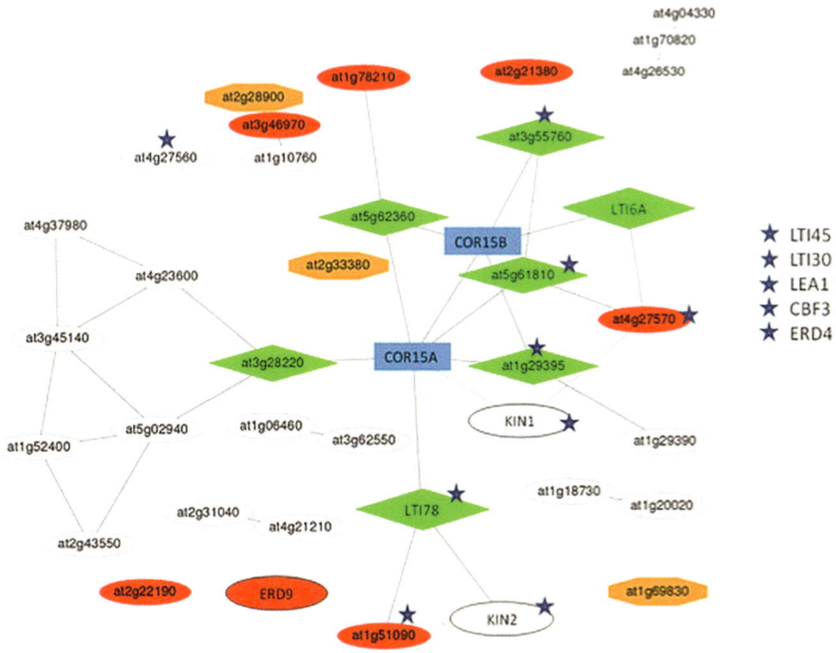

Fig. 4 Gene coexpression network using *COR15A* and *COR15B* as bait genes. The network was constructed using GeneCAT (Mutwil et al. 2008) with an *r*-value cut-off of 0.6. The *lines* connecting different gene identifiers indicate mutual coexpression ranks less than 10 after comparing the top 50 genes from the coexpression list in a pairwise manner. Genes showing mutual coexpression ranks greater than 10 are not connected to only show the strongest coexpression. *Arabidopsis* Genome Initiative (AGI) codes of known stress-regulated genes were replaced by conventional gene names for easier identification. Since the database linked to GeneCAT contains no stress-related expression data, the response to stress treatments is not mapped in this network. Therefore, genes indicated with a *star* were additionally detected using the Genevestigator coexpression tool (Hruz et al. 2008; Zimmermann et al. 2004) with the perturbation metaprofile. The different colors of the gene identifiers indicate bait genes (*blue*) and decreasing coexpression strength between node and bait gene (*green* > *orange* > *red*)

stress treatments is not mapped in Fig. 3a. A coexpression analysis, nevertheless, clearly showed that while the expression of the members of the two tandem repeat pairs was tightly correlated (*COR15A* vs. *COR15B*, $r=0.93$; *LEA11* vs. *LEA12*, $r=0.70$), there was no significant coexpression outside these gene pairs ($r<0.25$; Fig. 3b).

A GeneCAT coexpression network was created using *COR15A* and *COR15B* as bait genes to get a wider overview of the transcriptional coregulation of these genes. Both genes were found to be part of a closely correlated expression cluster (Fig. 4) even under unstressed conditions. A number of well-characterized stress-responsive genes were part of this cluster, for example, *KIN1* (Kurkela and Franck 1990), *KIN2* (Kurkela and Borg-Franck 1992), *ERD9* (Chen et al. 2012), and *COR78*

(Horvath et al. 1993). This finding is in good agreement with a recent experimental study on 54 *Arabidopsis* accessions (Zuther et al. 2012) that also showed tightly correlated expression of a number of cold-induced genes, including *COR15A* and *COR15B*, even under warm control conditions.

In a next step, the gene coexpression network under nonstressed conditions was compared to a network built by the Genevestigator tool (Hruz et al. 2008; Zimmermann et al. 2004) using the perturbation metaprofile in order to incorporate the expression profiles of genes under a wide range of stress conditions. Overall, the list of coexpressed genes was quite similar to that obtained using GeneCAT. Some additional stress-responsive genes were found, but most of them showed lower correlation coefficients ($0.64 > r > 0.51$) than those identified in the GeneCAT analysis. Among the genes whose expression was only correlated with the expression of *COR15A*/*COR15B* under stress conditions were *LTI45*, *LTI30* (Welin et al. 1994), *LEA1* (Kreps et al. 2002), *CBF3* (Gilmour et al. 1998), and *ERD4* (Zhang et al. 2004) as indicated in Fig. 4.

GeneCAT coexpression analysis of *LEA25* expression revealed a large network containing 267 genes ($r > 0.6$, not shown). Strikingly, of a total of 35 *Arabidopsis* seed-expressed *LEA* genes (Hundertmark and Hincha 2008), 63 % (22 genes) are part of this network (Fig. 5). Moreover, these coexpressed genes are not randomly distributed among the nine different LEA families. Genes from three families (Pv-LEA18, LEA_2, LEA_3) are not represented in the network, although these families contain seed-expressed genes. Of the members of the SMP and AtM families, approximately equal numbers of the seed-expressed genes are either represented in or outside the network. On the other hand, all or almost all seed-expressed members of the LEA_1, LEA_4, LEA_5, and dehydrin families are present in the coexpression network. This is insofar unexpected, as *LEA* gene classification is derived from sequence motifs (Hundertmark and Hincha 2008) and not from any regulatory properties of the genes. Since many *LEA* genes are regulated by abscisic acid (ABA), which also plays an important role in seed desiccation tolerance development (Ingram and Bartels 1996), regulation of the coexpressed *LEA* genes by ABA was an attractive hypothesis. However, when we compared ABA-regulated genes (Hundertmark and Hincha 2008) among the coexpressed or not coexpressed genes, we found that of the 22 coexpressed *LEA* genes 11 each were induced and not induced by ABA. Likewise, of those *LEA* genes that were not coexpressed, six were ABA-induced and seven were not. This distribution clearly excludes ABA as the decisive factor for the establishment of the seed *LEA* gene coexpression network.

To further explore the evolutionary relationships between the *COR15* genes and their closest homologues (Fig. 1), the gene duplication history of these five *LEA* genes was assessed using the number of synonymous substitutions per synonymous site (K_S) as a measure of evolutionary distance of genes. Such substitutions do not result in changes in the amino acid sequence of the encoded proteins and are, therefore, not under selection. Thus, their rate of fixation is relatively constant and they can be used as a measure for the overall mutation rate (Maere et al. 2005). It is widely accepted that the *Arabidopsis* genome underwent three whole genome duplication (WGD) events during its evolution (Barker et al. 2009; Maere et al. 2005),

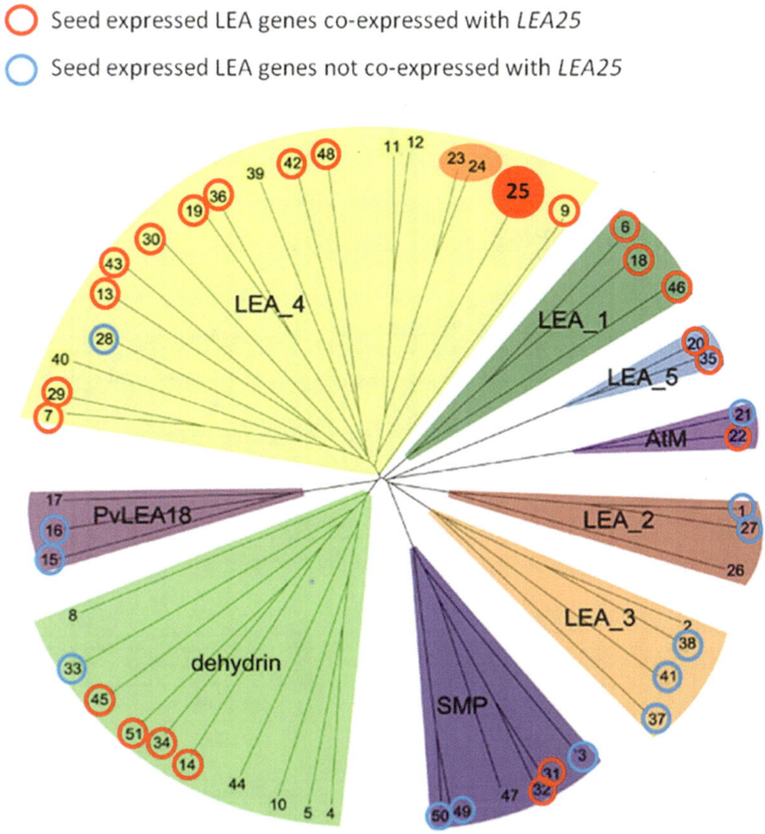

Fig. 5 Coexpression analysis of *LEA25* with other seed-expressed LEA genes. Coexpression analysis was performed with the GeneCAT tool as in Fig. 4 with an *r*-value cut-off of 0.65. The unrooted dendrogram of all 51 *Arabidopsis* LEA proteins was taken from Hundertmark and Hincha (2008)

referred to as 1R, 2R, and 3R (Fig. 6a), where the most recent 3R WGD event corresponds to the diversification of the Brassicaceae (Barker et al. 2009). Dissecting this most recent genome duplication event revealed, in addition to the whole genome duplication, a more recent tandem gene duplication event (Blanc and Wolfe 2004; Fig. 6b).

K_S-values of the *COR15* and *LEA11/12* gene duplications of about 0.3 indicate a duplication of these gene pairs during the 3R tandem gene duplication event. However, when the two LEA gene pairs were compared, an average K_S-value of 1.0 dates the separation of the progenitors of these tandem repeat pairs towards the earlier 3R WGD event (Fig. 6b, c). A common gene duplication past of *COR15A*, *COR15B*, *LEA11*, and *LEA12* with *LEA25*, on the other hand, is unlikely, since no consistent K_S-values could be calculated for the four possible comparisons.

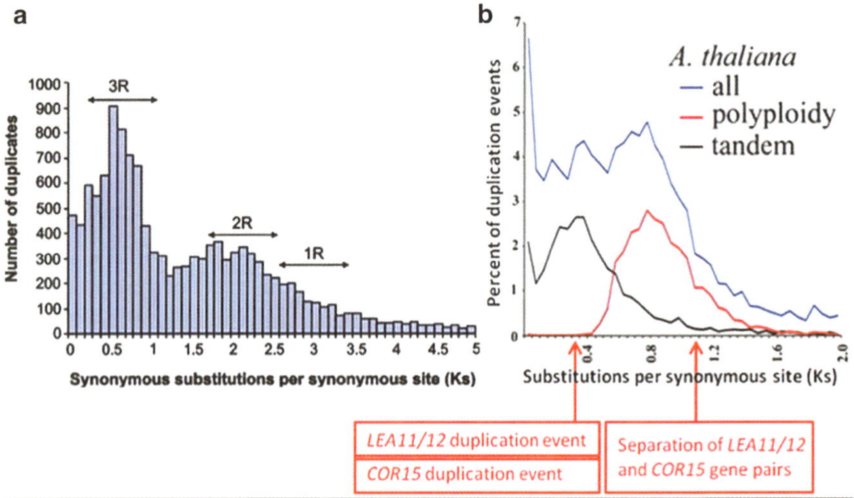

LEA gene pair	K_s	St.dev.	Estimated age of gene duplication	St.dev.
COR15A vs. *COR15B*	0.321	0.037	10.7 million years	1.250
LEA11 vs. *LEA12*	0.337	0.021	11.2 million years	0.728
COR15A+B vs. *LEA11+12*	1.019	0.189	32.3 million years	3.315

LEA gene pair	K_a/k_s	St.dev.	Selection
COR15A vs. *COR15B*	0.450	0.069	Predominantly purifiying
LEA11 vs. *LEA12*	0.506	0.046	Predominantly purifiying
COR15A+B vs. *LEA11+12*	0.368	0.106	Predominantly purifiying

Fig. 6 **a** Age distribution of the *Arabidopsis* paranome based on K_S (fraction of synonymous substitutions per synonymous site) values, calculated with the softwares KaKs_Calculator (Zhang et al. 2006) and DnaSP v5 (Librado and Rozas 2009) averaging over all available algorithms. *1R*, *2R*, and *3R* refer to the three genome-wide duplication events that have occurred in *Arabidopsis* or its predecessors (figure from Maere et al. 2005). **b** The 3R genome duplication event in higher resolution. Two underlying peaks can be resolved, one due to the genome duplication event, the other due to a tandem gene duplication event (figure modified from Blanc and Wolfe 2004). *COR15A/B* and *LEA11/12* duplication events were determined on the basis of K_S-values and are indicated with *red arrows*. **c** The age of all gene duplications was estimated based on K_S-values (see text for details). The parameter K_A/K_S was used to assess potential selective constraints of the duplicated *LEA* genes

The absolute ages of the gene duplication events were estimated using a mean substitution rate of 1.5 synonymous substitutions per 10^8 years (Koch et al. 2000). This leads to an approximate *COR15A/B* and *LEA11/12* tandem gene duplication age of about 11 million years, whereas the separation of both gene pairs is estimated at about 32 million years ago (Fig. 6c). These age estimates place the *COR15A/B* and *LEA11/12* duplications on a timescale matching both the divergence of *Brassica* and *Arabidopsis* (14–24 million years ago) and the separation of the *Arabidopsis* and *Arabis* (9–20 million years ago; Koch et al. 2000; Yang et al. 1999). Separation of the *LEA11/12* and *COR15A/B* ancestors is estimated to be as old as the family of Brassicaceae, which is dated to about 23–50 million years ago (Barker et al. 2009; Beilstein et al. 2010).

The vast majority of gene duplications is not maintained in the genome over evolutionary timescales. The average half-life of duplicate genes averaged over all species is approximately 4 million years (Lynch and Conery 2000). It is possible to assess the selective constraints on duplicated genes using the parameter K_A/K_S, where K_A is the number of nonsynonymous substitutions per synonymous site, i.e., the number of nucleotide exchanges that result in amino acid substitutions. In this context, a K_A/K_S-value greater than 1 implies positive directional selection with a high probability for neofunctionalization of one of the gene copies. K_A/K_S-values around 1 indicate an absence of selection, whereas K_A/K_S-values less than 1 point to a predominantly purifying selection where only one of the gene copies is maintained over a long time (Lynch and Conery 2000). In the case of all evaluated LEA gene duplications, the K_A/K_S-values were about 0.35–0.5 (Fig. 6c), making a long-term persistence or neofunctionalization of these gene duplicates unlikely. These findings point to a possible redundancy or dosage effect of all four LEA genes within the tandem repeats, a hypothesis that we are currently testing experimentally.

References

Artus NN, Uemura M, Steponkus PL, Gilmour SJ, Lin C, Thomashow MF (1996) Constitutive expression of the cold-regulated *Arabidopsis thaliana COR15a* gene affects both chloroplast and protoplast freezing tolerance. Proc Natl Acad Sci U S A 93:13404–13409

Barker MS, Vogel H, Schranz M (2009) Paleopolyploidy in the *Brassicales*: analyses of the cleome transcriptome elucidate the history of genome duplications in *Arabidopsis* and other *Brassicales*. Genome Biol Evol 1:391–399

Battaglia M, Olvera-Carrillo Y, Garciarrubio A, Campos F, Covarrubias AA (2008) The enigmatic LEA proteins and other hydrophilins. Plant Physiol 148:6–24

Beilstein MA, Nagalingum NS, Clements MD, Manchester SR, Mathews S (2010) Dated molecular phylogenies indicate a Miocene origin for *Arabidopsis thaliana*. Proc Natl Acad Sci U S A 107:18724–18728

Bies-Etheve N, Gaubier-Comella P, Debures A, Lasserre E, Jobet E, Raynal M, Cooke R, Delseny M (2008) Inventory, evolution and expression profiling diversity of the LEA (late embryogenesis abundant) protein gene family in *Arabidopsis thaliana*. Plant Mol Biol 67:107–124

Blanc G, Wolfe KH (2004) Widespread paleopolyploidy in model plant species inferred from age distributions of duplicate genes. Plant Cell 16:1667–1678

Boucher V, Buitink J, Lin X, Boudet J, Hoekstra FA, Hundertmark M, Renard D, Leprince O (2010) MtPM25 is an atypical hydrophobic late embryogenesis-abundant protein that dissociates cold and desiccation-aggregated proteins. Plant Cell Environ 33:418–430

Chakrabortee S, Boschetti C, Walton LJ, Sarkar S, Rubensztein DC, Tunnacliffe A (2007) Hydrophilic protein associated with desiccation tolerance exhibits broad protein stabilization function. Proc Natl Acad Sci U S A 104:18073–18078

Chakrabortee S, Tripathi R, Watson M, Kaminski Schierle GS, Kumiawan DP, Kaminski CF, Wise MJ, Tunnacliffe A (2012) Intrinsically disordered proteins as molecular shields. Mol BioSyst 8:210–219

Chen J-H, Jiang H-W, Hsieh E-J, Chen H-Y, Chien C-T, Hsieh H-L, Lin T-P (2012) Drought and salt stress tolerance of an *Arabidopsis* glutathione S-transferase U17 knockout mutant are attributed to the combined effect of glutathione and abscisic acid. Plant Physiol 158:340–351

Corpet F (1988) Multiple sequence alignment with hierarchical clustering. Nucl Acids Res 16:10881–10890

Dure L III, Greenway SC, Galau GA (1981) Developmental biochemistry of cottonseed embryogenesis and germination: changing messenger ribonucleic acid populations as shown by in vitro and in vivo protein synthesis. Biochemistry 20:4162–4168

Ehlert B, Hincha DK (2008) Chlorophyll fluorescence imaging accurately quantifies freezing damage and cold acclimation responses in *Arabidopsis* leaves. Plant Meth 4:12

Emanuelsson O, Brunak S, Heijne G von, Nielsen H (2007) Locating proteins in the cell using TargetP, SignalP, and related tools. Nat Protoc 2:953–971

Eriksson SK, Kutzer M, Procek J, Gröbner G, Harryson P (2011) Tunable membrane binding of the intrinsically disordered dehydrin Lti30, a cold-induced plant stress protein. Plant Cell 23:2391–2404

Fennell A, Li PH, Markhart III AH (1990) Influence of air and soil temperature on water relations and freezing tolerance of spinach (*Spinacia oleracea*). Physiol Plant 78:51–56

Gilmour SJ, Zarka DG, Stockinger EJ, Salazar MP, Houghton JM, Thomashow MF (1998) Low temperature regulation of the *Arabidopsis* CBF family of AP2 transcriptional activators as an early step in cold-induced *COR* gene expression. Plant J 16:433–442

Goujon M, McWilliam H, Li W, Valentin F, Squizzato S, Paern J, Lopez R (2010) A new bioinformatics analysis tools framework at EMBL-EBI. Nucl Acids Res 38:W695–W699

Goyal K, Walton LJ, Tunnacliffe A (2005) LEA proteins prevent protein aggregation due to water stress. Biochem J 388:151–157

Guy CL, Kaplan F, Kopka J, Selbig J, Hincha DK (2008) Metabolomics of temperature stress. Physiol Plant 132:220–235

Hand SC, Menze MA, Toner M, Boswell L, Moore D (2011) LEA proteins during water stress: not just for plants anymore. Annu Rev Physiol 73:115–134

Hincha DK, Höfner R, Schwab KB, Heber U, Schmitt JM (1987) Membrane rupture is the common cause of damage to chloroplast membranes in leaves injured by freezing or excessive wilting. Plant Physiol 83:251–253

Hincha DK, Thalhammer A (2012) LEA proteins: IDPs with versatile functions in cellular dehydration tolerance. Biochem Soc Trans 40:1000–1003

Horvath DP, McLarney BK, Thomashow MF (1993) Regulation of *Arabidopsis thaliana* L. (Heyn) cor78 in response to low temperature. Plant Physiol 103:1047–1053

Hruz T, Laule O, Szabo G, Wessendorp F, Bleuler S, Oertle L, Widmayer P, Gruissem W, Zimmermann P (2008) Genevestigator V3: A reference expression database for the meta-analysis of transcriptomes. Advances Bioinformatics 2008:420747. doi:10.1155/2008/420747

Hughes S, Graether SP (2011) Cryoprotective mechanism of a small intrinsically disordered dehydrin protein. Protein Sci 20:42–50

Hundertmark M, Dimova R, Lengefeld J, Seckler R, Hincha DK (2011) The intrinsically disordered late embryogenesis abundant protein LEA18 modulates membrane stability through binding and folding. Biochim Biophys Acta 1808:446–453

Hundertmark M, Hincha DK (2008) LEA (Late Embryogenesis Abundant) proteins and their encoding genes in *Arabidopsis thaliana*. BMC Genomics 9:118

Hundertmark M, Popova AV, Rausch S, Seckler R, Hincha DK (2012) Influence of drying on the secondary structure of intrinsically disordered and globular proteins. Biochem Biophys Res Commun 417:122–128

Huson DH, Bryant D (2006) Application of phylogenetic networks in evolutionary studies. Mol Biol Evol 23:254–267

Ingram J, Bartels D (1996) The molecular basis of dehydration tolerance in plants. Annu Rev Plant Physiol Plant Mol Biol 47:377–403

Koag M-C, Fenton RD, Wilkens S, Close TJ (2003) The binding of maize DHN1 to lipid vesicles. Gain of structure and lipid specificity. Plant Physiol 131:309–316

Koag M-C, Wilkens S, Fenton RD, Resnik J, Vo E, Close TJ (2009) The K-segment of maize DHN1 mediates binding to anionic phospholipid vesicles and concomitant structural changes. Plant Physiol 150:1503–1514

Koch MA, Haubold B, Mitchell-Olds T (2000) Comparative evolutionary analysis of chalcone synthase and alcohol dehydrogenase loci in *Arabidopsis*, *Arabis*, and related genera (*Brassicaceae*). Mol Biol Evol 17:1483–1498

Kovacs D, Kalmar E, Torok Z, Tompa P (2008) Chaperone activity of ERD10 and ERD14, two disordered stress-related plant proteins. Plant Physiol 147:381–390

Kreps JA, Wu Y, Chang HS, Zhu T, Wang X, Harper JF (2002) Transcriptome changes for *Arabidopsis* in response to salt, osmotic, and cold stress. Plant Physiol 130:2129–2141

Kurkela S, Borg-Franck M (1992) Structure and expression of *kin2*, one of two cold- and ABA-induced genes of *Arabidopsis thaliana*. Plant Mol Biol 19:689–692

Kurkela S, Franck M (1990) Cloning and characterization of a cold- and ABA-inducible *Arabidopsis* gene. Plant Mol Biol 15:137–144

Larkin MA, Blackshields G, Brown NP, Chenna R, McGettigan PA, McWilliam H, Valentin F, Wallace IM, Wilm A, Lopez R, Thompson JD, Gibson TJ, Higgins DG (2007) Clustal W and Clustal X version 2.0. Bioinformatics 23:2947–2948

Lee YP, Babakov A, Boer B de, Zuther E, Hincha DK (2012) Comparison of freezing tolerance, compatible solutes and polyamines in geographically diverse collections of *Thellungiella* sp. and *Arabidopsis thaliana* accessions. BMC Plant Biol 12:131

Librado P, Rozas J (2009) DnaSP v5: A software for comprehensive analysis of DNA polymorphism data. Bioinformatics 25:1451–1452

Lin C, Thomashow MF (1992a) A cold-regulated *Arabidopsis* gene encodes a polypeptide having potent cryoprotective activity. Biochem Biophys Res Comm 183:1103–1108

Lin C, Thomashow MF (1992b) DNA sequence analysis of a complementary DNA for cold-regulated *Arabidopsis* gene cor15 and characterization of the COR15 polypeptide. Plant Physiol 99:519–525

Liu Q, Kasuga M, Sakuma Y, Abe H, Miura S, Yamaguchi-Shinozaki K, Shinozaki K (1998) Two transcription factors, DREB1 and DREB2, with an EREBP/AP2 DNA binding domain separate two cellular signal transduction pathways in drought- and low-temperature-responsive gene expression, respectively, in *Arabidopsis*. Plant Cell 10:1391–1406

Liu Y, Chakrabortee S, Li R, Zheng Y, Tunnacliffe A (2011) Both plant and animal LEA proteins act as kinetic stabilisers of polyglutamine-dependent protein aggregation. FEBS Lett 585:630–634

Lynch M, Conery JS (2000) The evolutionary fate and consequences of duplicate genes. Science 290:1151–1155

Maere S, De Bodt S, Raes J, Casneuf T, Van Montagu M, Kuiper M, Van de Peer Y (2005) Modeling gene and genome duplications in eukaryotes. Proc Natl Acad Sci U S A 102:5454–5459

McKhann HI, Gery C, Berard A, Leveque S, Zuther E, Hincha DK, Mita S de, Brunel D, Teoule E (2008) Natural variation in CBF gene sequence, gene expression and freezing tolerance in the Versailles core collection of *Arabidopsis thaliana*. BMC Plant Biol 8:105

Mouillon J-M, Gustafsson P, Harryson P (2006) Structural investigation of disordered stress proteins. Comparison of full-length dehydrins with isolated peptides of their conserved segments. Plant Physiol 141:638–650

Mutwil M, Abro J, Willats WGT, Persson S (2008) GeneCAT—novel webtools that combine BLAST and co-expression analyses. Nucl Acids Res 36:W320–W326

Nakayama K, Okawa K, Kakizaki T, Honma T, Itoh H, Inaba T (2007) *Arabidopsis* Cor15am is a chloroplast stromal protein that has cryoprotective activity and forms oligomers. Plant Physiol 144:513–523

Nakayama K, Okawa K, Kakizaki T, Inaba T (2008) Evaluation of the protective activities of a late embryogenesis abundant (LEA) related protein, Cor15am, during various stresses i*n vitro*. Biosci Biotechnol Biochem 72:1642–1645

Popova AV, Hundertmark M, Seckler R, Hincha DK (2011) Structural transitions in the intrinsically disordered plant dehydration stress protein LEA7 upon drying are modulated by the presence of membranes. Biochim Biophys Acta 1808:1879–1887

Pouchkina-Stantcheva NN, McGee BM, Boschetti C, Tolleter D, Chakrabortee S, Popova AV, Meersman F, Macherel D, Hincha DK, Tunnacliffe A (2007) Functional divergence of former alleles in an ancient asexual invertebrate. Science 318:268–271

Rahman LN, Chen L, Nazim S, Bamm VV, Yaish MW, Moffatt BA, Dutcher JR, Harauz G (2010) Interactions of intrinsically disordered *Thellungiella salsuginea* dehydrins TsDHN-1 and TsDHN-2 with membranes – synergistic effects of lipid composition and temperature on secondary structure. Biochem Cell Biol 88:791–807

Rumich-Bayer S, Krause GH (1986) Freezing damage and frost tolerance of the photosynthetic apparatus studied with isolated mesophyll protoplasts of *Valerianella locusta* L. Photosynth Res 8:161–174

Rütten D, Santarius KA (1988) Cold acclimation of *Ilex aquifolium* under natural conditions with special regard to the photosynthetic apparatus. Physiol Plant 72:807–815

Schmid M, Davison TS, Henz SR, Pape UJ, Demar M, Vingron M, Schölkopf B, Weigel D, Lohmann JU (2005) A gene expression map of *Arabidopsis thaliana* development. Nat Genet 37:501–506

Segrest JP, Jones MK, Loof H de, Brouillette CG, Venkatachalapathi YV, Anantharamaiah GM (1992) The amphipathic helix in the exchangeable apolipoproteins: a review of secondary structure and function. J Lipid Res 35:141–166

Steponkus PL (1984) Role of the plasma membrane in freezing injury and cold acclimation. Annu Rev Plant Physiol 35:543–584

Steponkus PL, Uemura M, Joseph RA, Gilmour SJ, Thomashow MF (1998) Mode of action of the *COR15a* gene on the freezing tolerance of *Arabidopsis thaliana*. Proc Natl Acad Sci U S A 95:14570–14575

Steponkus PL, Uemura M, Webb MS (1993) A contrast of the cryostability of the plasma membrane of winter rye and spring oat: two species that widely differ in their freezing tolerance and plasma membrane lipid composition. In: Steponkus PL (ed) Advances in low-temperature biology, Vol. 2. JAI Press, London, pp 211–312

Stockinger EJ, Gilmour SJ, Thomashow MF (1997) *Arabidopsis thaliana* CBF1 encodes an AP2 domain-containing transcriptional activator that binds to the C-repeat/DRE, a cis-acting DNA regulatory element that stimulates transcription in response to low temperature and water deficit. Proc Natl Acad Sci U S A 94:1035–1040

Thalhammer A, Hundertmark M, Popova AV, Seckler R, Hincha DK (2010) Interaction of two intrinsically disordered plant stress proteins (COR15 A and COR15B) with lipid membranes in the dry state. Biochim Biophys Acta 1798:1812–1820

Thomashow MF (1999) Plant cold acclimation: freezing tolerance genes and regulatory mechanisms. Annu Rev Plant Physiol Plant Mol Biol 50:571–599

Tolleter D, Hincha DK, Macherel D (2010) A mitochondrial late embryogenesis abundant protein stabilizes model membranes in the dry state. Biochim Biophys Acta 1798:1926–1933

Tolleter D, Jaquinod M, Mangavel C, Passirani C, Saulnier P, Manon S, Teyssier E, Payet N, Avelange-Macherel M-H, Macherel D (2007) Structure and function of a mitochondrial late embryogenesis abundant protein are revealed by desiccation. Plant Cell 19:1580–1589

Tunnacliffe A, Hincha DK, Leprince O, Macherel D (2010) LEA proteins: versatility of form and function. In: Lubzens E, Cerda J, Clark M (eds) Sleeping beauties—dormancy and resistance in harsh environments. Springer, Berlin, pp 91–108

Tunnacliffe A, Wise MJ (2007) The continuing conundrum of LEA proteins. Naturwissenschaften 94:791–812

Welin BV, Olson A, Nylander M, Palva ET (1994) Characterization and differential expression of *dhn/lea/rab*-like genes during cold acclimation and drought stress in *Arabidopsis thaliana*. Plant Mol Biol 26:131–144

Wilhelm KS, Thomashow MF (1993) *Arabidopsis thaliana cor15b*, an apparent homologue of *cor15a*, is strongly responsive to cold and ABA, but not drought. Plant Mol Biol 23:1073–1077

Xin Z, Browse J (2000) Cold comfort farm: the acclimation of plants to freezing temperatures. Plant Cell Environ 23:893–902

Yang Y-W, Lai K-N, Tai P-Y, Li W-H (1999) Rates of nucleotide substitution in Angiosperm mitochondrial DNA sequences and dates of divergence between *Brassica* and other Angiosperm lineages. J Mol Evol 48:597–604

Yelenosky G, Guy CL (1989) Freezing tolerance of citrus, spinach, and petunia leaf tissue. Plant Physiol 89:444–451

Zhang X, Fowler SG, Cheng H, Lou Y, Rhee SY, Stockinger EJ, Thomashow MF (2004) Freezing-sensitive tomato has a functional CBF cold response pathway, but a CBF regulon that differs from that of freezing-tolerant *Arabidopsis*. Plant J 39:905–919

Zhang Z, Li J, Zhao X, Wang J, Wong G, Yu J (2006) KaKs Calculator: calculating Ka and Ks through model selection and model averaging. Genomics Proteomics Bioinformatics 4:259–263

Zimmermann P, Hirsch-Hoffmann M, Henning L, Gruissem W (2004) GENEVESTIGATOR. *Arabidopsis* microarray database and analysis toolbox. Plant Physiol 136:2621–2632

Zuther E, Schulz E, Childs LH, Hincha DK (2012) Natural variation in the non-acclimated and cold-acclimated freezing tolerance of *Arabidopsis thaliana* accessions. Plant Cell Environ 35:1860–1878

3D Reconstruction of Frozen Plant Tissue: A Unique Histological Analysis to Image Postfreeze Responses

David P. Livingston III and Tan Tuong

Introduction

Plant research has come a long way since Robert Hooke published images of cork tissue in 1664 and coined the term "cell." He described what he saw as "little boxes or cells distinct from one another" (Hooke 1664). As optics for the light microscope improved, the complexity of plant tissue became more and more apparent and by 1842 among other cellular structures, chromosomes were first observed (Sedgewick 2007). It would be about 40 years later that the nucleus, which contained the chromosomes, would be understood as the seat of heritability.

As various stains began to be used to distinguish plant tissues, a new complexity involving the biochemistry of chemicals interacting with molecules within plant tissues was realized. Our understanding of plant anatomy grew exponentially as a result of histological analysis with the light microscope. With the advent of the electron microscope, computer tomography, magnetic resonance imaging, confocal microscopy, and synchrotron radiation imaging one might wonder if the days of the usefulness of the light microscope are behind us. Is "merely" observational data with the light microscope to be regarded as irrelevant?

To underscore the importance of the light microscope in freezing tolerance research, we would like to present histological observations in the economically important small grains from the most winter hardy to the least hardy: rye (*Secale cereale*), wheat (*Triticum aestivum*), barley (*Hordeum vulgare*), and oat (*Avena sativa*). We would like to concentrate on a unique response to freezing in oat and describe a new technique to image this response in three dimensions.

D. P. Livingston III (✉) · T. Tuong
Crop Science, USDA ARS and North Carolina State University,
840 Method Rd, Unit 3, Raleigh, NC 27695, USA
e-mail: dpl@ncsu.edu

R. Imai et al. (eds.), *Plant and Microbe Adaptations to Cold in a Changing World*,
DOI 10.1007/978-1-4614-8253-6_9,

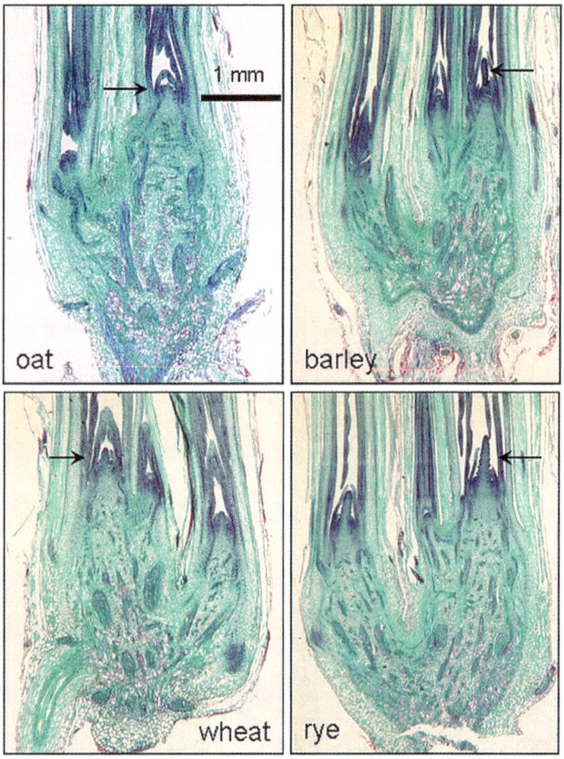

Fig. 1 Longitudinal sections of the crown of four winter cereals stained with Safranin and Fast Green. The *arrows* point to the shoot apex of the primary tiller of each species. (Used with permission from PLOS ONE, Livingston et al. 2013)

Understanding Crown Survival by Visualizing its Internal Anatomy

In its simplest portrayal, winter hardiness in plants consists of (1) cold-acclimation, (2) freeze-acclimation ("2nd phase hardening" or hardening at subzero temperatures), (3) the actual freezing process, and (4) recovery from freezing. Of these four stages, the period of recovery from freezing has probably received the least attention and it is to that aspect of winter hardiness we would like to turn our attention.

To understand how a winter cereal plant can survive freezing, we must first appreciate the anatomy of the below-ground portion of the stem called the crown. The crown is essentially the interface between roots and stems and is the region of the plant from which new stems (or tillers) are produced after freezing. At the bottom of the crown, xylem vessels enter the base of the stem through a complex series of fiber cells (Livingston et al. 2005). Here they intertwine with phloem and form bundles interspersed with mesophyll cells that appear to meander through the center of the crown (Fig. 1). As you progress up the crown, a transition zone subtending the shoot apex (Fig. 1, arrows) is encountered where most of the premeristem cells

Fig. 2 Oat crown 7 days after freezing at − 12 °C showing differences in survival of individual tillers. Note that the primary (1°) tiller is completely dead while both the secondary (2 °C) and tertiary (3 °C) tillers show some live and dead tissue. The secondary tiller will fully recover and eventually produce a head with seeds

are located. These regions give rise to new leaves and shoot apices which elongate and eventually form the flowering heads of the plant in spring and early summer.

Postfreeze Recovery: Differential Survival of Tillers

After freezing, survival of the plant is evaluated by observing regrowth of the tillers (Fig. 2). However, in winter cereals there does not appear to be a pattern in the survival of tillers. In fact, in an analysis of tiller survival in oat while the survival of the tertiary tiller (Table 1) was lower than the other three, the deviation from the mean was greater than the mean itself for all three tillers illustrating the seemingly random survival pattern.

Clearly, vital tissue is surviving in the secondary tiller of the plant in Fig. 2, which is not in the primary and tertiary tillers. In fact, from what we know by observing regrowth patterns and using vital stains, it is clear that much of the support tissue in the center of the crown can die, but the individual tiller may in fact survive.

This cannot be due to the presence or absence of genes since the genome is identical throughout the crown. The survival difference between the tillers must be due to differences in expression of one or more genes. This observation of differences in tiller survival with no apparent pattern led us to investigate the histological basis for differences in survival of the tillers.

Table 1 Percent survival of individual tillers of oat frozen at −12 °C and allowed to recover for 14 days. Zero is completely dead, 50 % survival is highly damaged and may or may not survive if left to recover for a longer period, 100 % indicates a healthy tiller with very little apparent freeze damage

Plant number	Tiller position		
	Primary	Secondary	Tertiary
	% Survival		
1	50	0	0
2	0	100	0
3	50	0	100
4	100	0	100
5	0	50	0
6	50	50	0
7	0	100	0
8	100	0	0
9	50	0	50
10	50	50	0
11	0	50	50
Average	41	36	27
Standard deviation	38	39	41

Delineating the Dead-Zone in Frozen Crown Tissue

For nearly 20 years, the median lethal tolerance (LT50) of cold-acclimated oats under controlled conditions was considered to be −12 °C (Livingston, 1996). At −14 or −15 °C the percent survival was zero. However, as is usually the case with plant systems, this conclusion is not as simple as we had assumed. An analysis of frozen tissue with the vital stain 2,3,5 triphenyl tetrazolium chloride caused us to reconsider much of what we thought we understood about freezing tolerance.

Tetrazolium chloride is reportedly reduced by electrons from the electron transport system of the mitochondria (Briggs 2009) forming a red-colored compound, formazan. This allows one to distinguish live (red) from dead (white) cells. The use of tetrazolium for identifying live tissue was first described by Parker (1953) in pine trees; it has also been used to identify live tissue within frozen wheat crowns (Tanino and Mckersie 1985). It is commonly used for identifying freeze damaged seeds primarily in the corn industry (AOSA 2000).

In cold-acclimated (unfrozen) crowns (Fig. 3a), all tissues were able to reduce tetratzolium. When frozen at −12 °C for 3 h a considerable portion of the center of the crown was unable to reduce tetrazolium and was presumed dead (Fig. 3b). At a freezing temperature of −20 °C there are no known oat genotypes that can survive. However, while the dead-zone extended upwards into the shoot apexes, there were still large areas of the crown able to reduce tetrazolium (Fig. 3c). Even when frozen at −80 °C overnight (nearly 14 h) there were still a few cells that were apparently able to reduce tetrazolium (Fig. 3d).

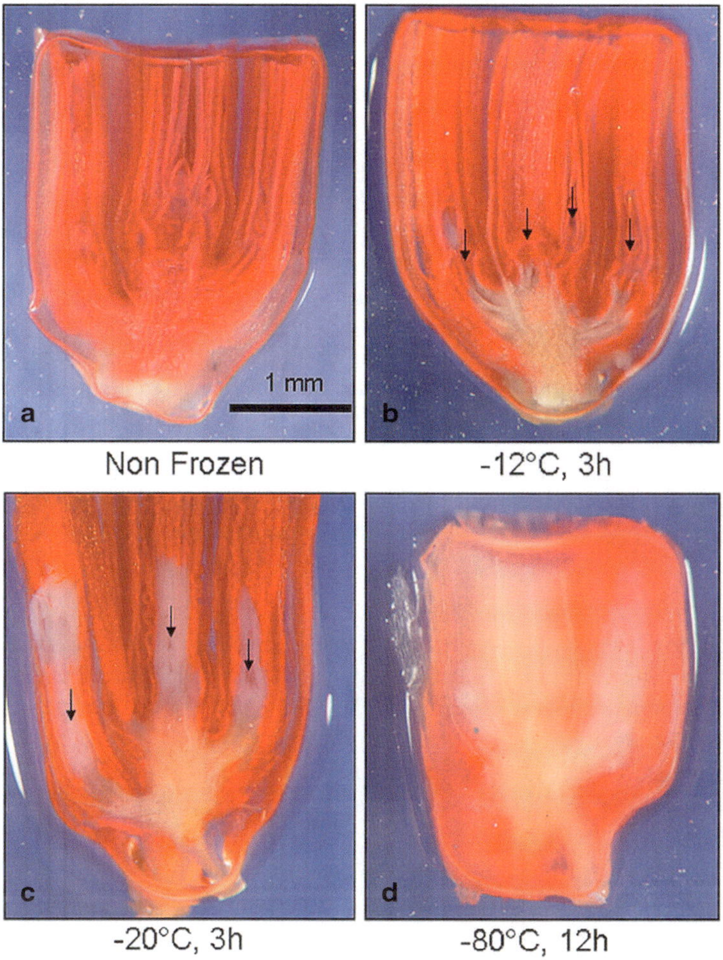

Fig. 3 Cold-acclimated Wintok oat crowns treated with tetrazolium chloride. **a** Unfrozen control. *Red* color indicates live tissue, or at least tissue that is able to reduce tetrazolium chloride to its red-colored derivative formazan. **b** The same oat cultivar frozen for 3 h at −12 °C, the LT50 for this cultivar. Note that nearly all the shoot-apices are still alive (*arrows*). **c** The same oat cultivar frozen at −20 °C for 3 h. No oat cultivar is known to survive this temperature. Note that the shoot-apices from all the tillers are dead (*arrows*) but tissue at the base of the leaves is alive (i.e., able to reduce tetrazolium). **d** Crown that were frozen at −80 °C for 12 h. Note extensive dead areas (*white*) but that a few cells were still able to reduce tetrazolium

One might wonder what mechanisms allow a particular group of tissues to survive freezing while another is killed by the same stress. Every cell in the plant has the same DNA so the genetics of both tissues is identical. The difference between tissues is determined by which DNA is expressed. Which compounds that protect from ice are synthesized in the leaf bases that are not in the center of the crown?

This can never be discovered by grinding and then extracting expressed genes from the whole plant. That would essentially be attempting to understand a four-dimensional system (three spatial dimensions and time) by studying data collected across one or possibly two dimensions.

It may be a relatively thin layer of cells that have the capacity to prevent ice from growing into the bases of the leaves, or there may be a few cells that have the capacity to supercool when the rest of the crown freezes. Indeed, Houde et al. (1995) localized several freezing tolerance associated proteins to the transition zone of wheat crowns. Without studying frozen tissue from a histological perspective, we would have no way to know that tissues may be expressing genes differently resulting in differences in survival of the tissues. As a consequence, we would never discover which expressed genes result in protection from freezing as opposed to genes that are simply correlated but have no cause and effect associated with their expression.

Postfreeze Recovery: Specific Visual Response Only in Oat

Palta et al. (1977) demonstrated that plants can recover from much of what is diagnosed as freeze-damage (which incidentally calls into question the use of electrolyte leakage to measure freezing tolerance). Those studies indicated that mechanisms of repair following freezing are a vital component of over all winter hardiness and may be least as important as mechanisms involved in cold acclimation (Palta et al. 1977). The question we began to consider is how do the plants recover from freezing and is there any part of the recovery process that we can visualize to understand it better?

Some of the more obvious visual responses during recovery from freezing were tissue disruption, pycnotic nuclei (not shown here, see Livingston et al. 2005), disintegrated shoot apices, and dark-red-staining regions that appeared to fill xylem vessels (Fig. 4). We noticed regions of dark-red-staining cells within the crown of winter cereals a few days after freezing. When we looked at oat, this area was quite pronounced and it looked like it formed a ring, especially when observed in cross section (Fig. 4b). We call this area a barrier because it separates live from dead tissue as shown in a tetrazolium analysis (Fig. 5).

Quantitative Analysis of Barrier

Using color recognition software we quantified the change in red-staining tissue from the base of the crown to the top of the shoot apex of the primary tiller (Fig. 6a). While Fig. 6a is the data only from a single crown, most of the images had a similar result (not shown). The graph indicates that the central part of the crown was the most variable between sections (each data point represents the area staining red in a single section) and that in general the central part of the crown had a greater red-staining area than the lower or upper portions of the crown (Fig. 6a).

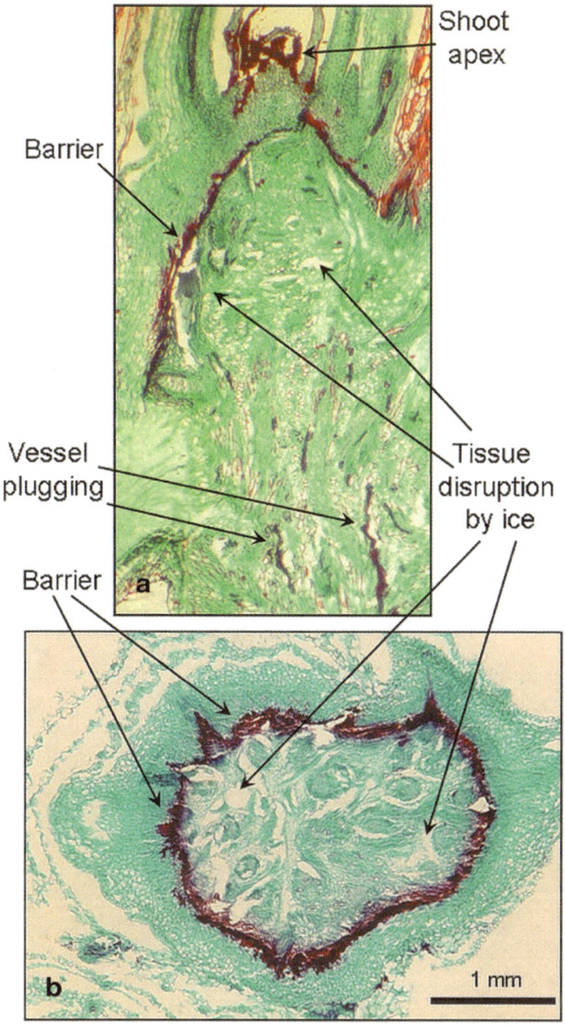

Fig. 4 Safranin and Fast Green-stained sections of oat crowns 7 days after freezing at −12 °C. Note the *dark red* barrier made visible by the Safranin stain. **a** Longitudinal section. **b** Cross section

In addition to a spatial analysis, we quantified the change in red color over time (Fig. 6b) and thus provided a basis for correlating metabolic changes over time. At day 0 (immediately after freezing and thawing) day 1 and day 3 differences between frozen and unfrozen controls were not significant. It was not until day 7 that significant differences between the frozen and unfrozen crowns were detected. Several metabolites were highly correlated with the change in red color (not shown) but it is not known at this time which of these may be related in a cause and effect manner.

Fig. 5 Tetrazolium analysis of oat crown 7 days after freezing showing the barrier separating live and dead tissue. The *lower* panel shows the area separating live from dead tissue (*arrows*) in closer detail

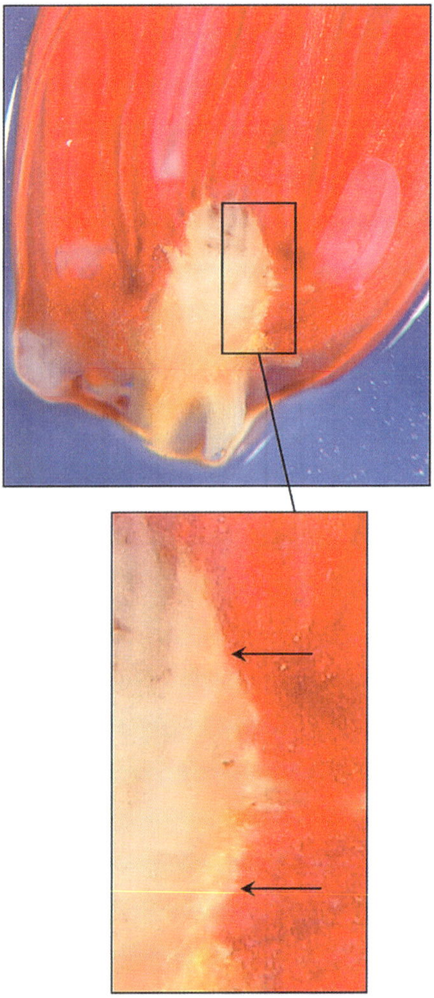

Three-Dimensional Reconstruction of Oat Crown after Freezing

The discovery of a unique freeze response in oat led us to consider the continuity of the barrier in three-dimensional (3D) space. After aligning and stacking over 150 images of individual sections (Fig. 7a, c), we applied a technique used extensively in the film industry called "color keying." This enabled us to digitally clear surrounding tissue and visualize the barrier within the crown in 3D (Fig. 7b, d; see Livingston et al. 2010 for details of the procedure; Please visit our YouTube site for a video animation of this freeze response: http://www.youtube.com/watch?v=HyNw0yFY14Y&feature=channel&list=UL).

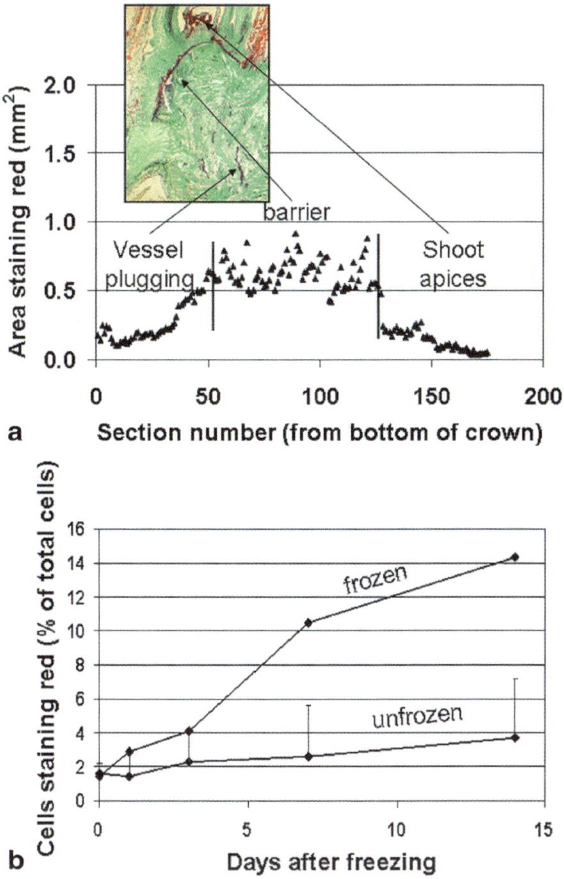

Fig. 6 Quantitative measurement of red-staining area within oat crowns using color recognition software. **a** Change in the area of red-staining cells over the length of a single crown. Each data point is the total area staining *red* in a single section. *Bars* on the graph provide an approximate separation of the three areas of the crown. *Inset* is a longitudinal section of a frozen crown and *arrows* identify the approximate regions of the crown indicated by the three regions of the graph. **b** Change in the percentage of red-staining cells over time after freezing. Each data point is the average percent red color in five crowns frozen at −12 °C. Total red color for each crown was determined by adding the percentage of red color in approximately 150 sections. *Bars* on the graph indicate the LSD at $p = 0.05$

Three days after freezing the beginnings of a barrier became visible (Fig. 7b) and by day 7 the barrier was fully developed (Fig. 7d). Once fully developed the barrier appeared to take on a somewhat spherical shape and extended upwards into apical regions of the crown (Fig. 7d).

Function of the Barrier: Speculation

While the barrier is clearly a response to freezing, its specific function is unclear. Safranin stains lignified tissue and phenolic compounds, so it is possible that the barrier is a form of lignin that was generated in phenol-producing cells and diffused into surrounding tissue (Beckman 2000). Beckman states that phenols can stabilize regions of tissue and "in the process create of the cells one large, durable inert macromolecule." The cells Beckman describes are stimulated to produce phenols by pathogens, so it is possible that the barrier is a reaction to secondary infection.

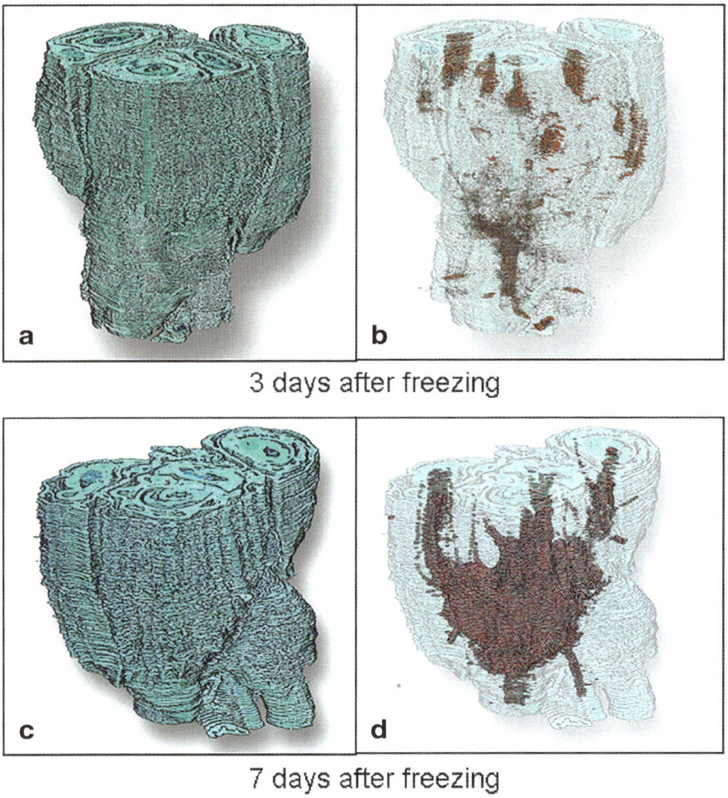

Fig. 7 Three-dimensional (3D) reconstruction of oat crowns after freezing. Over 150 individual images were aligned and stacked in z-space to produce the 3D volume. **a, c** Aligned and stacked images showing the outside of the crowns. **b, d** Sections from a and c that were digitally cleared to show the anatomical continuity between sections of the red-staining freeze response within the crown

In fact, postfreeze infection by *Pseudomonas* bacteria has been documented (Olien and Smith 1981; Marshall 1988). Olien and Smith (1981) reported that *Pseudomonas* moved from roots of barley into crowns after a freeze treatment and they credited some of the crown tissue death to bacterial infection.

Why oat was the only winter cereal to exhibit the freeze-induced barrier is not known. Even in the most freezing tolerant winter cereal, rye, when apical meristems were killed at −16 °C (not shown) no barrier was present at any time following freezing. When frozen wheat crowns recovering from freezing were treated with tetrazolium, a distinct demarcation of live from dead tissues was observed (not shown), similar to that in oat (Fig. 3) but without the safranin-staining barrier (Fig. 4). This implies that some kind of barrier may also exist in wheat as suggested by Tanino and Mckersie (1985) but it is likely of a different composition than oat. If the barrier in oat is a freeze-protection mechanism to limit postfreeze cell/tissue death, maybe rye, and wheat are able to withstand the freezing stresses causing the initial injury and do not need the barrier for recovery from freezing.

Concluding Remarks

The identification of a barrier between live and dead tissue provides a basis for studying differences in gene expression between the two regions. Combining 3D reconstruction with gene expression using in situ hybridization should allow an unprecedented understanding of the relationship between gene expression and the effect of freezing on specific tissue.

References

Association of Official Seed Analysts (AOSA) (2000) Tetrazolium Testing Handbook. In: Peters J (ed) Contribution 29. AOSA, Lincoln

Beckman CH (2000) Phenolic-storing cells: keys to programmed cell death and periderm formation in wilt disease resistance in general defense responses in plants? Physiol Mol Plant Pathol 57:101–110

Briggs DE, Sole SM, Latham P (2009) Tetrazolium staining, mitochondria and barley quality. J Ins Brew 115:41–48

Hooke R (1664) Micrographia. Some physiological descriptions of minute bodies made by magnifying glasses with observations and inquiries thereupon. Available through the Gutenberg eBook project: http://www.gutenberg.org/files/15491/15491-h/15491-h.htm

Houde M, Daniel C, Lachapelle M, Allard F, Laliberte S, Sarhan F (1995) Immunolocalization of freezing-tolerance-associated proteins in the cytoplasm and nucleoplasm of wheat crown tissues. Plant J 8:583–593

Livingston DP III (1996) The second phase of cold hardening: freezing tolerance and fructan isomer changes in winter cereal crowns. Crop Sci 36:1568–1573

Livingston DP III, Henson CA, Tuong TD, Wise ML, Tallury SP Duke SH (2013) Histological analysis and 3D reconstruction of winter cereal crowns recovering from freezing. PLoS one 8(1):e53468

Livingston DP III, Tallury SP, Premakumar R, Owens S, Olien CR (2005) Changes in the histology of cold hardened oat crowns during recovery from freezing. Crop Sci 45:1545–1558

Livingston DP III, Tuong TD, Gadi SRV, Haigler CH, Gelman RS, Cullen JM (2010) 3D volumes constructed from pixel-based images by digitally clearing plant and animal tissue. J Microsc 240:122–129

Marshall D (1988) A relationship between ice-nucleation-active bacteria, freeze damage, and genotype in oats. Phytopathology 78:952–957

Olien CR, Smith MN (1981) Extension of localized freeze injury in barley by acute post-thaw bacterial disease. Cryobiology 18:404–409

Parker J (1953) Some applications and limitations of tetrazolium chloride. Science 118:77–79

Palta JP, Levitt J, Stadelmann EJ (1977) Freezing injury in onion bulb cells. Plant Physiol 60:393–397

Sedgwick WT (2007) A short history of science (pp. 548). Stronck Press

Tanino KK, McKersie BD (1985) Injury within the crown of winter wheat seedlings after freezing and icing stress. Can J Bot 63:432–435

Post-transcriptional and Post-translational Modifications Controlling Cold Response

Elisabetta Mazzucotelli, Cristina Crosatti, Lorenzo Giusti, Davide Guerra and Luigi Cattivelli

Introduction

Plants become tolerant to environmental stresses by reprogramming metabolism and gene expression, to establish a new equilibrium between growth, development, and survival. In the past two decades, important advances have been made in the understanding of transcriptional changes induced by exposition to low temperature and in the identification of signaling proteins and transcription factors which regulate the cold-induced gene expression. Outcomes depicted a complex process where several pathways connecting stress perception and specific transcriptional changes play a main role. Typical and famous components of the cold response are the C-repeat (CRT) Binding Factors (CBFs), universally recognized as key regulators of response to low temperature, ICE1 (Inducer of CBF Expression 1) and the downstream *COR* (COld-Regulated) genes (Huang et al. 2012).

The characterization of new mutants as well as the integration of gene expression data and protein activities indicates that further levels of regulation based on post-transcriptional and post-translational mechanisms are responsible for the transcriptional changes related to the plant response to cold and other stresses. This chapter focuses on the most recent findings about these mechanisms—alternative splicing, RNA-mediated silencing, ubiquitination, and sumoylation—activated by plants after the perception of a cold stress. The network of such mechanisms targets transcription factors and other regulatory components of the stress signaling, resulting in either activation or repression of their activities (Fig. 1, Mazzucotelli et al. 2008). This ensures temporally and spatially appropriate patterns of downstream gene expression, thus the shaping of proteome and metabolome complexity, to ultimately switch on adaptive response.

E. Mazzucotelli (✉) · C. Crosatti · L. Giusti · D. Guerra · L. Cattivelli
Genomics Research Centre, Consiglio per la ricerca e la sperimentazione in agricoltura (CRA), Fiorenzuola d'Arda (PC), Italy
e-mail: elisabetta.mazzucotelli@entecra.it

R. Imai et al. (eds.), *Plant and Microbe Adaptations to Cold in a Changing World*,
DOI 10.1007/978-1-4614-8253-6_10,

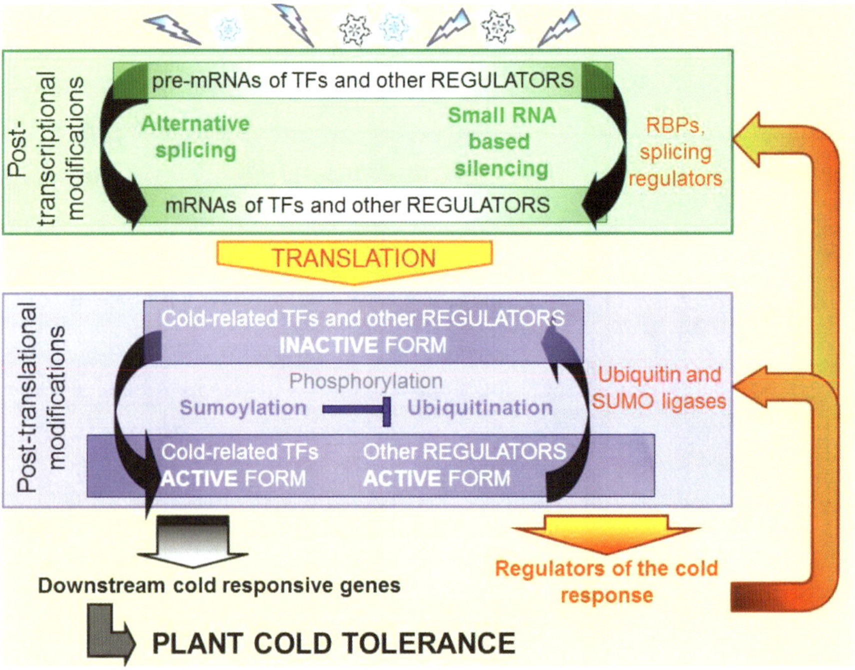

Fig. 1 Model describing the cross-talking among post-transcritpional and post-translational regulations involved in the control of the plant response to cold stress. See the text for details. (Modified from Mazzucotelli et al. 2008)

Post-transcriptional Processes Affecting mRNA Availability

Different mechanisms regulate synthesis, maturation, and stability of transcript, ranging from alternative splicing to association with polysomes, from the formation of RNA granules and mechanisms of RNA degradation, guided by noncoding RNAs and nonsense-mediated decay (NMD). All these steps are emerging as important point in the regulation of the plant stress responses.

RNA-Binding Proteins (RBP) are at the heart of all these post-transcriptional processes. They bind RNA molecules immediately after the transcription till the translation, in order to protect, organize, and prepare mRNAs for posttranscriptional processes. More than 200 RBP genes have been predicted in *Arabidopsis* and rice genomes (Ambrosone et al. 2012). RBPs contain different kinds of RNA-binding and auxiliary domains, which enable them to bind RNA in a versatile way, such as RNA recognition motif, glycine-rich motif, zinc finger motif, cold shock domain (CSD), and double-stranded RNA-binding motif. RBPs have a role in normal cellular functions particularly in regulating several developmental processes. Further, a large body of evidence is supporting their key function in plant adaptation to various environmental conditions when they regulate the stress-dependent export of mRNAs from the nucleus, selective translation of stress-associated genes, and increased stability of related transcripts (Ambrosone et al. 2012). RBPs execute this activity functioning as RNA chaperones (Kang et al. 2012). Indeed, when cells are exposed to low temperatures, misfolded RNA molecules become overstabilized and cannot assume a native conformation without the help of RNA chaperones. In *Arabidopsis*, a positive effect on cold and freezing tolerance associated with RNA chaperone activity has been shown for the Glycine-Rich Proteins GRP7 (Kim et al. 2008), GRP2 (Kim et al. 2007), and AtRZ-1a, b, c (Kim and Kang 2006), and for the CSD proteins CSP2 (Sasaki et al. 2007) and CSP3 (Kim et al. 2009). GRP7 acts in the export of mRNAs from the nucleus to the cytoplasm under cold stress conditions in the guard cells. GRP2 modulates the expression and activity of mitochondrial-encoded genes by exhibiting transcription antitermination activity. AtRZ-1a affects the translation of several genes involved in stress resistance and in RNA and protein metabolism.

Also RNA-helicases belong to the class of RBPs. The *Arabidopsis* gene *Los4* (Low expression of osmotically responsive genes 4) encodes an RNA-helicase acting on nuclear mRNA export, with particular effects in the response to temperature stress (Gong et al. 2005). The *los4-1* mutant has a reduced expression of *CBF3* and a delayed expression of *CBF1* and *CBF2* during cold acclimation resulting in chilling sensitivity, while cryophyte/los4-2 mutant (allelic to los4-1) promotes a superinduction of *CBF2* under cold stress and an enhanced freezing tolerance. Both mutants are heat sensitive. This divergent response is mediated through a differential effect on nuclear mRNA export: inhibited by los4-1 and enhanced by los4-2 at low temperatures, while both mutations disrupt the mRNA export at high temperatures.

Stress-Related Transcripts from Alternative Splicing Events

Alternative splicing (AS), a mechanism by which different RNA transcripts from a single gene are produced, occurs in up to 42 % of *Arabidopsis* genes containing introns. Alternative mRNA forms have different fates according to whether the protein still maintain an activity or not, which usually depends on the kind of splicing which takes place. Nonfunctional transcripts incur in degradation by NMD, which is both a surveillance mechanism avoiding accumulation of incomplete, potentially harmful proteins, and a fine modulator of gene expression (Mastrangelo et al. 2011).

A large number of gene networks related to abiotic stresses, involving, among others, protein kinases, transcription factors, splicing regulators have demonstrated the centrality of AS in the fine-tuning of plant responses to the environment (Mastrangelo et al. 2011). For example, the *Arabidopsis STA1* gene encodes a splicing regulator upregulated by cold. STA1 protein can regulate the stability and splicing pattern of a number of endogenous gene transcripts related to abiotic stress response (Lee et al. 2006).

AS can act to control the abundance of the active protein. For example, AS of the wheat transcription factor gene *Wdreb2* generates three transcripts in response to cold, only one of which still maintains its transcriptional activity (Egawa et al. 2006). Alternatively, AS can represent an interconnecting mechanism between the response to stress and other cellular processes. A very interesting example is provided by the interaction between circadian clock and cold tolerance, in which alternative splicing bridges the two pathways. CCA1 (CIRCADIAN CLOCK ASSOCIATED 1) and LHY (LATE ELONGATED HYPOCOTYL) are MYB transcription factors belonging to *Arabidopsis* circadian clock. Among other functions, they positively affect the expression of *CBF* pathway genes (Dong et al. 2011). *CCA1* transcript can be alternatively spliced into two variants: *CCA1α*, the fully spliced one, and *CCA1β*, which retains the fourth intron. The latter encodes a protein lacking a DNA-binding domain, but still able to form nonfunctional homodimers and heterodimers with CCA1α and LHY, suggesting a CCA1 self-regulation model in which CCA1β variant competitively inhibits CCA1α and LHY activity by sequestering them. Intriguingly, *CCA1* splicing is regulated by temperature, and the fully spliced *CCA1α* variant becomes dominant during cold stress, therefore promoting the expression of the *CBF* pathway genes (Seo et al. 2012). Interestingly, cold stress is responsible also for the alternative splicing of LHY, in this case leading to the downregulation of the functional transcript (James et al. 2012).

The alternative splicing of *IDD14* (INDETERMINATE DOMAIN14) generates a self-controlled regulatory loop that modulates starch accumulation in response to cold (Seo et al. 2011). *IDD14* binds to Qua-Quine Starch (QQS) promoter, which regulates starch accumulation. Under cold conditions, *IDD14* is alternatively spliced producing a nonfunctional form (IDD14β) lacking a DNA-binding domain, but able to form heterodimers with the fully spliced variant (IDD14α), thus preventing its binding to QQS.

Besides transcription factors, all levels of regulation of gene expression can be affected by AS in response to abiotic stresses. Splicing events are being described for genes coding for E3 ubiquitin ligases. For instance, the transcripts of the E3 ligase 6g2, later called TdRf1, undergoes cold-dependent alternative splicing (Mastrangelo et al. 2005). Similarly, the gene *HOS1* (High expression of Osmotically responsive gene1), encoding an E3 ligase with a negative regulative function on the cold response, produces two differential transcripts differentially regulated in both temporal, spatial manners, and at different temperatures (Lee et al. 2012).

Many splicing regulators, like *SR* (ser/arg) genes (Palusa et al. 2007) promote alternative splicing of their own transcript as well as of other gene products. An example is provided by the *Arabidopsis* genes encoding the RNA-binding proteins AtGRP8 and AtGRP7 (Schöning et al. 2008). During cold stress, AtGRP8 promotes the use of a 5' splice site of its own transcript, producing an inactive form (*as_AtGRP8*) that is immediately targeted by NMD machinery. At the same time, AtGRP8 promotes the alternative splicing of *AtGRP7* transcript, which leads to NMD of *as_AtGRP7*, and then to a decrease of AtGRP7 abundance. Furthermore, also AtGRP7 does the same with its own and *AtGRP8* mRNAs.

Degradation of Stress-Related Transcripts by miRNAs and tasiRNAs

The discovery of microRNAs (miRNAs) and tasiRNA (trans acting small interfering RNA) revealed their fundamental role in the regulatory network of the plant response to cold (Sunkar et al. 2012). The small noncoding RNAs post-transcriptionally silence target genes either by guiding degradation or repressing translation of target mRNAs.

Many plant cold-stress-regulated miRNAs have been identified by cloning, bioinformatic and high-throughput sequencing approaches in *Arabidopsis* (Zhou et al. 2008), *Populus* (Chen et al. 2012), *Brachypodium* (Zhang et al. 2009), rice (Lv et al. 2010). miRNAs are conserved among dicotyledonous and monocotyledonous plants, nevertheless, differences in their regulation suggest a significant level of species specificity. Comparative studies of miRNAs in *Brachypodium* and rice highlighted that the composition and location of miRNA families are different even in closely related plant species (Zhang et al. 2009). Even in sugarcane, the miR319 was shown to be differentially expressed among sugarcane genotypes contrasting for cold tolerance (Thiebaut et al. 2012).

Despite identification of many cold-related miRNAs, little is known about their target genes and their role in the response to cold. In general, miRNA targets are mainly represented by transcription factors, regulators of auxin signaling, RNA silencing, but also enzymes (Khraiwesh et al. 2012).

A cold-responsive miRNAs can act in different signaling pathways and simultaneously regulate genes with diverse functions linking together several pathways. For example, several cold-responsive miRNAs are also involved in the response of plant to biotic stress suggesting a cross-talk between pathways for biotic and abiotic stress responses (Lu et al. 2008).

Auxin signaling is crucial for plant growth and development, and several studies suggested that auxin is also involved in plant stress tolerance (Achard et al. 2006). The *Arabidopsis* miR393, upregulated in response to most stress conditions (including cold stress), is putatively directed against TIR1 and other closely related F-box proteins (Sunkar et al. 2004). Because TIR1 is the auxin receptor that targets repressors of the Auxin-Responsive Factor (ARF) for ubiquitin-mediated degradation in response to auxin, the miRNA inhibition of *TIR1* would downregulate auxin signaling and seedling growth.

Tang et al. (2012) sequenced a set of both miRNA and tasiRNAs from spike tissues of a wheat thermosensitive genic male-sterile line exposed to cold. tasiRNAs are noncoding RNA generated by miRNA processing of a *Tas* gene transcript resulting in the production of 21-nt RNAs phased with respect to the miRNA cleavage site. They found that miR167 and tasiRNA-ARF play roles in regulating the auxin-signaling pathway and possibly in the developmental response to cold stress.

Protein Degradation in Response to Stress: The Role of Ubiquitination

Ubiquitination is a unique, widely conserved molecular mechanism to degrade proteins. The covalent attachment of ubiquitin to a target substrate involves the sequential action of three enzymes: ubiquitin-activating enzyme (E1), ubiquitin-conjugating enzyme (E2), and ubiquitin ligase (E3). Usually once a target protein has been labeled with a polyubiquitin chain (at least four ubiquitin), it is recognized and then degraded by the 26 S proteasome. The E3 ubiquitin ligases are the enzymes which specifically recruit the target proteins ensuring the specificity of the entire process. More than 1,400 E3 enzymes are coded by *Arabidopsis* genome (Mazzucotelli et al. 2006). E3-mediated protein degradation regulates almost all cellular processes, including abiotic stress response (Lyzenga and Stone 2012).

One of the first E3 ligases identified in stress response is HOS1 which exerts a negative control on cold response being responsible for ICE1 degradation (Dong et al. 2006). Indeed, *hos1* mutation enhances the induction of *CBFs* and of the downstream cold-regulated genes by low temperatures.

Further studies identified other E3 ligases involved in low temperature response. In *Arabidopsis*, the overexpression of the U-Box E3 ubiquitin ligase AtCHIP (Carboxyl terminus of HSC70-Interacting Protein) causes higher sensitivity to low temperatures than wild type and also a growth-impaired phenotype suggesting its role as negative regulator of low temperature response (Yan et al. 2003).

Often E3 ligases represent a connecting point between different signaling pathways. The RING-finger E3 ligase TdRF1 (*Triticum durum* RING-finger protein 1) represents an interesting example (Guerra et al. 2012). *TdRF1* is induced upon exposure to low temperatures and dehydration. TdRF1 is phosphorylated by the kinase TdWNK5 (With No Lysine [K]5) a member of the *Arabidopsis* WNK family of MAP kinases involved in flowering time and circadian clock regulation (Wang et al. 2008). TdRF1 interacts with another E3 ligase, WVIP2 (Wheat Viviparus1 Interacting Protein2) strongly upregulated upon cold treatment and sharing high amino acid similarity with the wild oat VIP2 (Jones et al. 2000). Finally, it degrades the transcription factor WBLH1 (Wheat Bell-Type Homeodomain1), a previously described protein belonging to KNOX (Knotted1-like homeobox) gene family (Mizumoto et al. 2011). The overexpression of *TdRF1* increases tolerance of barley cells to dehydration, suggesting that it could protect plant under drought and freezing conditions.

Control of Stress Response by Sumoylation

Sumoylation is a post-translational transient modification of protein substrates based on the covalent conjugation of the SUMO (Small Ubiquitin-like Modifier) peptide. The biochemical steps are similar to those operating in the ubiquitination pathway, but sumoylation is specifically regulated at the target level, with phosphorylation accomplishing a critical role, besides the target consensus motif CKxE/D (C: hydrophobic amino acid; K: SUMO target lysine; x: any amino acid; D/E acidic amino acids), while the dynamic aspects are regulated by SUMO peptidase activities (Miura and Hasegawa 2010).

A wide variety of biological consequences of sumoylation have been observed, including changes in enzymatic activity, subcellular relocalization, and protection from ubiquitin-mediated degradation of regulatory proteins. Both loss and gain of function analyses, as well as the pattern of SUMO-conjugates revealed a key role of sumoylation in plants in response to environmental signals. *Arabidopsis siz1* (for SAP and Miz domain) mutants are hypersensitive to chilling and freezing stresses (Miura et al. 2007). Sumoylation is essential for freezing tolerance through the stabilization of ICE1, the inducer of *CBF* and repressor of *MYB15* expression. This modification blocks the ubiquitin-mediated degradation of ICE1 allowing it to activate *CBF* transcription. The sumoylated isoform of ICE1 also has a negative effect on the transcription of *MYB15*, which functions as repressor of *CBF* genes. The final effect of the AtSIZ1-mediated sumoylation is therefore the attenuation of a repressor system that in normal growing conditions blocks part of the transcriptional response to cold.

A general accumulation of SUMO conjugates is an early effect of the exposure to extreme temperatures, and other stress, thus many proteins are target of sumoylation upon exposure to stress (Miura et al. 2007). Studies on SUMO proteome list hundreds of potentially sumoylated proteins (Budhiraja et al. 2009; Elrouby and Coupland 2010; Miller et al. 2010; Park et al. 2011). The list is enriched in transcription factors and RNA-related factors regulating processing, splicing, export, and translation of mRNA. Thus many regulators of the plant cold response could represent the SUMO conjugates which accumulate under stress conditions. One example is STABILIZED1 (STA1), a splicing factor of the cold-induced gene *COR15 A* also involved in mRNA turnover. Its expression is induced by cold stress and the protein is sumoylated (Lee et al. 2006; Meier et al. 2012).

A Complex Network of Regulations

The overall regulations interacting at several levels represent a complex and dynamic network (Fig. 1). Factors responsible of post-transcriptional and post-translatrional mechanisms are often controlled at transcriptional level (i.e., many E3 ubiquitin ligases are stress induced), subjected to gene silencing by action of miRNA (Sunkar et al. 2012) or to alternative splicing (Mastrangelo et al. 2011). They might be modified by SUMO (many RNPs, RNA helicases, and components of the RNA export are substrates of sumoylation; Meier 2012), ubiquitin (ubiquitin conjugation controls transcription to mRNP assembly down to export through the nuclear pore complex; Babour 2012). Multiple signaling pathways may converge on the same target protein by multisite modifications, resulting in complex combinatorial regulatory patterns that dynamically and reversibly affect the activity of the target protein. Different post-translational mechanisms may act synergically or have antagonistic effects. The final output is a new shape of transcriptome, proteome, and metabolome which constitute the plant response to the stress.

The regulation of the activity of the transcription factor ICE1 offers a well-characterized example of the complexity of these regulatory systems. ICE1 is constitutively expressed, nevertheless, it activates the expression of CBF genes only upon cold treatment (Chinnusamy et al. 2012). Three different modifications are known, so far, to control the activity of ICE1 protein, and thus the expression of *CBFs*. At low temperature ICE1 is sumoylated through the action of AtSIZ1 (Miura et al. 2007), resulting in a fully active transcription factor. Alternatively, HOS1 ubiquitinates ICE1 leading to its proteasomal degradation (Dong et al. 2006). ICE1 may be more or less available for ubiquitination and sumoylation depending on its phosphorylation status, which is most likely temperature dependent (Zhu et al. 2007). The balance between activation and degradation allows a perfect tuning of ICE1 activity which in turn leads to the activation of the cold-induced molecular response. The function of HOS1, which is also subjected to alternative splicing, is activated by nuclear localization mediated by an unknown post-translational modification upon cold signal, in order to attenuate the stress response (Lee et al. 2012).

Epigenetic changes have been implicated in the acclimation process (Santos et al. 2011). In barley histone genes constitute an overrepresented functional category in the cold transcriptome of crown, the key organ for cereal overwintering (Janska et al. 2011), among them specific histone variants acting as temperature sensors in *Arabidopsis* (Kumar and Wigge 2010). Nontranscriptional regulations are implicated in epigenetic processes. Indeed, the final effect of RNA-mediated gene silencing is often the methylation of the genomic region producing the target RNA. Ubiquitination and sumoylation, beyond acetylation and methylation, act on nucleosome core histones and sumoylation regulates the activity of the chromatin-remodeling complexes. All together these modifications constitute a histone code which activates or silences gene expression by modifying chromatin structure and constituting a memory of the stress.

New Targets for Engineering Stress-Tolerant Plants?

A new generation of transgenic plants with improved performance under challenging environments could be developed using the increased knowledge on the network of post-transcriptional and post-translational regulations. Plants evolved this network to strictly regulate and perfectly fine-tune the molecular responses to abiotic stresses to achieve stress tolerance avoiding strongly physiological alteration and futile metabolic costs. Current transgenic strategies based on a rough manipulation of regulatory factors often produced plants with some increase of stress tolerance at the expense of growth and production. Indeed, many signaling pathways cross-talk with other developmental and metabolic pathways, thus increasing the risk of inadvertently affecting other routes producing yield penalties or developmental constraints. Future aims will be the development of plants with a finer and more specific regulation of upstream regulators of the stress response.

Acknowledgments This research was supported by Progetto Strategico MIPAF "OLEA – Genomica e Miglioramento genetico dell'olivo," D.M. 27011/7643/10.

References

Achard P, Cheng H, De Grauwe L, Decat J, Schoutteten H, Moritz T, Van Der Straeten D, Peng J, Harberd NP (2006) Integration of plant responses to environmentally activated phytohormonal signals. Science 311:91–94

Ambrosone A, Costa A, Leone A, Grillo S (2012) Beyond transcription: RNA-binding proteins as emerging regulators of plant response to environmental constraints. Plant Sci 182:12–8

Babour A, Dargemont C, Stutz F (2012) Ubiquitin and assembly of export competent mRNP. Biochim et Biophys Acta 1819:521–530

Budhiraja R, Hermkes R, Muller S, Schmidt J, Colby T, Panigrahi K, Coupland G, Bachmair A (2009) Substrates related to chromatin and to RNA-dependent processes are modified by *Arabidopsis* SUMO isoforms that differ in a conserved residue with influence on desumoylation. Plant Physiol 149:1529–1540

Chen L, Zhang Y, Ren Y, Xu J, Zhang Z, Wang Y (2012) Genome-wide identification of cold-responsive and new microRNAs in *Populus tomentosa* by high-throughput sequencing. Biochem Biophys Res Commun 417:892–896

Chinnusamy V, Ohta M, Kanrar S, Lee BH, Hong X, Agarwal M, Zhu JK (2003) ICE1: a regulator of cold-induced transcriptome and freezing tolerance in *Arabidopsis*. Genes Dev 17:1043–1054

Dong CH, Agarwal M, Zhang Y, Xie Q, Zhu JK (2006) The negative regulator of plant cold responses, HOS1, is a RING E3 ligase that mediates the ubiquitination and degradation of ICE1. Proc Natl Acad Sci U S A 103:8281–8286

Dong MA, Farre´ E, Thomashow MF (2011) Circadian clock-associated 1 and late elongated hypocotyl regulate expression of the C-repeat binding factor (CBF) pathway in *Arabidopsis*. Proc Natl Acad Sci U S A 108:7241–7246

Egawa C, Kobayashi F, Ishibashi M, Nakamura T, Nakamura C, Takumi S (2006) Differential regulation of transcript accumulation and alternative splicing of a DREB2 homolog under abiotic stress conditions in common wheat. Genes Genet Syst 81:77–91

Elrouby N, Coupland G (2010) Proteome-wide screens for small ubiquitin-like modifier (SUMO) substrates identify *Arabidopsis* proteins implicated in diverse biological processes. Proc Natl Acad Sci U S A 107:17415–17420

Gong Z, Dong CH, Lee H, Zhu J, Xiong L, Gong D, Stevenson B, Zhu JK (2005) A DEAD box RNA helicase is essential for mRNA export and important for development and stress responses in *Arabidopsis*. Plant Cell 17:256–267

Guerra D, Mastrangelo AM, Lopez-Torrejon G, Marzin S, Schweizer P, Stanca AM, del Pozo JC, Cattivelli L, Mazzucotelli E (2012) Identification of a protein network interacting with TdRF1, a wheat RING ubiquitin ligase with a protective role against cellular dehydration. Plant Physiol 158:777–89

Huang GT, Ma SL, Bai LP, Zhang L, Ma H, Jia P, Liu J, Zhong M, Guo ZF (2012) Signal transduction during cold, salt, and drought stresses in plants. Mol Biol Rep 39:969–87

James AB, Syed NH, Bordage S, Marshall J, Nimmo GA, Jenkins GI, Herzyk P, Brown JWS, Nimmo HG (2012) Alternative splicing mediates responses of the *Arabidopsis* circadian clock to temperature changes. Plant Cell 24:961–981

Janská A, Aprile A, Zámečnàk J, Cattivelli L, Ovesná J (2011) Transcriptional responses of winter barley to cold indicate nucleosome remodelling as a specific feature of crown tissues. Funct Integr Genomics 11:307–325

Jones HD, Kurup S, Peters NC, Holdsworth MJ (2000) Identification and analysis of proteins that interact with the Avena fatua homologue of the maize transcription factor VIVIPAROUS 1. Plant J 21:133–142

Kang H, Park SJ, Kwak KJ (2012) Plant RNA chaperones in stress response. Trends Plant Sci 4:130 (http://dx.doi.org/10.1016/j.tplants. 2012. 08.004)
Khraiwesh B, Zhu JH, Zhu J (2012) Role of miRNAs and siRNA in biotic and biotic and abiotic stress responses of plants. Biochim Biophys Acta 1819:137–148
Kim YO, Kang H (2006) The role of a zinc finger-containing glycine-rich RNA-binding protein during the cold adaptation process in *Arabidopsis thaliana*. Plant Cell Physiol 47:793–798
Kim JY, Park SJ, Jang B, Jung CH, Ahn SJ, Goh CH, Cho K, Han O, Kang H (2007) Functional characterization of a glycine-rich RNA-binding protein 2 in *Arabidopsis* thaliana under abiotic stress conditions. Plant J 50:439–451
Kim JS, Jung HJ, Lee HJ, Kim KA, Goh C-H, Woo Y, SH Oh, Han YS, Kang H (2008) Glycine-rich RNA-binding protein 7 affects abiotic stress responses by regulating stomata opening and closing in *Arabidopsis* thaliana. Plant J 55:455–466
Kim MH, Sasaki K, Imai R (2009) Cold shock domain protein 3 regulates freezing tolerance in Arabidopsis thaliana. J Biol Chem 284:23454–23460
Kumar SV, Wigge PA (2010) H2 A.Z-containing nucleosomes mediate the thermosensory response in *Arabidopsis*. Cell 140:136–147
Lee BH, Kapoor A, Zhu J, Zhu JK (2006) STABILIZED1, a stress-upregulated nuclear protein, is required for pre-mRNA splicing, mRNA turnover, and stress tolerance in *Arabidopsis*. Plant Cell 18:1736–1749
Lee JH, Kim SH, Kim JJ, Ahn JH (2012) Alternative splicing and expression analysis of high expression of osmotically responsive genes1 (HOS1) in Arabidopsis. BMB Rep 45:515–520
Lu S, Sun YH, Chiang VL (2008) Stress-responsive microRNAs in *Populus*. The Plant J 55:131–151
Lv DK, Bai X, Li Y, Ding XD, Ge Y, Cai H, Ji W, Wu N, Zhu YM (2010) Profiling of cold-stress-responsive miRNAs in rice by microarrays. Gene 459:39–47
Lyzenga WJ, Stone SL (2012) Abiotic stress tolerance mediated by protein ubiquitination. J Exp Bot 63:599–616
Mastrangelo AM, Belloni S, Barilli S, Ruperti B, Di Fonzo N, Stanca AM, Cattivelli L (2005) Low temperature promotes intron retention in two e-cor genes of durum wheat. Planta 221:705–715
Mastrangelo AM, Marone D, Laidò G, De Leonardis AM, De Vita P (2011) Alternative splicing: enhancing ability to cope with stress via transcriptome plasticity. Plant Sci 185–186:40–49
Mazzucotelli E, Belloni S, Marone D, De Leonardis AM, Guerra D, Di Fonzo N, Cattivelli L, Mastrangelo AM (2006) The E3 ubiquitin ligase gene family in plants: regulation by degradation. Curr Genomics 7:509–522
Mazzucotelli E, Mastrangelo AM, Crosatti C, Guerra D, Stanca AM, Cattivelli L (2008) Abiotic stress response in plants: when post-transcriptional and post-translational regulations control transcription. Plant Sci 174:420–431
Meier I (2012) mRNA export and sumoylation-Lessons from plants. Biochim Biophys Acta 1819:531–537
Miller MJ, Barrett-Wilt GA, Hua Z, Vierstra RD (2010) Proteomic analyses identify a diverse array of nuclear processes affected by small ubiquitin-like modifier conjugation in *Arabidopsis*. Proc Natl Acad Sci U S A 107:16512–16517
Miura K, Hasegawa PM (2010) Sumoylation and other ubiquitin-like post-translational modifications in plants. Trends Cell Biol 20:223–232
Miura K, Jin JB, Lee J, Yoo CY, Stirm V, Miura T, Ashworth EN, Bressan RA, Yun DJ, Hasegawa PM (2007) SIZ1-mediated sumoylation of ICE1 controls CBF3/DREB1 A expression and freezing tolerance in *Arabidopsis*. Plant Cell 19:1403–1414
Mizumoto K, Hatano H, Hirabayashi C, Murai K, Takumi S (2011) Characterization of wheat Bell1-type homeobox genes in floral organs of alloplasmic lines with Aegilops crassa cytoplasm. BMC Plant Biol 11:2
Palusa SG, Ali GS, Reddy ASN (2007) Alternative splicing of pre-mRNAs of *Arabidopsis* serine/arginine-rich proteins: regulation by hormones and stresses. Plant J 49:1091–1107
Park HC, Choi W, Park HJ et al (2011) Identification and molecular properties of SUMO-binding proteins in *Arabidopsis*. Mol Cells 32:143–151

Santos AP, Serra T, Figueiredo DD, Barros P, Lourenço T, Chander S, Oliveira MM, Saibo NJ (2011) Transcription regulation of abiotic stress responses in rice: a combined action of transcription factors and epigenetic mechanisms. OMICS 15:839–57

Sasaki K, Kim MH, Imai R (2007) Arabidopsis COLD SHOCK DOMAIN PROTEIN2 is a RNA chaperone that is regulated by cold and developmental signals. Biochem Biophys Res Commun 364:633–638

Schöning JC, Streitner C, Meyer IM, Gao Y, Staiger D (2008) Reciprocal regulation of glycine-rich RNA-binding proteins via an interlocked feedback loop coupling alternative splicing to nonsense-mediated decay in *Arabidopsis*. Nucleic Acids Res 36:6977–6987

Seo PJ, Kim MJ, Ryu JY, Jeong EY, Park CM (2011) Two splice variants of the IDD14 transcription factor competitively form nonfunctional heterodimers which may regulate starch me-tabolism. Nat Commun 2:303

Seo PJ, Park MJ, Lim MH, Kim SG, Lee M, Baldwin IT, Park CM (2012) A self-regulatory circuit of CIRCADIAN CLOCK-ASSOCIATED1 underlies the circadian clock regulation of temperature responses in *Arabidopsis*. Plant Cell 24:2427–2442

Sunkar R, Zhu J-K (2004) Novel and stress-regulated microRNAs and other small RNAs from *Arabidopsis*. Plant Cell 16:2001–2019

Sunkar R, Li YF, Jagadeeswaran G (2012) Functions of microRNA in plant stress responses. Trends in. Plant Sci 17:196–203

Tang Z, Zhang L, Xu C, Yuan S, Zhang F, Zheng Y, Zhao C (2012) Uncovering small RNA-mediated responses to cold stress in a wheat thermosensitive genic male-sterile line by deep sequencing. Plant Physiol 159:721–738

Thiebaut F, Rojas CA, Almeida KL, Grativol C, Domiciano GC, Regina C, Lamb C, De Almeida Engler J, Hemerly AS, Ferreira PCG (2012) Regulation of miR319 during cold stress in sugarcane. Plant Cell Envir 35:502–512

Wang Y, Liu K, Liao H, Zhuang C, Ma H, Yan X (2008) The plant WNK gene family and regulation of flowering time in *Arabidopsis*. Plant Biol 10:548–562

Yan J, Wang J, Li Q, Hwang JR, Patterson C, Zhang H (2003) AtCHIP, a U-Box-Containing E3 ubiquitin ligase, plays a critical role in temperature stress tolerance in *Arabidopsis*. Plant Physiol 132:861–869

Zhang J, Xu Y, Huan Q, Chong K (2009) Deep sequencing of *Brachypodium* small RNAs at the global genome level identifies microRNAs involved in cold stress response. BMC Genomics 10:449

Zhou X, Wang G, Sutoh K, Zhu JK, Zhang W (2008) Identification of cold-inducible microRNAs in plants by transcriptome analysis. Biochim Biophys Acta 1779:780–788

Zhu J, Dong CH, Zhu JK (2007) Interplay between cold-responsive gene regulation, metabolism and RNA processing during plant cold acclimation,Curr. Opin. Plant Biol. 10: 290–295

Cold Shock Domain Proteins in *Arabidopsis*: Functions in Stress Tolerance and Development

Ryozo Imai, Myung Hee Kim, Kentaro Sasaki, Shunya Sato and Yutaka Sonoda

Cold Shock Domain (CSD) Proteins: Backgrounds

The cold shock domain is one of the most conserved nucleic acid-binding domains, which is widely distributed from bacteria to mammals (Graumann and Marahiel 1996). CSD proteins are defined as they contain at least one copy of CSD. First report of CSD protein appeared in 1988, where a protein that binds Y-box of MHC class II gene promoter was reported (Didier et al. 1998). Independently, an approach for identifying proteins that are induced after cold shock in *Escherichia coli* discovered a class of cold-inducible proteins (Jones et al. 1987; Goldstein et al. 1990). One of them, F10.6 or CS7.4, was only detected in cold-treated cells and therefore designated as cold shock protein A (CspA; Goldstein et al. 1990). CspA solely consists of a CSD. Significant homology between CspA and CSDs within a wide range of eukaryotic proteins suggests an essential function of the domain through evolution (Fig. 1). Model structures of CSD have been solved for *E. coli* CspA and *Bacillus subtilis* CspB. They contain five stranded ß-barrel sheets with two juxtaposed consensus RNA-binding motifs (RNP1 and RNP2) that reside on ß2 and ß3 sheets (Schindelin et al. 1993, 1994).

Cold Shock Domains Proteins in Bacteria

In *E. coli*, nine members of the CSP gene family (*cspA* to *cspI*) have been identified. Four of them (*cspA*, *cspB*, *cspG*, and *cspI*) are induced by cold shock (Wang et al. 1999), while others are either constitutive or induced by other stresses. CspA, the most predominant cold shock protein in *E. coli*, may accumulate up to 10 % of total proteins subsequent to low temperature exposure. CSPs are essential for growth of

R. Imai (✉) · M. H. Kim · K. Sasaki · S. Sato · Y. Sonoda
Crop Breeding Research Division, NARO Hokkaido Agricultural Research Center, Sapporo, Hokkaido, Japan
e-mail: rzi@affrc.go.jp

R. Imai et al. (eds.), *Plant and Microbe Adaptations to Cold in a Changing World*, DOI 10.1007/978-1-4614-8253-6_11, © Springer Science+Business Media New York 2013

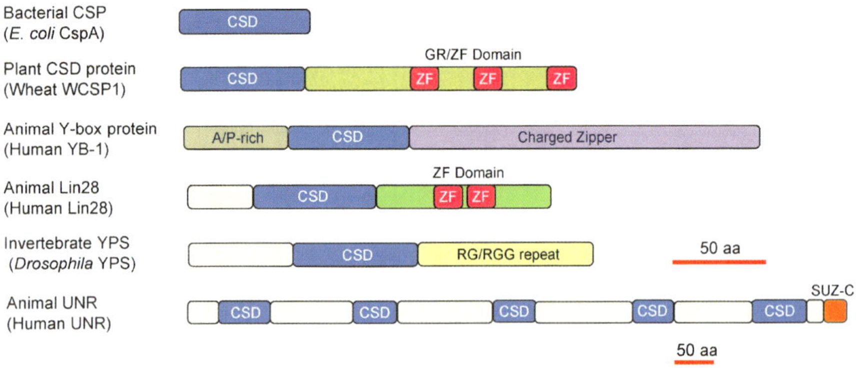

Fig. 1 Domain structures of CSD proteins from bacteria, plants, and animals. Schematic drawing of the domain organization of representative CSD proteins from bacteria, animals, and plants. Abbreviations used are: A/P-rich, Ala and Pro-rich domain; CSD, cold shock domain; GR, Gly-rich region; SUZ-C, SUZ-C domain (pfam12901); ZF, CCHC zinc finger motif.

E. coli under low temperature conditions. A quadruple deletion mutation in *cspA*, *cspB*, *cspG,* and *cspE* results in a growth defect at low temperatures (Xia et al. 2001). CSPs bind to ssRNA and ssDNA without an apparent sequence specificity (Jiang et al. 1997). However, CspA works on partially double-stranded RNA and unwinds the secondary structures (Jiang et al. 1997). RNA molecules typically form stable secondary structures in response to low temperature. With the prokaryotic transcriptional machinery, cold-induced RNA secondary structure may impose premature transcription termination. CspA, CspC, and CspE were confirmed to possess *in vivo* and *in vitro* transcription antitermination activity (Bae et al. 2000). CspA is also thought to enhance translation at low temperature through the elimination of stabilized RNA secondary structures (Jiang et al. 1997). These activities of CSPs established their function as RNA chaperones, according to the definition: "RNA chaperone" refers to proteins that prevent RNA misfolding and resolve misfolded RNAs, thereby ensuring their biological functions (Jiang et al. 1997).

Cold Shock Domains Proteins in Animals

CSD proteins are also extensively studied in eukaryotes. Among them Y-box proteins are the most extensively studied (Matsumoto and Wolffe 1998; Kohno et al. 2003; Eliseeva et al. 2011). The human Y-box protein, YB-1, was first identified as a transcription factor that binds to the Y-box of MHC class II promoters (Didier et al. 1988). Y-box proteins contain a single CSD with N-terminal and C-terminal auxiliary domains. The C-terminal domain is composed of alternating clusters of positively and negatively charged amino acid residues (also called charged zipper; Matsumoto and Wolffe 1998). Y-box proteins have been shown to be involved in transcriptional regulation of several genes (Matsumoto and Wolffe 1998;

Ohga et al. 1998). Y-box proteins also serve as a major component of cytoplasmic mRNP (Graumann and Marahiel 1998), where they play roles in repression of translation and stabilization of mRNAs in the early embryo and somatic cells (Matsumoto and Wolffe 1998). The frog Y-box protein, FRGY2, has been implicated in RNA masking to avoid translation of maternal mRNA in the fertilized egg (Ranjan et al. 1993). Although function of Y-box proteins in cold adaptation has not been explored in detail, disruption of YB-1 in chicken DT50 cells resulted in a cold-sensitive growth phenotype (Matsumoto et al. 2005).

Another class of CSD proteins in animals is the Lin28 family. Lin28 was first identified in the nematode *Caenorhabditis elegans* as a regulator of developmental timing and is required for cell fates to be expressed at the appropriate stage of larval development. Lin28 is a 25-kDa cytoplasmic protein that contains one CSD and a pair of retroviral-type Cys-Cys-His-Cys (CCHC) zinc fingers (Moss et al. 1997). Lin28 and its paralog, Lin28B, are highly expressed during embryogenesis and selectively down-regulated during differentiation (Moss and Tang 2003; Yang and Moss 2003). LIN28 is important for the maintenance of embryonic stem (ES) cell pluripotency and was used to produce induced pluripotent stem cell (iPSC) together with the three transcription factors, OCT4, SOX2, NANOG (Yu et al. 2009). Lin28 is also highly expressed in a range of different cancer cells and primary tumor tissues and implicated in oncogenesis (Thornton and Gregory 2012). Recent studies have revealed regulatory mechanisms of Lin28 and Lin28B on negative regulation of *let-7* miRNA biogenesis (Viswanathan et al. 2008). Nucleolar-localized Lin28B binds pri-*let-7* miRNAs and sequester them within the nucleus in order to block processing microprocessor, while cytoplasm-localized Lin28 binds miRNAs to inhibit pri-*let-7* processing (Piskounova et al. 2011; Thornton and Gregory 2012).

UNR (upstream of N-*ras*) is the third class of animal CSD protein that contains five tandem repeats of CSD (Fig. 1). UNR is an essential cytoplasmic protein. UNR binds to mRNA such as *c-myc* and *Apaf-1* at an internal ribosome entry site (IRES) or a 5′-untranslated region (UTR) and regulates translation and mRNA stability (Mihailovich et al. 2010). Binding of UNR to the IRES induces conformational rearrangement which promotes 40S ribosome to recognize the IRES, thereby stimulating translation independent of 5′-cap. Therefore, UNR has been proposed to function as an RNA chaperone (Mitchell et al. 2003).

The last class of animal CSD proteins is identified only in a range of invertebrates. Proteins of this class commonly contain one N-terminal CSD and C-terminal RG/RGG repeats. *Drosophila* Ypsilon Schachtel (YPS) is a protein of this class and involved in localization of specific mRNA to a subcellular site to regulate developmental patterning (Mansfield et al. 2002). YPS-like proteins, ApY1 from *Aplysia californica* (Skehel and Bartsch 1994) and DjY1 from planaria (Salvetti et al. 1998) are also reported as RNA-binding proteins but their detailed function is unknown. RBP16 from *Trypanosoma brucei* is involved in kinetoplastid RNA editing and translation (Thieringer et al. 1997; Pelletier et al. 2000). Interestingly, a recent report showed that the scallop YPS-like protein, CfCSP, is cold-inducible and able to complement the *E. coli csp* quadruple mutant (Yang et al. 2012).

Cold Shock Domain Proteins in Plants

Plant CSD proteins typically contain a single N-terminal CSD and variable copies of C-terminal retroviral-like CCHC zinc fingers that are interspersed by glycine-rich regions (Fig. 1). The first plant CSD protein that was functionally characterized is wheat cold shock protein 1 (WCSP1; Karlson et al. 2002). Both *WCSP1* mRNA and protein levels steadily increase in crown tissue during cold acclimation. WCSP1 protein binds both single-stranded and double-stranded nucleic acids and melts double-stranded RNA/DNA. *WCSP1* complements a cold-sensitive phenotype of the *E. coli csp* quadruple mutant (*cspA*, *cspB*, *cspG* and *cspE*), suggesting that WCSP1 shares a function with *E. coli* CSPs for cold adaptation (Nakaminami et al. 2006). According to the public sequence databases, plant CSD proteins can be detected in virtually all higher plant species. However, the number of paralogous genes and the copy numbers of CCHC zinc fingers within the proteins are markedly diverged among genera (Karlson and Imai 2003; Sasaki and Imai 2011). A genome-wide sequence analysis revealed that plant genome generally contains two classes of CSD proteins with fewer CCHC repeats (Class I) and more repeats (Class II; Sasaki and Imai 2011).

Four CSD proteins are found in Arabidopsis genome (AtCSP1/CSDP1: At4g36020; AtCSP2/CSDP2: At4g38680; AtCSP3: At2g17870; AtCSP4: At2g21060). Class I AtCSPs, AtCSP2 and AtCSP4 contain one CSD in the N-terminal region and two CCHC zinc fingers interspersed by Gly-rich sequences. Class II AtCSPs, AtCSP1 and AtCSP3, contain one N-terminal CSD and C-terminal seven CCHC zinc fingers with Gly-rich sequences.

AtCSP1

AtCSP1 contains one CSD and seven zinc fingers (ZFs). AtCSP1 binds to RNA and ss/dsDNA and melt partially double-stranded nucleic acids (Kim et al. 2007). Expression of *AtCSP1* in the *E. coli csp* quadruple mutant complemented the cold-sensitive growth phenotype of the mutant (Fig. 2a), RNA chaperone activity was also demonstrated with RNase susceptibility assay (Kim et al. 2007) and molecular beacon melting (Fig. 2b; Kim et al. 2007). It is interesting to note that either of the domains, CSD or ZFs, alone exhibits ability to complement *E. coli csp* mutant and melt an RNA duplex (Park et al. 2010). Subcellular localization of AtCSP1-GFP was observed in the nucleus and cytoplasm, which is similar to the pattern of AtCSP3-GUS localization (Fig. 3). Expression of *AtCSP1* is upregulated in response to cold, but down-regulated by salt and drought stresses (Kim et al. 2007; Park et al. 2009). High levels of *AtCSP1* mRNA accumulation were observed in shoot apical tissue and siliques. Expression of AtCSP2 in shoot apical tissue increases during floral transition induced by long day treatment (Nakaminami et al. 2009). Currently, it is not known if AtCSP1 is involved in the regulation of flowering time or silique/seed development. Overexpression of *AtCSP1* delayed seed germination under dehydration and salinity stress (Park et al. 2009).

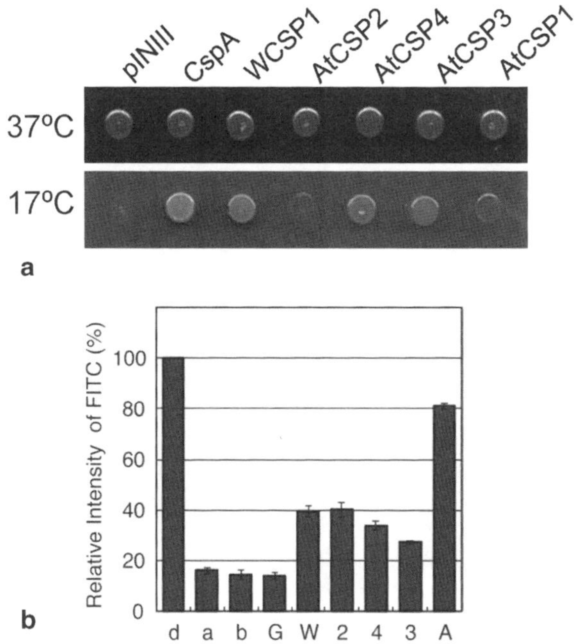

Fig. 2 RNA chaperone activity of AtCSPs. **a** Complementation of a low temperature sensitivity of an *E. coli cspA*, *cspB*, *cspE*, *cspG* quadruple mutant (BX04) with *AtCSPs*. Overnight liquid cultures were adjusted for OD, uniformly spotted onto LB-ampicillin plates, and grown on at either 37 °C (overnight) or 17 °C (5 days). CspA and WCSP1 are positive controls. **b** Ability of AtCSPs to melt DNA. A fluorescence molecular beacon system developed in the previous study was used for an *in vitro* dsDNA-melting assay (Nakaminami et al. 2006). Two partially complementing oligonucleotides, one of which was fluorescein isothiocyanate (FITC)-labeled at 5′ terminus and the other labeled with a quencher at the 3′ terminus, were used as a substrate. Relative fluorescence with each protein is shown in comparison to completely heat-denatured molecular beacons (100 % relative intensity equals to fluorescence from denatured beacons). d, denatured; a, annealed; b, buffer; G, GST; W, WCSP1; 2, AtCSP2; 4, AtCSP4; 3, AtCSP3; A, CspA

AtCSP2

AtCSP2 is a Class I CSP containing one CSD and two zinc fingers. AtCSP2 complements the cold-sensitive phenotype of *E. coli csp* quadruple mutant but to a lesser extent as compared with other AtCSPs (Fig. 2a). However, AtCSP2 displays comparable nucleic acid melting activity *in vitro*, indicating that it holds RNA chaperone activity (Fig. 2b; Sasaki et al. 2007). Immunological detection identified the nucleus and cytoplasm localization of AtCSP2 (Fusaro et al. 2006; Nakaminami et al. 2009). Fluorescent images for AtCSP2-GFP protein expressed in Arabidopsis root cells detected GFP signals in the nucleus and cytoplasm (Fusaro et al. 2006; Sasaki et al. 2007). Within the nucleus, preferential accumulation of the GFP-fused protein was observed within the nucleolus (Fig. 3; Sasaki et al. 2007). *AtCSP2* shows the highest expression level among the AtCSPs in seedlings (Fusaro et al. 2006; Nakaminami et al. 2009). *AtCSP2* is predominantly expressed in shoot apical

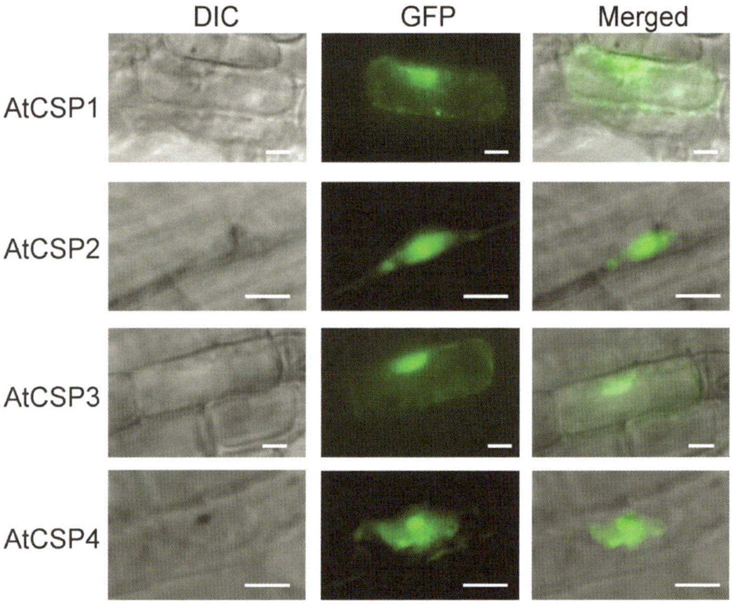

Fig. 3 Subcellular localization of AtCSPs in *Arabidopsis* root cells. For green fluorescent protein (GFP)-fused constructs, amplified AtCSP cDNAs were cloned into sGFP (S65T) vector (Niwa et al. 1999). For transient expression, gold particles (1.0 μm) coated with plasmid DNA (2.5 μg) were introduced into *Arabidopsis* root cells using a PDS1000/He particle gun (Bio-Rad, USA) according to the manufacturer's instructions. Arabidopsis root tissues were placed on MS agar medium and used for bombardment with a rupture setting of 1,100 psi. The bombarded samples were incubated for 16 h at 22 °C and were observed by a Leica FW 4000 microscope. Bar scale: 10 μm

meristem and root tip during the vegitative growth phase (Fusaro et al. 2006; Sasaki et al. 2007). The transcript levels increase in response to cold and salt but not by dehydration (Fusaro et al. 2006; Kim et al. 2007; Sasaki et al. 2007; Park et al. 2009).

A knockdown mutant of *AtCSP2* (*atcsp2–3*), where *AtCSP2* expression levels are reduced to about 40 % of the wild type, showed no change in freezing tolerance. However, crossed lines with *AtCSP4* knockout mutant (*atcsp2–3 atcsp4–1*) exhibited increased freezing tolerance (Sasaki et al. 2013). Interestingly, the increased freezing tolerance of the double mutant was only observed when the plants were cold-acclimated, and nonacclimated plants showed the same level of freezing tolerance as wild type. By contrast, overexpression of *AtCSP2* resulted in decreased freezing tolerance when cold-acclimated. These data suggested an overlapping function between the two Class I AtCSPs, AtCSP2 and AtCSP4, in the negative regulation of cold-acclimation (Sasaki et al. 2013). AtCSP2-mediated regulation of cold acclimation involves the C-repeat-binding factor (CBF)-dependent signaling pathway, which is known as the main regulatory pathway for cold acclimation. Expression levels of *CBF*s and its regulon genes are down-regulated by AtCSP2 (Sasaki et al. 2013).

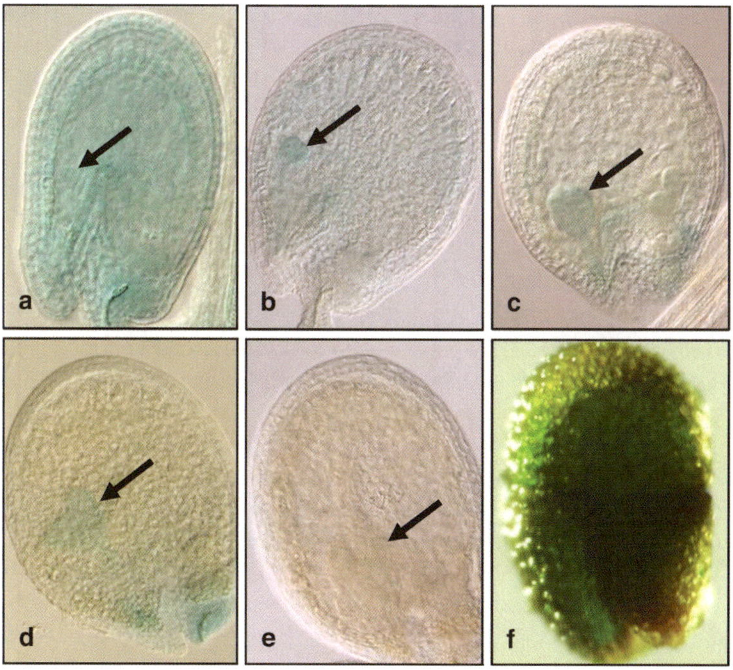

Fig. 4 GUS expression patterns during embryogenesis of *Arabidopsis* plants transformed with AtCSP2 promoter-GUS construct. AtCSP2 promoter-GUS transgenic plant was generated in the previous study (Sasaki et al. 2007). GUS staining was carried out with seeds of the following developmental stages: a, dermatogen stage embryo; b, globular stage embryo; c, triangular stage embryo; d, heart stage embryo; e, torpedo stage embryo; f, dry seed. *Arrow* indicates embryo

Expression of *AtCSP2* is strictly regulated during development processes. *AtCSP2* is expressed in anthers and pistils especially, in ovules and transmitting tissues during floral development (Fusaro et al. 2006; Sasaki et al. 2013). At the determatogen stage of seed development, *AtCSP2* promoter-GUS expression is detected in whole seeds (Fig. 4). As embryo development proceeds, GUS expression in embryo is maintained while tissues outside embryo drastically reduce the GUS expression. At the torpedo stage, GUS expression in embryo also becomes undetectable. GUS expression in embryo is again detectable in mature seeds. The GUS expression data correspond well with *AtCSP2* expression levels determined by qRT-PCR (Nakaminami et al. 2009). Over/ectopic expression of *AtCSP2* resulted in morphological and developmental alterations in the C24 ecotype of Arabidopsis (Fusaro et al. 2006). C24 plants overexpressing *AtCSP2* displayed late flowering, while RNAi lines with reduced *AtCSP2* expression displayed early flowering phenotype. Similar results were observed with Col-0 ecotype but the effects were rather small (Sasaki et al. 2013). In addition, AtCSP2-overexpressors show reduced number of stamens and high rates of abnormal development of seeds/embryos in C24 ecotype (Fusaro et al. 2006), although this phenotype is not observed with Col-0 (Sasaki et al. 2013).

It is also found that AtCSP2 negatively regulates silique length without affecting the seed number per silique and viability of the seeds (Sasaki et al. 2013). The mechanism that AtCSP2 regulates developmental processes is currently unknown but it is possible to speculate that plant hormones such as gibberellins are involved (Sasaki et al. 2013).

AtCSP3

AtCSP3 contains one CSD and seven ZnFs (Class II). AtCSP3 is currently the most extensively characterized CSD protein in Arabidopsis. AtCSP3 shows nucleic acid melting activity and complements the cold-sensitive phenotype of the *E. coli csp* quadruple mutant (Kim et al. 2009). AtCSP3 is exclusively expressed in meristematic tissues of roots and shoots of the seedlings (Kim et al. 2009). Transcript levels of AtCSP3 are up-regulated during cold acclimation in root and shoot of seedlings. Upon cold treatment, expression area expands from root meristematic tissue to a wider region of the root (Kim et al. 2009).

A loss of function mutant of AtCSP3 (*atcsp3–2*) is sensitive to freezing as compared to the wild-type plant under non-acclimated or cold-acclimated conditions. Overexpression of *AtCSP3* in transgenic plants confers enhanced freezing tolerance over wild-type plants (Kim et al. 2009). Expression levels of *CBF* genes and representative CBF regulon genes are not altered in the mutant. Microarray analysis revealed that AtCSP3 regulates expression of several stress-inducible genes that are not classified as CBF regulon genes. It was therefore suggested that AtCSP3 regulates freezing tolerance in *Arabidopsis* during cold acclimation independent of the CBF/DREB1 pathway. Analysis of promoter-GUS expression revealed that AtCSP3 is expressed in anthers during floral development. However, no phenotype regarding anther development was observed with the *atcsp3–2* mutant and *AtCSP3*-overexpressors. Interestingly, a recent report noticed that mutations in *AtCSP3* also result in change in leaf shape (Yang and Karlson 2012). It is therefore possible that AtCSP3 regulates developmental processes as well.

In order to understand the mechanisms through which AtCSP3 regulates gene expression, an interactome analysis was carried out (Kim et al. 2013). Sixteen proteins have been identified as interactors of AtCSP3 after a screening using yeast two-hybrid assay and bi-molecular fluorescence complementation (BiFC) assay. BiFC assay revealed that AtCSP3 interacts with different class of proteins in nucleolus, nuclear speckles, and cytoplasmic P-body (Kim et al. 2013). For example, AtCSP3 interacts with nuclear poly (A)-binding proteins within the speckles. AtCSP3 also interacts with decapping 5 (DCP5) within the cytoplasmic P-body (Kim et al. 2013). On the basis of these observations, we therefore propose a model in which AtCSP3 has versatile functions in RNA processing and metabolism by forming a variety of mRNP complexes. AtCSP3 may regulate pre-mRNA splicing, polyadenylation, RNA stability, RNA export, and translation by influencing mRNA processing in response to stress and developmental cues (Fig. 5).

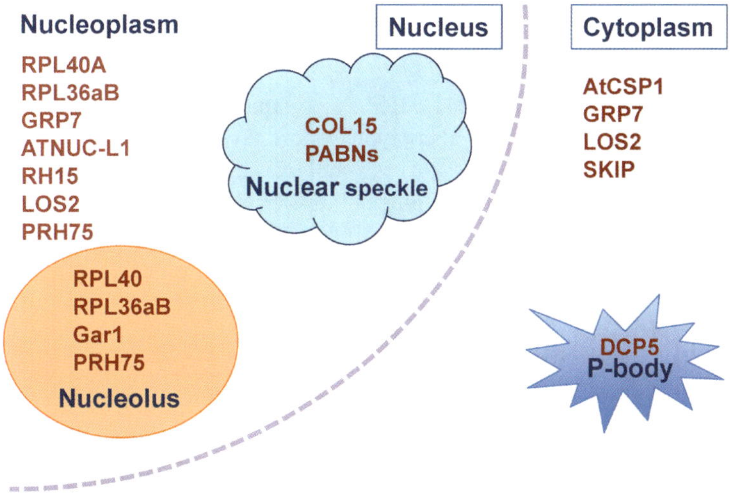

Fig. 5 Possible functional interactions of AtCSP3 in the nucleus and cytoplasm. AtCSP3-interacting proteins identified through a yeast two-hybrid screening and BiFC analysis are grouped by their subcellular localizations

AtCSP4

AtCSP4 is a Class I CSP containing one CSD and two zinc fingers. AtCSP4 shows RNA chaperon activity by complementing the *E. coli csp* quadruple mutant and *in vitro* melting assay (Fig. 2). Subcellular localization of AtCSP4-GFP was detected most intensively in nucleolus but the GFP signals were also observed in nucleoplasm and very weakly in cytoplasm (Fig. 3). The subcellular localization pattern of AtCSP4-GFP is very similar to that of AtCSP2-GFP (Fig. 3, Sasaki et al. 2007). Tissue-specific expression pattern of *AtCSP4* has been investigated using qRT-PCR and promoter-GUS expression (Nakaminami et al. 2009; Yang and Karlson 2011). In contrast to *AtCSP2* and *AtCSP3*, expression of *AtCSP4* is detectable in a wider range of plant tissues. During the vegetative stage, preferential expression of *AtCSP4* was detected in apical meristems, vascular tissues of young leaves and root tips (Yang and Karlson 2011). Higher levels of expression were detected in reproductive organs, such as flowers and siliques (Yang and Karlson 2011). A knockout mutant of *AtCSP4* did not show growth defects or morphological changes. In addition, *AtCSP4* is 1000-fold down-regulated in the Ler ecotype relative to Col-0 (Nakaminami et al. 2009). These data imply that *AtCSP4* alone is not an essential gene for normal growth or is functionally redundant with other AtCSPs. Recently, it was demonstrated that AtCSP2 and AtCSP4 are functionally redundant at least in the negative regulation of cold acclimation and flowering time (Sasaki et al. 2013).

Over/ectopic expression of *AtCSP4* resulted in morphological alterations in silique and seed development (Yang and Karlson 2011). Silique lengths of the homogygous overexpressors are slightly reduced and ~50 % of the seeds set on the heterozygous plants show a shrunken phonotype. The shrunken-seed phenotype is very similar to that of mutants in endosperm development (Yang and Karlson 2011). Accordingly, expression levels of *MEA* and *FIS2* are affected in the overexpressor lines (Yang and Karlson 2011). Reduced silique length is also observed with *AtCSP2* overexpressors but the phenotype is more severe with the *AtCSP2* overexpressors. In addition, embryo lethality is not observed with *AtCSP2* overexpressing lines. These data suggest that functions of AtCSP2 and AtCSP4 do not completely overlap.

Possible Mode of Action of Plant CSD Proteins

Given that plant CSD proteins share RNA chaperone activity with bacterial CSPs, it was initially postulated that plant CSD proteins enhance translation under low temperature conditions. However, interactome analysis of AtCSP3 suggested possible functions of AtCSP3 in multiple steps in RNA processing and stability control. It is reasonable to speculate that RNA chaperone activity can be required for a variety of RNA processing events within the eukaryotic cells. Next research on AtCSPs should be directed toward identifying the targets of AtCSPs. In the case of AtCSP3, several genes that are regulated by AtCSP3 have been identified. However, we do not know if these mRNAs are direct targets of AtCSP3. The two functional elements within the plant CSD proteins, CSD and CCHC zinc finger, are common to animal Lin28 proteins. As recent studies have revealed, Lin28 binds to *let-7* miRNA precursors and repress maturation of *let-7*. It is plausible to hypothesize that plant CSDs also interact with miRNAs.

References

Bae W, Xia B, Inouye M, Severinov K (2000) *Escherichia coli* CspA-family RNA chaperones are transcription antiterminators. Proc Natl Acad Sci U S A 97:7784–7789

Didier DK, Schiffenbauer J, Woulfe SL, Zacheis M, Schwartz BD (1988) Characterization of the cDNA encoding a protein binding to the major histocompatibility complex class II Y box. Proc Natl Acad Sci U S A 85:7322–7326

Eliseeva IA, Kim ER, Guryanov SG, Ovchinnikov LP, Lyabin DN (2011) Y-box-binding protein 1 (YB-1) and its functions. Biochemistry Mosc 76:1402–1433

Fusaro AF, Bocca SN, Ramos RLB, Barrôco RM, Magioli C, Jorge VC, Coutinho TC, Rangel-Lima CM, Rycke R, Inzé D et al (2006) AtGRP2, a cold-induced nucleo-cytoplasmic RNA-binding protein, has a role in flower and seed development. Planta 225:1339–1351

Goldstein J, Pollitt NS, Inouye M (1990) Major cold shock protein of *Escherichia coli*. Proc Natl Acad Sci U S A 87:283–287

Graumann P, Marahiel MA (1996) A case of convergent evolution of nucleic acid binding modules. BioEssays 18:309–315

Graumann PL, Marahiel MA (1998) A superfamily of proteins that contain the cold-shock domain. Trends Biochem Sci 23:286–290

Jiang W, Hou Y, Inouye M (1997) CspA, the major cold-shock protein of *Escherichia coli*, is an RNA chaperone. J Biol Chem 272:196–202

Jones PG, VanBogelen RA, Neidhardt FC (1987) Induction of proteins in response to low temperature in *Escherichia coli*. J Bacteriol 169:2092–2095

Karlson D, Imai R (2003) Conservation of the cold shock domain protein family in plants. Plant Physiol 131:12–15

Karlson D, Nakaminami K, Toyomasu T, Imai R (2002) A cold-regulated nucleic acid-binding protein of winter wheat shares a domain with bacterial cold shock proteins. J Biol Chem 277:35248–35256

Kim JS, Park SJ, Kwak KJ, Kim YO, Kim JY, Song J, Jang B, Jung C-H, Kang H (2007) Cold shock domain proteins and glycine-rich RNA-binding proteins from *Arabidopsis thaliana* can promote the cold adaptation process in *Escherichia coli*. Nucleic Acids Res 35:506–516

Kim MH, Sasaki K, Imai R (2009) Cold shock domain protein 3 regulates freezing tolerance in *Arabidopsis thaliana*. J Biol Chem 284:23454–23460

Kim M-H, Sonoda Y, Sasaki K, Kaminaka H, Imai R (2013) Interactome analysis reveals versatile functions of Arabidopsis COLD SHOCK DOMAIN PROTEIN 3 in RNA processing within the nucleus and cytoplasm. Cell Stress Chaperones. 18:517–525

Kohno K, Izumi H, Uchiumi T, Ashizuka M, Kuwano M (2003) The pleiotropic functions of the Y-box-binding protein, YB-1. BioEssays 25:691–698

Mansfield JH, Wilhelm JE, Hazelrigg T (2002) Ypsilon Schachtel, a *Drosophila* Y-box protein, acts antagonistically to Orb in the *oskar* mRNA localization and translation pathway. Development 129:197–209

Matsumoto K, Wolffe AP (1998) Gene regulation by Y-box proteins: coupling control of transcription and translation. Trends Cell Biol 8:318–323

Matsumoto K, Tanaka KJ, Tsujimoto M (2005) An acidic protein, YBAP1, mediates the release of YB-1 from mRNA and relieves the translational repression activity of YB-1. Mol Cell Biol 25:1779–1792

Mihailovich M, Militti C, Gabaldón T, Gebauer F (2010) Eukaryotic cold shock domain proteins: highly versatile regulators of gene expression. BioEssays 32:109–118

Mitchell SA, Spriggs KA, Coldwell MJ, Jackson RJ (2003) The Apaf-1 internal ribosome entry segment attains the correct structural conformation for function via interactions with PTB and unr. Mol Cell 11:757–771

Moss EG, Tang L (2003) Conservation of the heterochronic regulator Lin-28, its developmental expression and microRNA complementary sites. Dev Biol 258:432–442

Moss EG, Lee RC, Ambros V (1997) The cold shock domain protein LIN-28 controls developmental timing in C. elegans and is regulated by the *lin-4* RNA. Cell 88:637–646

Nakaminami K, Karlson DT, Imai R (2006) Functional conservation of cold shock domains in bacteria and higher plants. Proc Natl Acad Sci U S A 103:10122–10127

Nakaminami K, Hill K, Perry SE, Sentoku N, Long JA, Karlson DT (2009) *Arabidopsis* cold shock domain proteins: relationships to floral and silique development. J Exp Bot 60:1047–1062

Niwa Y, Hirano T, Yoshimoto K, Shimizu M, Kobayashi H (1999) Non-invasive quantitative detection and applications of non-toxic, S65T-type green fluorescent protein in living plants. Plant J 18:455–463

Ohga T, Uchiumi T, Makino Y, Koike K, Wada M, Kuwano M, Kohno K (1998) Direct involvement of the Y-box binding protein YB-1 in genotoxic stress-induced activation of the human multidrug resistance 1 gene. J Biol Chem 273:5997–6000

Park SJ, Kwak KJ, TR Oh, Kim YO, Kang H (2009) Cold shock domain proteins affect seed germination and growth of *Arabidopsis thaliana* under abiotic stress conditions. Plant Cell Physiol 50:869–878

Park SJ, Kwak KJ, Jung HJ, Lee HJ, Kang H (2010) The C-terminal zinc finger domain of *Arabidopsis* cold shock domain proteins is important for RNA chaperone activity during cold adaptation. Phytochemistry 71:543–547

Pelletier M, Miller MM, Read LK (2000) RNA-binding properties of the mitochondrial Y-box protein RBP16. Nucleic Acids Res 28:1266–1275

Piskounova E, Polytarchou C, Thornton JE, LaPierre RJ, Pothoulakis C, Hagan JP, Iliopoulos D, Gregory RI (2011) Lin28A and Lin28B inhibit let-7 microRNA biogenesis by distinct mechanisms. Cell 147:1066–1079

Ranjan M, Tafuri SR, Wolffe AP (1993) Masking mRNA from translation in somatic cells. Genes Dev 7:1725–1736

Salvetti A, Batistoni R, Deri P, Rossi L, Sommerville J (1998) Expression of DjY1, a protein containing a cold shock domain and RG repeat motifs, is targeted to sites of regeneration in planarians. Dev Biol 201:217–229

Sasaki K, Imai R (2011) Pleiotropic roles of cold shock domain proteins in plants. Front Plant Sci 2:116

Sasaki K, Kim M-H, Imai R (2007) *Arabidopsis* COLD SHOCK DOMAIN PROTEIN2 is a RNA chaperone that is regulated by cold and developmental signals. Biochem Biophys Res Commun 364:633–638

Sasaki K, Kim M-H, Imai R (2013) Arabidopsis COLD SHOCK DOMAIN PROTEIN 2 is a negative regulator of cold acclimation. New Phytol 198:95–102

Schindelin H, Marahiel MA, Heinemann U (1993) Universal nucleic acid-binding domain revealed by crystal structure of the *B. subtilis* major cold-shock protein. Nature 364:164–168

Schindelin H, Jiang W, Inouye M, Heinemann U (1994) Crystal structure of CspA, the major cold shock protein of *Escherichia coli*. Proc Natl Acad Sci U S A 91:5119–5123

Skehel PA, Bartsch D (1994) Characterization of a Y-Box factor from Aplysia californica. Gene 145:231–235

Thieringer HA, Singh K, Trivedi H, Inouye M (1997) Identification and developmental characterization of a novel Y-box protein from *Drosophila melanogaster*. Nucleic Acids Res 25:4764–4770

Thornton JE, Gregory RI (2012) How does Lin28 let-7 control development and disease? Trends Cell Biol 22:474–482

Viswanathan SR, Daley GQ, Gregory RI (2008) Selective blockade of microRNA processing by Lin28. Science 320:97–100

Wang N, Yamanaka K, Inouye M (1999) CspI, the ninth member of the CspA family of *Escherichia coli*, is induced upon cold shock. J Bacteriol 181:1603–1609

Xia B, Ke H, Inouye M (2001) Acquirement of cold sensitivity by quadruple deletion of the cspA family and its suppression by PNPase S1 domain in *Escherichia coli*. Mol Microbiol 40:179–188

Yang D-H, Moss EG (2003) Temporally regulated expression of Lin-28 in diverse tissues of the developing mouse. Gene Expr Patterns 3:719–726

Yang Y, Karlson DT (2011) Overexpression of *AtCSP4* affects late stages of embryo development in *Arabidopsis*. J Exp Bot 62:2079–2091

Yang Y, Karlson D (2012) Effects of mutations in the *Arabidopsis* cold shock domain protein 3 (*AtCSP3*) gene on leaf cell expansion. J Exp Bot 63:4861–4873

Yang C, Wang L, Siva VS, Shi X, Jiang Q, Wang J, Zhang H, Song L (2012) A novel cold-regulated cold shock domain containing protein from scallop *Chlamys farreri* with nucleic acid-binding activity. PLoS ONE 7:e32012

Yu J, Hu K, Smuga-Otto K, Tian S, Stewart R, Slukvin II, Thomson JA (2009) Human induced pluripotent stem cells free of vector and transgene sequences. Science 324:797–801

Regulation of RNA Metabolism in Plant Adaptation to Cold

Hunseung Kang and Su Jung Park

Abbreviations

CSD	Cold shock domain
CSP	Cold shock protein
GRP	Glycine-rich RNA-binding protein
RBP	RNA-binding protein
RH	RNA helicase
RZ	Zinc finger GRP

Introduction

The adaptation of plants to cold requires a regulated expression of multiple genes involved in the stress adaptation process. The regulation of gene expression occurs at the posttranscriptional level as well as transcriptional level, which are of crucial importance for plant stress responses. Posttranscriptional regulation of RNA metabolism includes precursor-RNA processing, splicing, capping, polyadenylation, transport, turnover, and translational control (Simpson and Filipowicz 1996; Floris et al. 2009). These posttranscriptional events are controlled by direct binding of RNA-binding proteins (RBPs) to RNA targets or indirect interactions of RBPs with other regulatory factors. How specific RBPs recognize RNA targets and promote RNA metabolism is thus a central question to better understand the cold adaptation process in cells.

H. Kang (✉)
Department of Plant Biotechnology, Chonnam National University,
300 Yongbong-dong, Buk-gu, Gwangju 500-757, South Korea
e-mail: hskang@jnu.ac.kr

S. J. Park
Department of Plant Biotechnology, College of Agriculture and Life Sciences,
Chonnam National University, Gwangju, South Korea

R. Imai et al. (eds.), *Plant and Microbe Adaptations to Cold in a Changing World,*
DOI 10.1007/978-1-4614-8253-6_12,

Recent progress in genome sequence analysis has led to the discovery of different types of RBPs harboring several conserved structural motifs, such as RNA-recognition motifs (RRMs), arginine-rich domains, glycine-rich domains, RD-repeats, SR-repeats, and zinc finger motifs (Burd and Dreyfuss 1994; Albà and Pagès 1998; Lorković and Barta 2002). The typical RBPs contain one or more RRMs at their N-terminus and auxiliary motifs, such as glycine-rich domains, arginine-rich domains, RD-repeats, SR-repeats, and zinc finger motifs, at their C-terminus. The RRM is the best characterized RNA-binding motif and comprises 80–90 amino acids that are capable of binding to RNA substrates. The most conserved RRM sequences are ribonucleoprotein 1 (RNP1) (RGFGFVTF) and RNP2 (CFVGGL), which normally determines the affinity for RNA substrates. The auxiliary domains present at the C-terminus are the major determinants of RNA-binding specificity (Kenan et al. 1991; Burd and Dreyfuss 1994; Nagai et al. 1995). Among the diverse RBPs found in plants, RBPs containing RRMs at their N-terminus and a glycine-rich region at their C-terminus, thus referred to as glycine-rich RNA-binding proteins (GRPs), cold shock domain proteins (CSPs), and RNA helicases (RHs) have been extensively demonstrated for their roles in the cold adaptation process in plants. The RBPs whose functions have been demonstrated during cold stress response in plants are summarized in Table 1, and we will discuss the recent progress of our understanding of the roles of these RBPs during the cold adaptation process in plants.

Glycine-Rich RNA-Binding Proteins in Cold Adaptation

Plant GRPs discussed in this chapter are small in size (approximately 16–20 kDa) and contain a single RRM at the N-terminal half and a glycine-rich region at the C-terminal half. The genes encoding GRPs have been found in diverse plant species, including maize, alfalfa, *Arabidopsis*, barley, Brasicca, rice, and tobacco (Gómez et al. 1988; van Nocker and Vierstra 1993; Hirose et al. 1994; Ferullo et al. 1997; Aneeta et al. 2002; Stephen et al. 2003; Nomata et al. 2004; Shinozuka et al. 2006). The *Arabidopsis* and rice (*Oryza sativa*) genomes contain eight and at least six GRPs, respectively (Lorković and Barta 2002; Kim et al. 2010a). Increasing numbers of reports have demonstrated that GRPs play important roles in the response of plants to changing environmental conditions (Sachetto-Martins 2000; Lorković 2009; Mangeon et al. 2010). In particular, the functional roles of several GRPs in plants under cold stress conditions have been recently characterized (Table 1). Comparative and systematic analysis of GRPs present in *Arabidopsis* has revealed the different roles of AtGRP family members in seed germination, seedling growth, and stress tolerance of *Arabidopsis* plants under cold stress conditions (Kwak et al. 2005; Kim et al. 2007a, 2008a). It has been demonstrated that AtGRP2 and AtGRP7 but not AtGRP4 accelerate seed germination and seedling growth under low temperatures and confer freezing tolerance to *Arabidopsis* plants. Although our understanding of the functional roles of GRPs in monocotyledonous plants is far less than that in dicotyledonous plants, several recent studies also showed that rice OsGRP1

Table 1 Overview of plant RNA-binding proteins (RBPs) and their expression and roles in cold stress response

Protein name	Plant source	Gene ID	Expression/function	Reference
Glycine-rich RNA-binding protein (GRP)				
AtGRP2	Arabidopsis	At4g13850	Cold/freezing tolerance	Kim et al. (2007a)
AtGRP4	Arabidopsis	At3g23830	Responds to cold	Kwak et al. (2005)
AtGRP7	Arabidopsis	At2g21660	Cold/freezing tolerance	Kim et al. (2008a)
OsGRP1	Rice	Os01g68790	Cold/freezing tolerance	Kim et al. (2010a)
OsGRP4	Rice	Os04g33810	Cold/freezing tolerance	Kim et al. (2010a)
OsGRP6	Rice	Os12g31800	Cold/freezing tolerance	Kim et al. (2010a)
AtRZ-1a	Arabidopsis	At3g26420	Cold/freezing tolerance	Kim et al. (2005)
OsRZ-2	Rice	Os07g08960	Cold/freezing tolerance	Kim et al. (2010b)
NtGRP	Tobacco	D16204	Responds to cold	Chen et al. (2010)
Cold shock domain protein (CSP)				
AtCSP1	Arabidopsis	At4g36020	Cold/freezing tolerance	Park et al. (2010)
AtCSP2	Arabidopsis	At4g38680	Responds to cold	Sasaki et al. (2007)
AtCSP3	Arabidopsis	At2g17870	Cold/freezing tolerance	Kim et al. (2009)
AtCSP4	Arabidopsis	At2g21060	Responds to cold	Karlson and Imai (2003)
OsCSP1	Rice	Os02g02870	Responds to cold	Chaikam and Karlson (2008)
OsCSP2	Rice	Os08g03520	Responds to cold	Chaikam and Karlson (2008)
WCSP1	Wheat	AB066265	Responds to cold	Karlson et al. (2002)

Table 1 (continued)

Protein name	Plant source	Gene ID	Expression/function	Reference
DEAD-box RNA helicase (RH)				
AtRH9	Arabidopsis	At3g22310	Cold/freezing tolerance	Kim et al. (2008b)
AtRH22	Arabidopsis	At1g59990	Responds to cold	Tripurani et al. (2011)
AtRH25	Arabidopsis	At5g08620	Cold/freezing tolerance	Kim et al. (2008b)
AtRH38	Arabidopsis	At3g53110	Cold/freezing tolerance	Gong et al. (2005)
AtRH52	Arabidopsis	At3g58570	Responds to cold	Tripurani et al. (2011)
GmRH	Soybean	FJ462142	Responds to cold	Chung et al. (2009)

and OsGRP4 confer cold and freezing tolerance in *Arabidopsis* plants (Kim et al. 2010a). Complementation analysis of rice GRPs in GRP-deficient *Arabidopsis* mutants has further demonstrated that GRPs in rice and *Arabidopsis* are functionally conserved during the cold adaptation process (Kim et al. 2010a). Recently, it has been shown that four GRP genes in tobacco (NtGRPs) were highly induced by cold stress but not by ABA, which suggests that these NtGRPs play a role in abiotic stress response via an ABA-independent pathway (Chen et al. 2010).

In addition to these GRPs harboring an RRM and a glycine-rich region, plants contain a different type of GRPs that harbor an RRM at the N-terminal half and a glycine-rich region interspersed with a CCHC-type zinc finger at the C-terminal half, designated as RZs. The *Arabidopsis* and rice genomes contain a gene family encoding three RZ proteins (Lorković and Barta 2002; Kim et al. 2010b), which share approximately 40 % of sequence similarity with each other and have a highly conserved CCHC-type zinc finger motif (Kim et al. 2010b). Although the functional roles of RZ proteins have been determined in limited cases, their roles in stress response are emerging. Among the three RZs present in *Arabidopsis*, it has been determined that AtRZ-1a affects seed germination and seedling growth under cold stress conditions and enhances freezing tolerance in *Arabidopsis* (Kim et al. 2005; Kim and Kang 2006), while AtRZ-1b and AtRZ-1c do not affect seed germination or seedling growth of *Arabidopsis* under cold stress conditions (Kim et al. 2010c). These results demonstrate that the three AtRZ-1 family members contribute differently to the cold adaptation process in plants. Interestingly, it has also been demonstrated that, among the three RZ proteins in rice, only OsRZ2 has the ability to confer cold and freezing tolerance in plants (Kim et al. 2010b).

Cold Shock Domain Proteins in Cold Adaptation

Cold shock proteins found in bacteria are homologous to a domain called the cold shock domain (CSD) of eukaryotic Y-box proteins (Didier et al. 1988), which is capable of binding RNA, single-stranded DNA, and double-stranded DNA (Landsman 1992; Graumann and Marahiel 1998; Manival et al. 2001). Cold shock proteins are highly induced during the cold acclimation phase in bacteria and fungi, and are involved in the regulation of gene expression under cold stress conditions (Goldstein et al. 1990; Wolffe 1994; Yamanaka et al. 1998; Phadtare et al. 1999; Fang and Leger 2010). It has been demonstrated that certain CSPs in bacteria function as RNA chaperones that facilitate translation at low temperatures via the destabilization of the overstabilized secondary structures in mRNA (Jiang et al. 1997). Contrary to the small size (about 7–10 kDa) of CSPs found in prokaryotes, the CSPs present in eukaryotes including plants contain additional glycine-rich regions interspersed by CCHC-type zinc fingers at the C-terminal half as well as a CSD at the N-terminal half (Karlson et al. 2002; Karlson and Imai 2003; Chaikam and Karlson 2010). It has been determined that *Arabidopsis* and rice genomes contain four CSPs that harbor highly homologous CSDs at the N-terminal half but variable glycine-rich regions interspersed with different numbers of CCHC zinc fingers at the C-terminal half (Karlson and Imai 2003; Chaikam and Karlson 2008).

Plant CSPs have been implicated in cold stress response based on highly induced expression levels and their structural similarity to prokaryotic CSPs, which function as RNA chaperones to enhance translation at low temperatures (Jiang et al. 1997; Bae et al. 2000). The domain structures and functions of plant CSPs have been reviewed recently (Sasaki and Imai, 2012). *Arabidopsis* AtCSP2 was cold-regulated and complemented the cold-sensitive phenotype of *E. coli* mutant at low temperatures, demonstrating its role as an RNA chaperone during the cold adaptation process (Sasaki et al. 2007). It has been demonstrated that *Arabidopsis* AtCSP3 contributes to enhance freezing tolerance (Kim et al. 2009). In addition, the CSPs isolated from winter wheat and rice were shown to complement the cold-sensitive phenotype of *E. coli* mutant at low temperatures (Nakaminami et al. 2006; Chaikam and Karlson 2008), which suggests that they play a role as RNA chaperones during the cold adaptation process in *E. coli.* All of these results clearly indicate that CSPs in plants as well as in bacteria are important players in the response of living organisms to cold stress.

RNA Helicases in Cold Adaptation

RNA helicases are ubiquitous enzymes that catalyze the unwinding of duplex RNA secondary structures in an ATP-dependent manner. Among the six superfamily (SF) of RHs, the DEAD-box family belongs to the SF2 helicase superfamily, which contains a core region of highly conserved helicase and Q motifs (Gorbalenya and Koonin 1993; Tanner et al. 2003; Fairman-Williams et al. 2010). The *Arabidopsis* and rice genomes harbor 58 and 50 DEAD-box RHs, respectively (Aubourg et al. 1999; Mingam et al. 2004; Umate et al. 2010). DEAD-box RHs have been implicated to play roles during stress adaptation process in plants. RNA helicases have the ability to unwind the overstabilized RNAs at low temperatures, which is required for proper RNA metabolism, including ribosome biogenesis, pre-mRNA splicing, transport, decay, and translational control (Jones et al. 1996; Jankowsky 2011). Despite the potential roles of RHs in the response of plants to environmental stimuli, the reports demonstrating the functional roles of RHs during stress adaptation process in plants are severely limited.

The cold-regulated expressions of DEAD-box RHs have been investigated in diverse systems, including csdA in *E. coli*, crhC in cyanobacteria, deaD in *Methanococcoides burtonii,* and several RHs in plants (Jones et al. 1996; Chamot and Owttrim 2000; Lim et al. 2000; Chung et al. 2009) The potential role of RHs in plant cold stress response was reported at the first time for the low expression of an osmotically responsive gene (los4) encoding *Arabidopsis* AtRH38 that conferred freezing tolerance in *Arabidopsis* (Gong et al. 2002; 2005; Zhu et al. 2007). The two DEAD-box RHs, STRESS RESPONSE SUPPRESSOR1 and STRESS RESPONSE SUPPRESSOR2, have been determined to play a role in *Arabidopsis* response to multiple abiotic stresses (Kant et al. 2007). It has been demonstrated that *Arabidopsis* AtRH25 has an ability to enhance freezing tolerance, while AtRH9 does not have such ability (Kim et al. 2008b). In a recent study, it has been demonstrated that the cold-responsive AtRH22 and AtRH52 genes are essential during *Arabidopsis* embryogenesis (Tripurani et al. 2011). All of these reports clearly indicate that RHs should play an important role in plant response to cold stress.

Cellular Roles of RNA-Binding Proteins in the Cold Adaptation Process

Despite the increase in understanding of the functional roles of GRPs, CSPs, and RHs in the response of plants to cold stress, the cellular roles and molecular mechanisms of RBP's action during cold adaptation have not yet been firmly established. Several recent studies have suggested that RNA chaperone activity of these RBPs is closely correlated with their roles in plants under cold stress conditions. RNA molecules are prone to fold into nonfunctional alternative secondary structures (Herschlag 1995). It has been suggested that formation of misfolded structures is prevented or reversed by the action of proteins generally known as RNA chaperones. RNA chaperones are nonspecific RBPs found in all living organisms and function by aiding the correct folding of RNA molecules during RNA metabolism (Herschlag 1995; Kang et al. 2013). The roles of bacterial CSPs as RNA chaperones have been well characterized, in which CSPs destabilize the overstabilized secondary structures in mRNAs for efficient translation at low temperatures (Graumann et al. 1997; Jiang et al. 1997; Phadtare et al. 1999; 2002). The lack of CSPs in cyanobacteria and the structural similarity of GRPs to RRM proteins found in cyanobacteria suggested that GRP proteins may actually substitute for the function of CSPs in cyanobacteria (Maruyama et al. 1999).

The roles of GRPs as RNA chaperones have been demonstrated under cold stress conditions. The *Arabidopsis* AtGRP2 and AtGRP7, which conferred cold and freezing tolerance in plants (Kim et al. 2007a; 2008a), were found to harbor RNA chaperone activity as evidenced by their complementation ability in Csp-deficient *E. coli* mutant cells and the nucleic acid-melting activities (Kim et al. 2007b). By contrast, AtGRP4, which did not increase the cold or freezing resistance of *Arabidopsis* plants (Kwak et al. 2005), did not harbor RNA chaperone activity (Kim et al. 2007b). Similarly, rice OsGRP1 and OsGRP4 have the ability to confer cold and freezing tolerance in *Arabidopsis* and possess RNA chaperone activity (Kim et al. 2010a). In addition, the ability of *Arabidopsis* AtRZ-1a and rice OsRZ2 to enhance freezing tolerance was closely correlated with their RNA chaperone activity (Kim et al. 2010b). Through domain-swapping and deletion analysis of *Arabidopsis* CSPs, it was demonstrated that the number of zinc finger motif and the length of glycine-rich domains of CSPs are crucial for the RNA chaperone activity (Park et al. 2010). The CSPs isolated from rice and wheat also harbor RNA chaperone activity during the cold adaptation process (Nakaminami et al. 2006;

Chaikam and Karlson 2008). The *Arabidopsis* AtRH25 conferred freezing tolerance of the plant and had the ability to complement the cold-sensitive phenotype of Csp-deficient *E. coli* mutant cells at low temperatures (Kim et al. 2008b). All of these findings strongly suggest that GRPs, CSPs, and RHs function as RNA chaperones in plants during the cold adaptation process.

The cellular roles of RBPs in plant cold adaptation are just emerging. In order to regulate RNA metabolism in cells, RBPs should be localized to the nucleus and/or the cytoplasm, where RNA processing and translation occurs. It has been demonstrated that the nuclear-localized AtRH38 is involved in the regulation of mRNA export from the nucleus to the cytoplasm under cold stress conditions (Gong et al. 2005; Chinnusamy et al. 2008). The *Arabidopsis* AtGRP7, which is localized to the nucleus and cytoplasm, has also been demonstrated to be involved in mRNA export from the nucleus to the cytoplasm under cold stress conditions (Kim et al. 2008a). Ectopic expression of rice GRP or RZ in *Arabidopsis grp7* mutant restored the defect in mRNA export from the nucleus to the cytoplasm, demonstrating a conserved role of GRPs and RZs in mRNA export of dicots and monocots (Kim et al. 2010a; 2010b). All of these results support the emerging idea that certain GRPs, RZs, and RHs regulate mRNA export from the nucleus to the cytoplasm and thereby play a key role during the cold adaptation process in plants.

Concluding Remarks

Although our understanding of the functional roles of RBPs in the plant cold adaptation process is steadily increasing, the cellular roles of most of the RBPs are still poorly understood. In particular, GRPs, CSPs, and RHs have been determined to play important roles in the plant response to cold stress, yet our knowledge of the cellular functions of these RBPs in cold stress response remains far from sufficient. Recent advances in characterizing the conserved functions of GRPs and CSPs in *Arabidopsis* and rice have provided opportunities for the mechanistic examination of their cellular roles during the stress adaptation processes. Further studies should focus on the identification of target RNAs for RBPs and their interactions during the regulation of RNA metabolism, which is an indispensible step to get more comprehensive picture of the cellular roles of RBPs and posttranscriptional regulation of RNA metabolism during the cold adaptation process in plants.

Acknowledgments The authors were supported, in part, by a grant from the Next-Generation BioGreen21 Program (PJ00820303), Rural Development Administration, Republic of Korea, and by the Mid-career Researcher Program (2011–0017357) through the National Research Foundation of Korea grant funded by the Ministry of Education, Science and Technology, Republic of Korea.

References

Albà MM, Pagès M (1998) Plant proteins containing the RNA-recognition motif. Trends Plant Sci 3:15–21

Aneeta NS-M, Tuteja N, Sopory SK (2002) Salinity- and ABA-induced up-regulation and light-mediated modulation of mRNA encoding glycine-rich RNA-binding protein from *Sorghum bicolor*. Biochem Biophys Res Commun 296:1063–1068

Aubourg S, Kreis M, Lecharny A (1999) The DEAD box RNA helicase family in *Arabidopsis thaliana*. Nucleic Acids Res 27:628–636

Bae W, Xia B, Inouye M et al (2000) *Escherichia coli* CspA-family RNA chaperones are transcription antiterminators. Proc Natl Acad Sci USA 97:7784–7789

Burd CG, Dreyfuss G (1994) Conserved structures and diversity of functions of RNA-binding proteins. Science 265:615–621

Chaikam V, Karlson D (2008) Functional characterization of two cold shock domain proteins from *Oryza sativa*. Plant Cell Environ 31:995–1006

Chaikam V, Karlson D (2010) Comparison of structure, function and regulation of plant cold shock domain proteins to bacterial and animal cold shock domain proteins. BMB Rep 43:1–8

Chamot D, Owttrim GW (2000) Regulation of cold shock-induced RNA helicase gene expression in the *Cyanobacterium Anabaena* sp. Strain PCC 7120. J Bacteriol 182:1251–1256

Chen X, Zeng Q, Lu X, Yu D, Li W (2010) Characterization and expression analysis of four glycine-rich RNA-binding proteins involved in osmotic response in tobacco (*Nicotiana tabacum cv*. Xanthi). Agricultural Sciences in China 9:1577–1587

Chinnusamy V, Gong Z, Zhu J-K (2008) Nuclear RNA export and its importance in abiotic stress responses of plants. In: Reddy ASN, Golovkin M (eds) Nuclear pre-mRNA processing in plants. Berlin, Springer-Verlag, pp. 235–255

Chung E, Cho CW, Yun BH, Choi HK, So HA, Lee SW, Lee JH (2009) Molecular cloning and characterization of the soybean DEAD-box RNA helicase gene induced by low temperature and high salinity stress. Gene 443:91–99

Didier DK, Schiffenbauer J, Woulfe SL et al (1988) Characterization of the cDNA encoding a protein binding to the major histocompatibility complex class II Y box. Proc Natl Acad Sci USA 85:7322–7326

Fairman-Williams ME, Guenther UP, Jankowsky E (2010) SF1 and SF2 helicases: family matters. Curr Opin Struct Biol 20:313–324

Fang W, Leger RJ (2010) RNA binding proteins mediate the ability of a fungus to adapt to the cold. Environ Microbiol 12:810–820

Ferullo J-M, Vézina LP, Rail J et al (1997) Differential accumulation of two glycine-rich proteins during cold-acclimation alfalfa. Plant Mol Biol 33:625–633

Floris M, Mahgoub H, Lanet L et al (2009) Post-transcriptional regulation of gene expression in plants during abiotic stress. Int J Mol Sci 10:3168–3185

Goldstein J, Pollitt NS, Inouye M (1990) Major cold shock protein of *Escherichia coli*. Proc Natl Acad Sci USA 87:283–287

Gómez J, Sánchez-Martànez D, Stiefel V et al (1988) A gene induced by the plant hormone abscisic acid in response to water stress encodes a glycine-rich protein. Nature 344:262–264

Gong Z, Lee H, Xiong L et al (2002) RNA helicase-like protein as an early regulator of transcription factors for plant chilling and freezing tolerance. Proc Natl Acad Sci USA 99:11507–11512

Gong Z, Dong C-H, Lee H et al (2005) A dead box RNA helicase is essential for mRNA export and important for development and stress responses in Arabidopsis. Plant Cell 17:256–267

Gorbalenya AE, Koonin EV (1993) Helicases: amino acids sequence comparisons and structure–function relationship. Curr Opin Struct Biol 3:419–429

Graumann PL, Marahiel MA (1998) A superfamily of proteins that contain the cold-shock domain. Trends Biochem Sci 23:286–290

Graumann PL, Wendrich TM, Weber MH et al (1997) A family of cold shock proteins in *Bacillus subtilis* is essential for cellular growth and for efficient protein synthesis at optimal and low temperatures. Mol Microbiol 25:741–756

Herschlag D (1995) RNA chaperones and the RNA folding problem. J Biol Chem 270:20871–20874

Hirose T, Sugita M, Sugiura M (1994) Characterization of a cDNA encoding a novel type of RNA-binding protein in tobacco: its expression and nucleic acid-binding properties. Mol Gen Genet 244:360–366

Jankowsky E (2011) RNA helicases at work: binding and rearranging. Trends Biochem Sci 36:19–29

Jiang W, Hon Y, Inouye M (1997) CspA, the major cold-shock protein of *Escherichia coli*, is an RNA chaperone. J Biol Chem 272:196–202

Jones PG, Mitta M, Kim Y et al (1996) Cold shock induces a major ribosomal-associated protein that unwinds double-stranded RNA in *Escherichia coli*. Proc Natl Acad Sci USA 93:76–80

Kang H, Park SJ, Kwak KJ (2013) Plant RNA chaperones in stress response. Trends Plant Sci 18(2):100–106. doi:10.1016/j.tplants.2012.08.004

Kant P, Kant S, Gordon M et al (2007) Stress response suppressor1 and stress response suppressor2, two DEAD-box RNA helicases that attenuate *Arabidopsis* responses to multiple abiotic stresses. Plant Physiol 145:814–830

Karlson D, Imai R (2003) Conservation of the cold shock domain protein family in plants. Plant Physiol 131:12–15

Karlson D, Nakaminami K, Yoyomasu T et al (2002) A cold-regulated nucleic acid-binding protein of winter wheat shares a domain with bacterial cold shock proteins. J Biol Chem 277:35248–35256

Kenan DJ, Query CC, Keene JD (1991) RNA recognition: towards identifying determinants of specificity. Trends Biochem Sci 16:214–220

Kim YO, Kang H (2006) The role of a zinc finger-containing glycine-rich RNA-binding protein during the cold adaptation process in *Arabidopsis thaliana*. Plant Cell Physiol 47:793–798

Kim YO, Kim JS, Kang H (2005) Cold-inducible zinc finger-containing glycine-rich RNA-binding protein contributes to the enhancement of freezing tolerance in *Arabidopsis thaliana*. Plant J 42:890–900

Kim JY, Park SJ, Jang B et al (2007a) Functional characterization of a glycine-rich RNA-binding protein2 in *Arabidopsis thaliana* under abiotic stress conditions. Plant J 50:439–451

Kim JS, Park SJ, Kwak KJ et al (2007b) Cold shock domain proteins and glycine-rich RNA-binding proteins from *Arabidopsis thaliana* can promote the cold adaptation process in *Escherichia coli*. Nucleic Acids Res 35:506–516

Kim JS, Jung HJ, Lee HJ et al (2008a) Glycine-rich RNA-binding protein7 affects abiotic stress responses by regulating stomata opening and closing in *Arabidopsis thaliana*. Plant J 55:455–466

Kim JS, Kim KA, Oh TR et al (2008b) Functional characterization of DEAD-box RNA helicases in *Arabidopsis thaliana* under abiotic stress conditions. Plant Cell Physiol 49:1563–1571

Kim M-H, Sasaki K, Imai R (2009) Cold shock domain protein 3 regulates freezing tolerance in *Arabidopsis thaliana*. J Biol Chem 284:23454–23460

Kim JY, Kim WY, Kwak KJ et al (2010a) Glycine-rich RNA-binding proteins are functionally conserved in *Arabidopsis thaliana* and *Oryza sativa* during cold adaptation process. J Exp Bot 61:2317–2325

Kim JY, Kim WY, Kwak KJ et al (2010b) Zinc finger-containing glycine-rich RNA-binding protein in *Oryza sativa* has an RNA chaperone activity under cold stress conditions. Plant Cell Environ 33:759–768

Kim WY, Kim JY, Jung HJ et al (2010c) Comparative analysis of *Arabidopsis* zinc finger-containing glycine-rich RNA-binding proteins during cold adaptation. Plant Physiol Biochem 48:866–872

Kwak KJ, Kim YO, Kang H (2005) Characterization of transgenic *Arabidopsis* plants overexpressing GR-RBP4 under high salinity, dehydration, or cold stress. J Exp Bot 56:3007–3016

Landsman D (1992) Rnp-1, an RNA-binding motif is conserved in the DNA-binding cold shock domain. Nucleic Acids Res 20:2861–2864

Lim J, Thomas T, Cavicchioli R (2000) Low temperature regulated DEAD-box RNA helicase from the Antarctic archaeon, *Methanococcoides burtonii*. J Mol Biol 297:553–567

Lorković ZJ (2009) Role of plant RNA-binding proteins in development, stress response and genome organization. Trends Plant Sci 14:229–236

Lorković ZJ, Barta A (2002) Genomic analysis: RNA recognition motif (RRM) and K homology (KH) domain RNA-binding proteins from the flowering plant *Arabidopsis thaliana*. Nucleic Acids Res 30:623–635

Mangeon A, Junqueira RM, Sachetto-Martins G (2010) Functional diversity of the plant glycine-rich proteins superfamily. Plant Signal Behav 5:99–104

Manival X, Ghisolfi-Nieto L, Joseph G et al (2001) RNA-binding strategies common to cold-shock domain and RNA recognition motif-containing proteins. Nucleic Acids Res 29:2223–2233

Maruyama K, Sato N, Ohta N (1999) Conservation of structure and cold-regulation of RNA-binding proteins in cyanobacteria: probable convergent evolution with eukaryotic glycine-rich RNA-binding proteins. Nucleic Acid Res 27:2029–2036

Mingam A, Toffano-Nioche C, Brunaud V et al (2004) DEAD-box RNA helicases in *Arabidopsis thaliana*: establishing a link between quantitative expression, gene structure and evolution of a family of genes. Plant Biotechnol J 2:401–415

Nagai K, Oubridge C, Ito N et al (1995) The RNP domain: a sequence specific RNA-binding domain involved in processing and transport of RNA. Trends Biochem Sci 20:235–240

Nakaminami K, Karlson D, Imai R (2006) Functional conservation of cold shock domains in bacteria and higher plants. Proc Natl Acad Sci USA 103:10122–10127

Nocker S van, Vierstra RD (1993) Two cDNAs from Arabidopsis thaliana encode putative RNA binding proteins containing glycine-rich domains. Plant Mol Biol 21:695–699

Nomata T, Kabeya Y, Sato N (2004) Cloning and characterization of glycine-rich RNA-binding protein cDNAs in the moss *Physcomitrella patens*. Plant Cell Physiol 45:48–56

Park SJ, Kwak KJ, Jung HJ et al (2010) The C-terminal zinc finger domain of *Arabidopsis* cold shock domain proteins is important for RNA chaperone activity during cold adaptation. Phytochem 71:543–547

Phadtare S, Alsina J, Inouye M (1999) Cold-shock response and cold-shock proteins. Curr Opin Microbiol 2:175–180

Phadtare S, Inouye M, Severinov K (2002) The nucleic acid melting activity of *Escherichia coli* CspE is critical for transcription antitermination and cold acclimation of cells. J Biol Chem 277:7239–7245

Sachetto-Martins G, Franco LO, Oliveira DE de (2000) Plant glycine-rich proteins: a family or just proteins with a common motif? Biochim Biophys Acta 1492:1–14

Sasaki K, Imai R (2012) Pleiotropic roles of cold shock domain proteins in plants. Front Plant Sci 2:116

Sasaki K, Kim MH, Imai R (2007) *Arabidopsis* cold shock domain protein2 is a RNA chaperone that is regulated by cold and developmental signals. Biochem Biophys Res Commun 364:633–638

Shinozuka H, Hisano H, Yoneyama S et al (2006) Gene expression and genetic mapping analyses of a perennial ryegrass glycine-rich RNA-binding protein gene suggest a role in cold adaptation. Mol Genet Genom 275:399–408

Simpson GG, Filipowicz W (1996) Splicing of precursors to mRNA in higher plants: mechanism, regulation and sub-nuclear organization of the spliceosomal machinery. Plant Mol Biol 32:1–41

Stephen JR, Dent KC, Finch-Savage WE (2003) A cDNA encoding a cold-induced glycine-rich RNA binding protein from *Prunus avium* expressed in embryonic axes. Gene 320:177–183

Tripurani SK, Nakaminami K, Thompson KB, Crowell SV, Guy CL, Karlson DT (2011) Spatial and temporal expression of cold-responsive DEAD-box RNA helicases reveals their functional roles during embryogenesis in *Arabidopsis thaliana*. Plant Mol Biol Rep 29:761–768

Tanner NK, Cordin O, Banroques J et al (2003) The Q motif: a newly identified motif in DEAD box helicases may regulate ATP binding and hydrolysis. Mol Cell 11:127–138
Umate P, Tuteja R, Tuteja N (2010) Genome-wide analysis of helicase gene family from rice and *Arabidopsis*: a comparison with yeast and human. Plant Mol Biol 73:449–465
Wolffe AP (1994) Structural and functional properties of the evolutionarily ancient Y-box family of nucleic acid binding proteins. Bioessays 16:245–251
Yamanaka K, Fang L, Inouye M (1998) The CspA family in *Escherichia coli*: multiple gene duplication for stress adaptation. Mol Micobiol 27:247–255
Zhu J, Dong C-H, Zhu J-K (2007) Interplay between cold-responsive gene regulation, metabolism and RNA processing during plant cold acclimation. Curr Opin Plant Biol 10:290–295

Protein Phosphorylation Network in Abscisic Acid Signaling

Taishi Umezawa, Naoyuki Sugiyama, Jeffrey C. Anderson, Fuminori Takahashi, Yasushi Ishihama, Scott C. Peck and Kazuo Shinozaki

Introduction

Abscisic acid (ABA), a phytohormone discovered in the 1960s, was originally thought to be involved in abscission, but it is now known to play a critical role in abiotic and biotic stress responses. It is also important for multiple developmental processes in plants such as seed development, flowering, and bud dormancy (see reviews by Yamaguchi-Shinozaki and Shinozaki 2006, Cutler et al. 2010). Because many agronomical traits are tightly linked to ABA, it is important to understand how its responses are regulated in plants. ABA responses are thought to be controlled by multiple phases including the endogenous ABA levels, short/long-distance transport, and intracellular signal transduction. Recently, several studies have reviewed how ABA responses are regulated in plants.

ABA is a sesquiterpenoid (C15) and its precursor is cleaved from C40 carotenoids in chloroplasts (reviewed by Nambara and Marion-Poll 2005). This reaction is thought to be a key step in the ABA biosynthetic pathway and is catalyzed by

T. Umezawa (✉)
Faculty of Agriculture, Tokyo University of Agriculture and Technology,
2-24-16 Nakacho, Koganei, Tokyo 184-8588, Japan
e-mail: taishi@cc.tuat.ac.jp

N. Sugiyama · Y. Ishihama
Institute of Advanced Life Sciences, Keio University, Yamagata, Japan

J. C. Anderson · S. C. Peck
Department of Biochemistry and Bond Life Sciences Center,
University of Missouri, Columbia, MO, USA

F. Takahashi · K. Shinozaki
Gene Discovery Research Group, RIKEN Plant Science Center, Ibaraki, Japan

Y. Ishihama
Graduate School of Pharmaceutical Sciences, Kyoto University, Kyoto, Japan

S. C. Peck
Division of Biochemistry, University of Missouri-Columbia, Columbia, MO, USA

R. Imai et al. (eds.), *Plant and Microbe Adaptations to Cold in a Changing World*,
DOI 10.1007/978-1-4614-8253-6_13,

9-*cis*-epoxycarotenoid dioxygenase (NCED). On the other hand, ABA degradation is initiated by 8'-hydroxylation catalyzed by the CYP707A family, which is also important in the ABA catalytic pathway. These enzymes are thought to contribute to the rapid fluctuations of endogenous ABA levels in response to drought stress or during seed maturation. Other biosynthetic or catabolic enzymes play significant roles and individual members in ABA biosynthesis and catabolism have been identified and characterized. In addition, the major site of ABA biosynthesis is in vascular tissues, and ABA is transported to each tissue to carry out a specific function (Koiwai et al. 2004; Endo et al. 2008). ABA transport is mediated by two types of transporters: the ABC transporter family and the NRT1/PTR family (Kang et al. 2010; Kuromori et al. 2010; Kanno et al. 2012). After transport, ABA is recognized by plant cells in which various responses are induced via intracellular signal transduction pathways.

In addition to ABA biosynthesis/catabolism and transport pathways, recent studies have increased our understanding of ABA signaling mechanisms (Cutler et al. 2010; Hubbard et al. 2010; Klingler et al. 2010; Raghavendra et al. 2010; Umezawa et al. 2010). A recent model of ABA signaling strongly suggested that protein phosphorylation generated by SnRK2s plays an important role (Fujii and Zhu 2009; Fujita et al. 2009; Nakashima et al. 2009; Umezawa et al. 2009). Thus, we investigated the protein phosphorylation network in ABA signaling using a phosphoproteomics approach.

PYR/PYL/RCAR Receptors and PP2C in ABA Signaling

Although intracellular ABA signaling is complex, recent studies have revealed a simple mechanism consisting of three major components (Cutler et al. 2010; Umezawa et al. 2010). Initially, two research groups identified soluble ABA receptors (the PYR/PYL/RCAR family) using chemical genetics or biochemical approaches (Ma et al. 2009; Park et al. 2009).

Dr. Sean Cutler's group at the University of California Riverside screened chemical libraries to identify ABA agonists/antagonists for seed germination, and identified "pyrabactin" as an agonist (Park et al. 2009). They then screened Arabidopsis mutants using pyrabactin and isolated the *pyrabactin resistant 1* (*pyr1*) mutant. Map-based cloning identified *PYR1*, which encodes a START family protein, and demonstrated that PYR1 can bind to ABA. They also found 14 members of PYR1 and PYR1-like (PYL) proteins in the Arabidopsis genome. On the other hand, Dr. Erwin Grill's group identified the Regulatory Component of ABA Receptor 1 (RCAR1/PYL9) from yeast two-hybrid screening while searching for proteins that associate with ABA-Insensitive 1 (ABI1), a 2C-type protein phosphatase (PP2C) (Ma et al. 2009). ABI1 was identified from a genetic screen in the mid-1990s (Leung et al. 1994; Meyer et al. 1994) and belongs to the group A subfamily of plant PP2C, consisting of nine members in the Arabidopsis genome (Schweighofer et al. 2004). Other studies have revealed that all group A PP2Cs negatively regulate ABA responses (Gosti et al. 1999; Merlot et al. 2001). However, the upstream or downstream factors of PP2C remained unknown.

Therefore, it was surprising that both groups demonstrated that PYR/PYL/RCAR proteins interact with PP2Cs in an ABA-dependent manner to inhibit PP2C activity (Ma et al. 2009; Park et al. 2009). These results clearly indicated that PYR/PYL/RCAR ABA receptors directly and negatively regulate PP2C. Thus, it was important to identify the downstream factor(s) of PP2C.

Subclass III SnRK2s are Major Positive Regulators

Protein phosphorylation/dephosphorylation plays a critical role(s) in ABA signaling, because the importance of PP2C was determined prior to the discovery of PYR/PYL/RCARs, as described earlier. Further studies have attempted to identify protein kinases in ABA signaling using various techniques. Our group performed a biochemical study to identify ABA-activated protein kinases in Arabidopsis. We identified the SNF1-related protein kinase 2 (SnRK2) family as ABA-activated protein kinases based on its molecular weight and PI value determined by 2D-PAGE (Yoshida et al. 2002). There are ten members of the SnRK2 family in the Arabidopsis genome, designated as SRK2A-J or SnRK2.1–10 (Yoshida et al. 2002; Hrabak et al. 2003). Among them, we found that SRK2E/SnRK2.6 is an ABA-activated protein kinase that regulates stomatal closure and ABA-responsive gene expression. On the other hand, Mustilli et al. (2002) performed a genetic screen for stomatal response to ABA using thermography and isolated the *open stomata 1* (*ost1*) mutant. OST1 encodes SnRK2, corresponding to SRK2E/SnRK2.6. The SnRK2 family can be classified into subclasses I–III, and subclass III consists of three ABA-activated protein kinases: SRK2E/OST1/SnRK2.6, SRK2D/SnRK2.2, and SRK2I/SnRK2.3. SRK2D/SnRK2.2 and SRK2I/SnRK2.3 are also involved in ABA responses, especially for seed germination or ABA-responsive gene expression in plants, suggesting that subclass III SnRK2s have distinctive roles with tissue specificities (Fujii et al. 2007). Two research groups generated a triple knockout mutant of subclass III SnRK2s (*srk2dei* or *snrk2.2/2.3/2.6*), both of which showed impaired global ABA responses (Fujii and Zhu 2009; Fujita et al. 2009; Nakashima et al. 2009; Umezawa et al. 2009). For example, *srk2dei* showed a viviparous phenotype, suggesting that its seed dormancy was absent, as well as a wilted and dehydration-sensitive phenotype associated with fully opened stomata and a reduction in ABA-responsive gene expression. These results indicated that subclass III SnRK2s are global and positive regulators of ABA signaling. However, the upstream or downstream factors of subclass III SnRK2s remained unknown (Hubbard et al. 2010; Umezawa 2011).

Establishing a Basic ABA Signaling Model

Protein–protein interactions can be used to study upstream or downstream signaling events. To identify proteins that interact with SnRK2s, we used yeast two-hybrid (Y2H) analysis and examined interactions between SRK2E/OST1 and several known ABA signaling factors. We found that ABI1 interacts with the C-terminal region of SRK2E/OST1 in yeast cells (Yoshida et al. 2006). This was important because PP2C and SnRK2 were thought to be major negative and positive regulators in ABA signaling, respectively. Next, we extended the Y2H analysis to three subclass III SnRK2s and nine group A PP2Cs and found that they interact in various combinations (Umezawa et al. 2009). The SnRK2-PP2C interactions were also detected using co-immunoprecipitation or BiFC assays. Therefore, subclass III SnRK2s and group A PP2Cs can specifically interact in vivo. However, the biochemical relationship between PP2C and SnRK2, and how they function together in ABA signaling, remained unclear.

To address these questions, we performed a series of biochemical analyses for SnRK2s and PP2Cs (Umezawa et al. 2009). We first purified SnRK2 proteins by immunoprecipitation from Arabidopsis seedlings treated with/without ABA, and monitored SnRK2 activity using an in-gel kinase assay after in vitro incubation in the presence or absence of recombinant PP2C proteins. We found that PP2C rapidly inactivated SnRK2 in vitro. Next, we examined whether PP2C can directly dephosphorylate SnRK2. We obtained ^{32}P-labeled SnRK2 by in vivo labeling of Arabidopsis cultured cells, and then incubating with recombinant PP2C in vitro. We found that PP2C can directly dephosphorylate SnRK2, and MS/MS analysis revealed phosphorylation/dephosphorylation sites in SnRK2. Taken together, these results suggest that PP2C negatively regulates SnRK2 by direct dephosphorylation (Umezawa et al. 2009). This conclusion was also supported by an in-gel kinase assay in which ABA-dependent SnRK2 activation was increased in *ahg1* or *ahg3* PP2C-disrupted mutants (Umezawa et al. 2009). Other previous studies have also reported similar results, demonstrating that PP2C can dephosphorylate OST1 kinase (Vlad et al. 2009).

The results of the studies described previously suggest that ABA signal transduction is associated with receptors and protein phosphatases/kinases, such as PYR/PYL/RCARs, PP2C, and SnRK2s (Fujii et al. 2009; Umezawa et al. 2009). This is a typical double negative regulation system because PYR/PYL/RCARs negatively regulate PP2Cs while PP2Cs negatively regulate SnRK2s. Those three proteins are currently thought to be core components of a mainframe of ABA signaling in plants.

SnRK2-Dependent Protein Phosphorylation Networks in ABA Signaling

A recent ABA signaling model highlights the importance of protein phosphorylation/dephosphorylation in ABA signaling (Fig. 1). SnRK2 plays an important role because it transmits ABA signals to downstream factors (substrates) by phosphorylation (Umezawa et al. 2010; Umezawa 2011). Several SnRK2 substrate candidates have already been identified such as the basic leucine-zipper (bZIP) transcription factors AREB/ABF, which are major regulators of ABA-responsive gene expression

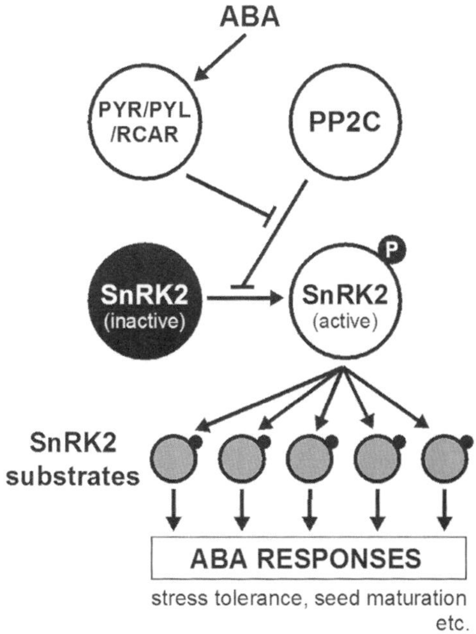

Fig. 1 Current model of the core ABA signaling pathway. Under normal conditions, ABA signals are shut down by the negative regulation of SnRK2 activity by PP2C. SnRK2 and PP2C interact and PP2C directly dephosphorylates SnRK2. Some environmental or developmental cues trigger the elevation of endogenous ABA levels, after which ABA is recognized by PYR/PYL/RCAR receptors. ABA-bound PYR/PYL/RCARs inhibit PP2C activity through physical interaction, resulting in ABA-responsive internal phosphorylation and activation of SnRK2. ABA-activated SnRK2 seems to activate ABA signals and induce various ABA responses in plants by phosphorylation (*black circles*) of multiple downstream factors (substrates)

(Yamaguchi-Shinozaki and Shinozaki 2006). Previous studies have demonstrated that AREB/ABFs are phosphorylated in response to ABA, and SnRK2 may be an upstream kinase of AREB/ABFs (Uno et al. 2000; Furihata et al. 2006). There are nine members of the AREB/ABF family in the Arabidopsis genome including four conserved regions that contain five putative phosphorylation sites consisting of the typical motif R-x-x-S/T. All five phosphorylation sites may be involved in the regulation of transcriptional activity of AREB/ABFs because Asp mutations in all five motifs enhance the activity of AREB/ABFs to the maximum level (Furihata et al. 2006). Therefore, SnRK2 may entirely or partially contribute to the phosphorylation of AREB/ABFs.

On the other hand, SnRK2 localization has been found not only in the nucleus but also in the cytosol (Fujita et al. 2009; Umezawa et al. 2009; Mizoguchi et al. 2010). This suggests that SnRK2 targets are not limited to nuclear proteins. Slow anion channel 1 (SLAC1) may be phosphorylated and regulated by SnRK2 (Geiger et al. 2009; Lee et al. 2009). SLAC1 is a critical component of ABA-dependent stomatal closure (Negi et al. 2008; Vahisalu et al. 2008) and explains how SnRK2 regulates stomatal closure in response to ABA. SnRK2 may phosphorylate a potassium channel, KAT1, which is also an important component of the stomatal response to ABA (Sato et al. 2009). Furthermore, SnRK2 might be involved in reactive oxygen species (ROS) signaling because an NADPH oxidase, RBOHF, is phosphorylated by SnRK2 in vitro (Sirichandra et al. 2009). Although several factors have been identified as SnRK2 substrates, this may account for only part of the SnRK2 regulatory network because a triple knockout mutant of SnRK2 (*srk2dei*) shows a reduced

ABA response, as described earlier, suggesting that a number of downstream factors could be regulated by SnRK2. To further understand ABA signaling, it is important to identify the SnRK2-dependent phosphorylation network in plants.

Application of Phosphoproteomics to ABA Signaling Studies

Phosphoproteomics is a targeted proteomic technology for specifically analyzing phosphorylated proteins/peptides (Macek et al. 2009; Eyrich et al. 2011). Although there are several methods used in phosphoproteomics, there are two key steps: enrichment/purification of phosphoproteins/peptides from crude extracts and detection and quantification of each phosphoprotein/peptide. To this end, phosphoproteins are often separated on 2D gels and selectively detected/quantified by ^{32}P-labeling. Currently, 2D-PAGE-based phosphoproteomics often applies fluorescent dyes (e.g., Pro-Q Diamond) instead of ^{32}P (Peck 2006). On the other hand, a high-performance liquid chromatography and a high-accuracy mass spectrometer (LC-MS)-based shotgun proteomics is a powerful method for analyzing a large number of proteins/peptides. A combination of LC-MS analysis and the recent development of enrichment techniques for phosphopeptides allowed us to analyze thousands of in vivo protein phosphorylations in a single analysis. Several techniques are already available for phosphopeptide enrichment such as immobilized metal affinity chromatography, metal oxide chromatography (MOC), and cation exchange chromatography. However, each technique has advantages and disadvantages, and a common problem is significant contamination of acidic peptides. Sugiyama et al. (2007) found that hydroxy acids can dramatically reduce contamination in MOC, and reported their method as a hydroxy acid-modified MOC (HAMMOC) system. Thus, we employed this technology to examine ABA signaling and SnRK2.

Our study was based on comparative and differential analysis of phosphoproteomic data to screen protein kinase substrates. We used seedlings of Arabidopsis wild-type (WT) and the *srk2dei* triple mutant treated with ABA or subjected to dehydration stress for 0, 15, 30, and 90 min. This experimental design allowed us to compare data between WT and *srk2dei*, as well as between ABA and dehydration stress. After the enrichment of phosphopeptides by HAMMOC from each sample, over 5,000 phosphopeptides were analyzed using an LC-MS system. We quantified each phosphopeptide by extracted ion-current chromatography and identified amino acid sequences and phosphorylation sites from the MS/MS data. Then we classified phosphopeptides as upregulated or downregulated based on ABA-treated and dehydration-stressed samples.

We first compared phosphoproteomic data between ABA and dehydration stress. It is typically thought that dehydration stress signaling consists of ABA-dependent and ABA-independent pathways (Yamaguchi-Shinozaki and Shinozaki 2006). Previous studies have evaluated the contribution of the two pathways to plant responses to dehydration at the transcriptomic or metabolomics levels, but our approach can provide a novel view of the protein phosphorylation level. For example, we identified 53 and 152 phosphopeptides upregulated by ABA and dehydration, respectively, with 27 in common (Umezawa et al. 2013). This suggests that a large part of the dehydration-responsive phosphopeptides is ABA-independent. Therefore, multiple protein phosphorylation pathways could be activated in response to dehydration.

As described earlier, our major goal was to identify SnRK2 substrates in ABA signaling. Thus, we compared phosphoproteomic profiles between WT and *srk2dei* (Umezawa et al. 2013). Because *srk2dei* has a reduced ABA response, ABA-responsive phosphorylation of SnRK2 substrates should be reduced or absent. Based on this hypothesis, we identified 32 SnRK2 substrate candidates after examining the phosphoproteomic data. We believe that some of these candidates are true SnRK2 substrates with significant roles in ABA signaling. For example, we identified one protein, SnRK2 Substrate 1 (SNS1), which was phosphorylated in response to ABA but showed decreased phosphorylation in *srk2dei* (Umezawa et al. 2013). SNS1 is an unknown protein, but the knockout mutant *sns1-1* showed a clear ABA-hypersensitive phenotype during a germination test in the presence of 0.5 or 1 μM ABA. This clearly indicates that SNS1 acts as a negative regulator of ABA signaling. This is interesting because SnRK2 may regulate a negative regulator even though previous known substrates are all positive factors, such as AREB/ABFs or SLAC channels. Therefore, it is likely that SnRK2 and SNS1 are involved in a negative feedback mechanism of ABA signaling, mainly for ABA-responsive gene expression. Such negative feedback mechanism may contribute to keep ABA responses of plant at a certain level.

Our approach, integrating genetics and phosphoproteomics to analyze a large number of phosphorylated proteins regulated by ABA or dehydration stress in Arabidopsis, increases our understanding of ABA-activated SnRK2-dependent phosphorylation networks (Umezawa et al. 2013). We believe that this approach can be used to analyze protein phosphorylation networks in ABA signaling, as well as other signaling pathways in plants.

Acknowledgments The authors were supported by a grant from the Bio-Oriented Technology Research Advancement Institution of Japan, Grants-in-Aid for Scientific Research from MEXT, research funds from Yamagata Prefecture and Tsuruoka City, and from National Science Foundation Grant IOS-1025837. We apologize to our colleagues who could not be cited in this review due to space restrictions.

References

Cutler SR, Rodriguez PL, Finkelstein RR, Abrams SR (2010) Abscisic acid: emergence of a core signaling network. Annu Rev Plant Biol 61:651–679

Endo A, Sawada Y, Takahashi H, Okamoto M, Ikegami K, Koiwai H, Seo M, Toyomasu T, Mitsuhashi W, Shinozaki K et al (2008) Drought induction of Arabidopsis 9-cis-epoxycarotenoid dioxygenase occurs in vascular parenchyma cells. Plant Physiol 147:1984–1993

Eyrich B, Sickmann A, Zahedi RP (2011) Catch me if you can: mass spectrometry-based phosphoproteomics and quantification strategies. Proteomics 11(4):554–570
Fujii H, Chinnusamy V, Rodrigues A, Rubio S, Antoni R, Park S-Y, Cutler SR, Sheen J, Rodriguez PL, Zhu J-K (2009) In vitro reconstitution of an abscisic acid signaling pathway. Nature 462:660–664
Fujii H, Verslues PE, Zhu J-K (2007) Identification of two protein kinases required for abscisic acid regulation of seed germination, root growth, and gene expression in Arabidopsis. Plant Cell 19:485–494
Fujii H, Zhu J-K (2009) Arabidopsis mutant deficient in 3 abscisic acid-activated protein kinases reveals critical roles in growth, reproduction, and stress. Proc Natl Acad Sci USA 106:8380–8385
Fujita Y, Nakashima K, Yoshida T, Katagiri T, Kidokoro S, Kanamori N, Umezawa T, Fujita M, Maruyama K, Ishiyama K et al (2009) Three SnRK2 protein kinases are the main positive regulators of abscisic acid signaling in response to water stress in Arabidopsis. Plant Cell Physiol 50:2123–2132
Furihata T, Maruyama K, Fujita Y, Umezawa T, Yoshida R, Shinozaki K, Yamaguchi-Shinozaki K (2006) Abscisic acid-dependent multisite phosphorylation regulates the activity of a transcription activator AREB1. Proc Natl Acad Sci USA 103:1988–1993
Geiger D, Scherzer S, Mumm P, Stange A, Marten I, Bauer H, Ache P, Matschi S, Liese A, Al-Rasheid KAS et al (2009) Activity of guard cell anion channel SLAC1 is controlled by drought-stress signaling kinase-phosphatase pair. Proc Natl Acad Sci USA 106:21425–21430
Gosti F, Beaudoin N, Serizet C, Webb AAR, Vartanian N, Giraudat J (1999) ABI1 protein phosphatase 2C is a negative regulator of abscisic acid signaling. Plant Cell 11:1897–1910
Hrabak EM, Chan CWM, Gribskov M, Harper JF, Choi JH, Halford N, Kudla J, Luan S, Nimmo HG, Sussman MR et al (2003) The Arabidopsis CDPK-SnRK superfamily of protein kinases. Plant Physiol 132:666–680
Hubbard KE, Nishimura N, Hitomi K, Getzoff ED, Schroeder JI (2010) Early abscisic acid signal transduction mechanisms: newly discovered components and newly emerging questions. Genes Dev 24:1695–1708
Kang J, Hwang J-U, Lee M, Kim Y-Y, Assmann SM, Martinoia E, Lee Y (2010) PDR-type ABC transporter mediates cellular uptake of the phytohormone abscisic acid. Proc Natl Acad Sci USA 107:2355–2360
Kanno Y, Hanada A, Chiba Y, Ichikawa T, Nakazawa M, Matsui M, Koshiba T, Kamiya Y, Seo M (2012) Identification of an abscisic acid transporter by functional screening using the receptor complex as a sensor. PNAS 109:9653–9658
Klingler JP, Batelli G, Zhu J-K (2010) ABA receptors: the START of a new paradigm in phytohormone signaling. J Exp Bot 61:3199–3210
Koiwai H, Nakaminami K, Seo M, Mitsuhashi W, Toyomasu T, Koshiba T (2004) Tissue-specific localization of an abscisic acid biosynthetic enzyme, AAO3, in Arabidopsis. Plant Physiol 134:1697–1707
Kuromori T, Miyaji T, Yabuuchi H, Shimizu H, Sugimoto E, Kamiya A, Moriyama Y, Shinozaki K (2010) ABC transporter AtABCG25 is involved in abscisic acid transport and responses. Proc Natl Acad Sci USA 107:2361–2366
Lee SC, Lan W, Buchanan BB, Luan S (2009) A protein kinase-phosphatase pair interacts with an ion channel to regulate ABA signaling in plant guard cells. Proc Natl Acad Sci USA 106:21419–21424
Leung J, Bouvier-Durand M, Morris PC, Guerrier D, Chefdor F, Giraudat J (1994) Arabidopsis ABA response gene ABI1: features of a calcium-modulated protein phosphatase. Science 264:1448–1452
Ma Y, Szostkiewicz I, Korte A, Moes D, Yang Y, Christmann A, Grill E (2009) Regulators of PP2 C phosphatase activity function as abscisic acid sensors. Science 324:1064–1068
Macek B, Mann M, Olsen JV (2009) Global and site-specific quantitative phosphoproteomics: principles and applications. Annu Rev Pharmacol Toxicol 49:199–221

Merlot S, Gosti F, Guerrier D, Vavasseur A, Giraudat J (2001) The ABI1 and ABI2 protein phosphatases 2 C act in a negative feedback regulatory loop of the abscisic acid signaling pathway. Plant J 25:295–303

Meyer K, Leube MP, Grill E (1994) A protein phosphatase 2 C involved in ABA signal transduction in Arabidopsis thaliana. Science 264:1452–1455

Mizoguchi M, Umezawa T, Nakashima K, Kidokoro S, Takasaki H, Fujita Y, Yamaguchi-Shinozaki K, Shinozaki K (2010) Two closely related subclass II SnRK2 protein kinases cooperatively regulate drought-inducible gene expression. Plant Cell Physiol 51:842–847

Mustilli A-C, Merlot S, Vavasseur A, Fenzi F, Giraudat J (2002) Arabidopsis OST1 protein kinase mediates the regulation of stomatal aperture by abscisic acid and acts upstream of reactive oxygen species production. Plant Cell 14:3089–3099

Nakashima K, Fujita Y, Kanamori N, Katagiri T, Umezawa T, Kidokoro S, Maruyama K, Yoshida T, Ishiyama K, Kobayashi M et al (2009) Three Arabidopsis SnRK2 protein kinases, SRK2D/SnRK2.2, SRK2E/SnRK2.6/OST1 and SRK2I/SnRK2.3, involved in ABA signaling are essential for the control of seed development and dormancy. Plant Cell Physiol 50:1345–1363

Nambara E, Marion-Poll A (2005) Abscisic acid biosynthesis and catabolism. Annu Rev Plant Biol 56:165–185

Negi J, Matsuda O, Nagasawa T, Oba Y, Takahashi H, Kawai-Yamada M, Uchimiya H, Hashimoto M, Iba K (2008) CO2 regulator SLAC1 and its homologues are essential for anion homeostasis in plant cells. Nature 452:483–486

Park S-Y, Fung P, Nishimura N, Jensen DR, Fujii H, Zhao Y, Lumba S, Santiago J, Rodrigues A, Chow TF et al (2009) Abscisic acid inhibits type 2 C protein phosphatases via the PYR/PYL family of START proteins. Science 324:1068–1071

Peck SC (2006) Phosphoproteomics in Arabidopsis: moving from empirical to predictive science. J Exp Bot 57:1523–1527

Raghavendra AS, Gonugunta VK, Christmann A, Grill E (2010) ABA perception and signaling. Trends Plant Sci 15:395–401

Sato A, Sato Y, Fukao Y, Fujiwara M, Umezawa T, Shinozaki K, Hibi T, Taniguchi M, Miyake H, Goto DB et al (2009) Threonine at position 306 of the KAT1 potassium channel is essential for channel activity and is a target site for ABA-activated SnRK2/OST1/SnRK2.6 protein kinase. Biochem J 424:439–448

Schweighofer A, Hirt H, Meskiene I (2004) Plant PP2 C phosphatases: emerging functions in stress signaling. Trends Plant Sci 9:236–243

Sirichandra C, Gu D, Hu H-C, Davanture M, Lee S, Djaoui M, Valot B, Zivy M, Leung J, Merlot S et al (2009) Phosphorylation of the Arabidopsis AtrbohF NADPH oxidase by OST1 protein kinase. Febs Lett 583:2982–2986

Sugiyama N, Masuda T, Shinoda K, Nakamura A, Tomita M, Ishihama Y (2007) Phosphopeptide enrichment by aliphatic hydroxy acid-modified metal oxide chromatography for nano-LC-MS/MS in proteomics applications. Mol Cell Proteomics 6:1103–1109

Umezawa T (2011) Systems biology approaches to abscisic acid signaling. J Plant Res 124:539–548

Umezawa T, Nakashima K, Miyakawa T, Kuromori T, Tanokura M, Shinozaki K, Yamaguchi-Shinozaki K (2010) Molecular basis of the core regulatory network in abscisic acid responses: sensing, signaling, and transport. Plant Cell Physiol 51:1821–1839

Umezawa T, Sugiyama N, Mizoguchi M, Hayashi S, Myouga F, Yamaguchi-Shinozaki K, Ishihama Y, Hirayama T, Shinozaki K (2009) Type 2C protein phosphatases directly regulate abscisic acid-activated protein kinases in Arabidopsis. Proc Natl Acad Sci USA 106:17588–17593

Uno Y, Furihata T, Abe H, Yoshida R, Shinozaki K, Yamaguchi-Shinozaki K (2000) Arabidopsis basic leucine zipper transcription factors involved in an abscisic acid-dependent signal transduction pathway under drought and high-salinity conditions. Proc Natl Acad Sci USA 97:11632–11637

Umezawa T, Sugiyama N, Takahashi F, Anderson JC, Ishihama Y, Peck SC, Shinozaki K (2013) Genetics and phosphoproteomics reveal a protein phosphorylation network in the abscisic acid pathway in Arabidopsis thaliana. Sci Signal 6:rs8

Vahisalu T, Kollist H, Wang Y-F, Nishimura N, Chan W-Y, Valerio G, Lamminmaki A, Brosche M, Moldau H, Desikan R et al (2008) SLAC1 is required for plant guard cell S-type anion channel function in stomatal signaling. Nature 452:487–491

Vlad F, Rubio S, Rodrigues A, Sirichandra C, Belin C, Robert N, Leung J, Rodriguez PL, Lauriere C, Merlot S (2009) Protein phosphatases 2 C regulate the activation of the Snf1-related kinase OST1 by abscisic acid in Arabidopsis. Plant Cell 21:3170–3184

Yamaguchi-Shinozaki K, Shinozaki K (2006) Transcriptional regulatory networks in cellular responses and tolerance to dehydration and cold stresses. Annu Rev Plant Biol 57:781–803

Yoshida R, Hobo T, Ichimura K, Mizoguchi T, Takahashi F, Aronso J, Ecker JR, Shinozaki K (2002) ABA-activated SnRK2 protein kinase is required for dehydration stress signaling in Arabidopsis. Plant Cell Physiol 43:1473–1483

Yoshida R, Umezawa T, Mizoguchi T, Takahashi S, Takahashi F, Shinozaki K (2006) The regulatory domain of SRK2E/OST1/SnRK2.6 interacts with ABI1 and integrates abscisic acid (ABA) and osmotic stress signals controlling stomatal closure in. Arabidopsis J Biol Chem 281:5310–5318

Using Synchrotron FTIR and Confocal Cryomicroscopy to Explore Mechanisms of Cold Acclimation and Freezing Resistance Using a Single Cell Layer of *Allium fistulosum* L

Karen Tanino, Jun Liu, Shion Kobayashi, Yukio Kawamura, Ferenc Borondics and Matsuo Uemura

Synchrotron Fourier Transform Infrared Microspectroscopy, its Use in Biology and Cold Acclimation Studies

Infrared (IR) spectroscopy is a valuable technique in plant sciences to identify and characterize biological molecules. The IR region is partitioned into near (14,000–4,000/cm), mid (4,000–400/cm), and far (400–10/cm) infrared based on relative proximity to the visible part of the electromagnetic spectrum. The mid-IR region is further divided into X-H stretching regions (4,000–2500/cm), C≡C triple bond (2,500–2,000/cm), C=C double bond (2,000–1,500/cm), and the so-called fingerprint region (1,500–600/cm), since molecules may show distinct absorption features at these ranges (Stuart 2012). To date, Fourier Transform Infrared (FTIR) microspectroscopy has played increasingly critical roles in the field of scientific research.

By connecting the FTIR with a synchrotron source (SR-FTIR), much higher resolution is achieved (Williams 1989; Nasse et al. 2011). Compared with a thermal source like a Globar, a synchrotron source has numerous advantages. A synchrotron source by nature has a small effective size because of the narrow emission angles, thereby producing a much brighter (2–3 orders of magnitude) source (Williams 1989; Carr et al. 1995; Dumas et al. 2004; Miller and Dumas 2006). As a consequence diffraction-limited spot size is available at synchrotron sources; the signal-to-noise ratio (S/N) is dramatically better compared to the conventional

K. Tanino (✉) · J. Liu
Department of Plant Sciences, University of Saskatchewan,
Saskatoon, SK, Canada
e-mail: karen.tanino@usask.ca

S. Kobayashi · Y. Kawamura · M. Uemura
Cryobiofrontier Research Center, Faculty of Agriculture,
Iwate University, Morioka, Japan

F. Borondics
Canadian Light Source, Saskatoon, SK, Canada

R. Imai et al. (eds.), *Plant and Microbe Adaptations to Cold in a Changing World,*
DOI 10.1007/978-1-4614-8253-6_14, © Springer Science+Business Media New York 2013

thermal sources, thus enabling the use of small apertures (Carr et al. 1995; Reffner et al. 1995; Carr 2001; Dumas et al. 2004; Miller and Dumas 2006). Thus, synchrotrons have been attractive tools due to their high brightness of light, allowing superior resolution in comparison to thermal sources (Nanba et al. 2003).

In plant sciences, synchrotron-based FTIR microspectroscopy is a powerful approach for understanding structure and composition in a range of plant tissues including roots, stems, leaves, and reproductive structures (Dokken and Davis 2005; Heraud et al. 2007). Here, we review how the synchrotron and infrared spectroscopy can be used to examine the biological mechanisms in plants and present some of our own research results.

Molecules have unique absorbance peaks based on their functional groups in the infrared region and represent a fingerprint of the substance (Yu et al. 2004). The fundamental infrared bands of biomolecules, including proteins, lipids, and carbohydrates appear in the mid-IR range (Sacksteder et al. 2001; Duygu et al. 2009). There are well-known absorption frequencies associated with particular functional groups. However, since organic molecules are composed of more than one functional group, combining the integration across all functional groups for a specific molecule would be ideal. For instance, the most regularly followed vibrational modes for proteins are the Amide I at 1650/cm containing 80% C=O stretching vibration, but, there is also the C–N and the Amide II at 1550/cm including 60% N–H stretching vibration, bound with 40% C–N stretching vibration (Susi and Byler 1983; Byler and Susi 1986). The IR region of carbohydrates is within the double bond and fingerprint region from 1800 to 950/cm (Michael and Mantsch 1996; Himmelsbach et al. 1998; Wetzel et al. 1998; Miller 2000; Wilson et al. 2000; Marinkovic et al. 2002). The distribution of diverse spectral characteristics in biological samples, coupled with the information on biological processes of plants and the quantitative variations and structures of protein, lipid, nucleic acid, and carbohydrates can contribute important details about the mechanisms in plant biology (Miller and Dumas 2006). The various published response measurements are summarized in Table 1. Because of the advantages of SR-FTIR, single living cells can be used to study biomolecules in vivo and measure subcellular locations, the orientation of molecular components, and concentration gradients (Quaroni and Zlateva 2011).

A further development is the ability to produce two-dimensional images with Focal Plane Array (FPA) detectors (see Stuart 2012 for a review). FTIR spectroscopy is normally based on the data of one detector but FPA detectors enable a recording of thousands of interferograms at a time to build three-dimensional data blocks (two spatial and one spectral dimension).

In our study, we explored the mechanism of freezing resistance in *Allium fistulosum* L. after cold acclimation with both Globar and synchrotron-based FTIR microspectroscopy using an FPA detector. The distinct spectra of the functional groups (carbohydrates and ester group, α-helix and β-pleated sheet proteins) surrounding the plasma membrane and the apoplastic region (including the cell wall and middle lamella) after 0, 1, and 2 weeks cold acclimated treatments were examined (Fig. 1a, b). The cell walls and middle lamella after 1 and 2 weeks cold acclimation were enriched in the α–helix conformations integrated with reduced β–sheet secondary structure,

Table 1 Summary of plant responses using synchrotron-based Fourier Transform Infrared (FTIR) microspectroscopy

Responses	Species	References
Analyzed the chemical (lipid, lignin, protein, and simple sugars) differences and similarities within the root zone with or without phosphorus	Mung bean (*Vigna radiata* L.)	Raab and Martin 2001
Localized the nucleus and the chloroplasts in living, nutrient-replete cells according to the spectrum of lipids and proteins	*Micrasterias hardyi*	Heraud et al. 2005
	Green flagellate *Euglena gracilis*	Hirschmugl et al. 2006
The detection and identification of pathogens, including Rhizoctonia, Colletotrichum, Verticillium, and Fusarium oxysporum; and brown-rot and white-rot	Scot pine (*Pinus sylvestris* L.) and beech (*Fagus sylvatica* L.)	Pandey and Pitman 2003
	Potato (*Solanum tuberosum* L.)	Salman et al. 2010
The structure imaging of proteins, lipids, and carbohydrates of various plant tissues	Wheat endosperm (*Triticum aestivum* L.)	Wetzel et al. 2003
	Corn (*Zea mays* L.)	Yu et al. 2004
	Soybean (*Glycine max* L.)	Pietrzak and Miller 2005
	Corn (*Zea mays* L.), Barley (*Hordeum vulgare* L.), Canola (*Brassica napus* L.), Flax (*Linum usitatissimum* L.), Oat (*Avena sativa* L.)	Yu 2005a
	Wheat (*Triticum aestivum* L.)	Yu et al. 2007
	Sorghum seed (*Sorghum bicolor* L.)	Yu 2011
The fate and transport of organic contaminants (such as benzotriazole and 2,6-dinitrotoluene) and their damage to lignin and root structure	Corn (*Zea mays* L.)	Dokken et al. 2005a
	Sunflower (*Helianthus annuus* L.)	Dokken et al. 2005b
Localized the nucleus and the chloroplasts in living, nutrient-replete cells according to the spectrum of lipids and proteins	*Micrasterias hardyi*	Heraud et al. 2005
	Green flagellate *Euglena gracilis*	Hirschmugl et al. 2006
Explored ultrastructural–chemical features within cellular dimensions of yellow- (*Brassica rapa*) and brown-seeded (*Brassica napus*) canola	Canola (*Brassica napus* L.)	Yu et al. 2005a

Table 1 (continued)

Responses	Species	References
Showed physiological differences in low-temperature tolerances according to protein molecular structural–chemical differences and protein secondary structures between two types of winterfat seeds	Winterfat (*Krascheninnikovia lanata*)	Yu et al. 2005c
Heat-induced changes to food/feed protein secondary structure based on the protein α-helix to β-sheet ratio	Flaxseed (*Linum usitatissimum*)	Yu 2005b; Yu et al. 2005b; Doiron et al. 2009
Reported plant cell wall architecture at the cellular level in situ and biopolymers, such as cellulose, lignin, and proteins in roots	Sunflower (*Helianthus annuus*. L) and maize	Dokken and Davis 2007
Identified the differences in proteins and carbohydrate molecular structure between barley varieties	Barley (*Hordeum vulgare* L.)	Yu 2007; Yu et al. 2008
Observed in situ chemical imaging and spatial arrangement of plant cell wall components	Green seaweed *Codium fragile*	Estevez et al. 2009
Examined the evolution of anaerobic metabolic products and the formation of ethanol in single, living cells	*Chlamydomonas reinhardtii*	Goff et al. 2009
Compared the protein structure of Lc-transgenic and nontransgenic alfalfa within cellular and subcellular dimensions	Alfalfa (*Medicago sativa* L.)	Yu et al. 2009
Characterized the organic and amino acid standards and the extracted xylem sap. Determined the Ni speciation in leaf, shoot, and extracted sap	*Alyssum murale*	McNear et al. 2010
Used individual flattened starch granules to determine which granules were esterified in the processing step	Corn (*Zea mays* L.)	Wetzel et al. 2010
Examined the role of phosphorus in response to manganese toxicity	Douglas fir (*Pseudotsuga menziesii*)	Dučić et al. 2012
Localized carbohydrate, esters, secondary protein structure in intact single cell layers during cold acclimation	Japanese or Welsh bunching onion (*Allium fistulosum* L.)	Tanino et al. 2013

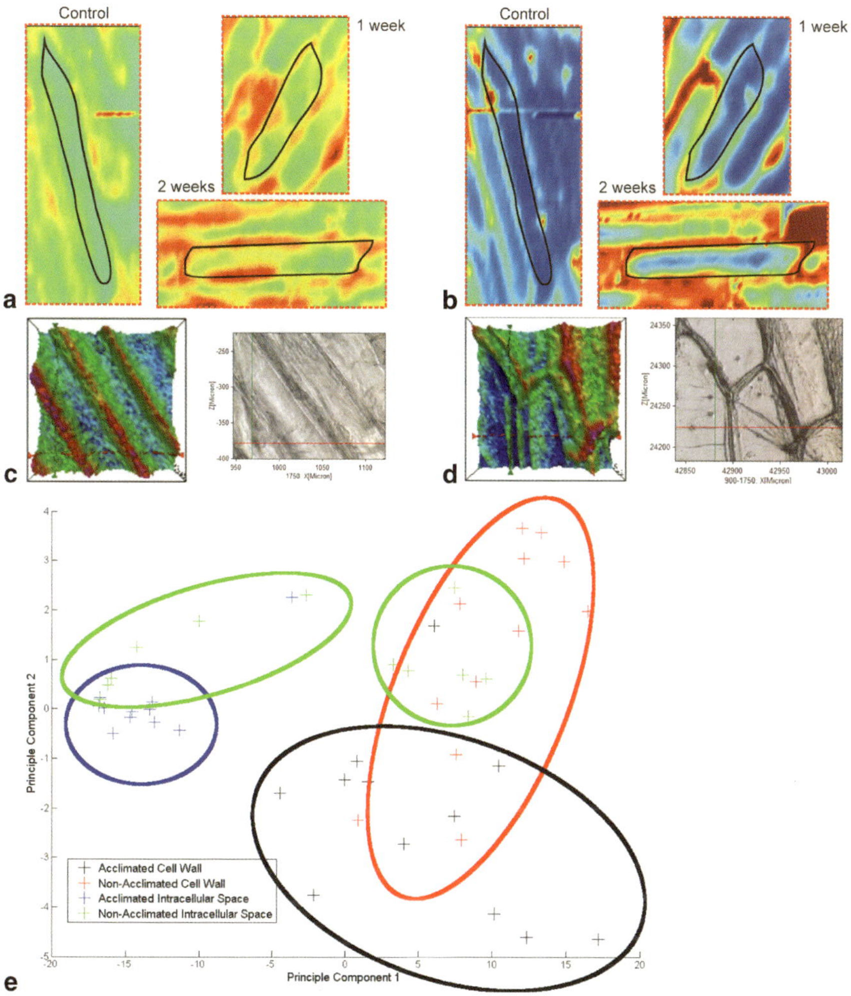

Fig. 1 a) Functional group map over 0, 1, 2 weeks of cold acclimation of the ratio of the α-helix secondary structures to β-sheet secondary structures, the false colour ranges from low (green) to high ratios (red) and shows the relative concentrations of the two types of secondary structures. The cell outlines from the light micrographs are outlined to highlight the position of at least one of the cells from the tissue for reference. b) Functional group map of the ratio of the total carbohydrate band to the total ester band, the false colour ranges from low (dark blue) to high ratios (red). c) and d) Representative three dimensional FTIR spectral plots integrated over 963 - 1745/cm corresponding to spectra which include pectin and cellulose vibrational modes in Allium cepa (Wilson et al. 2000). c) non-acclimated cells, d) four week acclimated cells. Corresponding cells displayed under visible light are shown beside. e) Principle component analysis of acclimated apoplast (black), acclimated symplast (blue), non-acclimated apoplast (red) and non-acclimated symplast (green). (Source: Tanino et al. 2013)

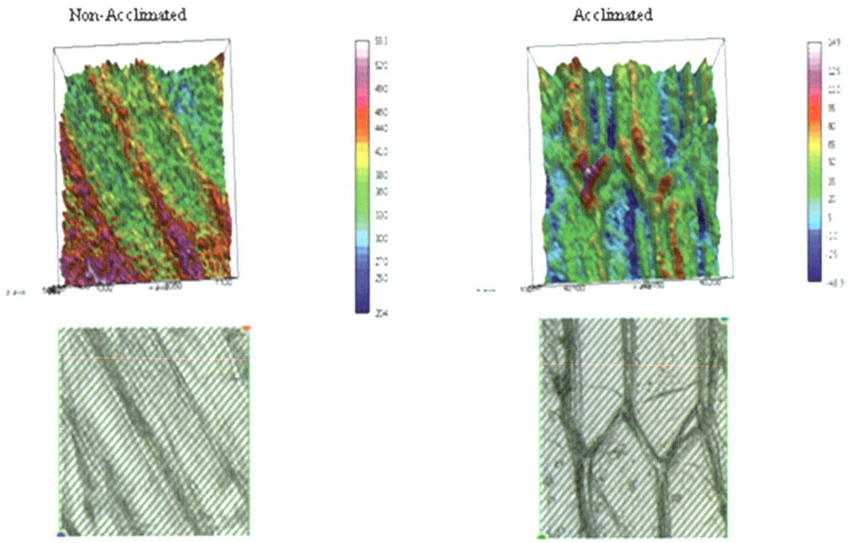

Fig. 2 Surface plots of nonacclimated and cold acclimated *A. fistulosum* live epidermal peel samples based on FPA data integrated over the 963–1,745/cm range. (Source: Tanino et al. 2013)

and increased carbohydrate concentration coupled with decreased ester group. The differences of cell wall and intracellular space regions between nonacclimated and cold acclimated samples can be distinguished based on the spectra of nonacclimated and cold acclimated treatments combined over the 963–1745/cm range (Figs. 1c–e and 2). The distinct quantitative increases in the detected spectrum range encompassing cell walls and middle lamella were assigned to pectins and cellulose according to another onion species, *Allium cepa* (Wilson et al. 2000). Three-dimensional spectra of cold acclimated treatments of Fig. 2 were then integrated over one of the published *A. cepa* pectin peaks (Wilson et al. 2000) and integrated over the 1700–1778/cm wavenumber region (Fig. 3a). It is apparent that while this particular region can account for some of the variation, it does not account for all of the responses observed in Fig. 2. This is supported by the two-dimensional Principle Component Analysis (PCA) of the spectral data of Fig. 3a (shown in Fig. 3b). PCA enables a statistical analysis of multivariate data such as that generated by FPA (Heraud et al. 2007). Figure 3b indicates darker red regions (increased concentration across the integrated region) in the cell wall compared to the blue regions (decreased concentration across the integrated region) in the intercellular spaces. Since pectins are not only present in the cell walls, but also in the intercellular spaces, it is apparent that more validation of the methods is required. Nevertheless, it indicates the potential to localize specific components on a single cell basis. Pectin has several peaks and as mentioned earlier, one of the next steps is to combine all of the individual peaks for pectin in the FPA. If this can be accomplished, then the molecules associated with the observed distinct differences between nonacclimated and cold acclimated samples can be better identified.

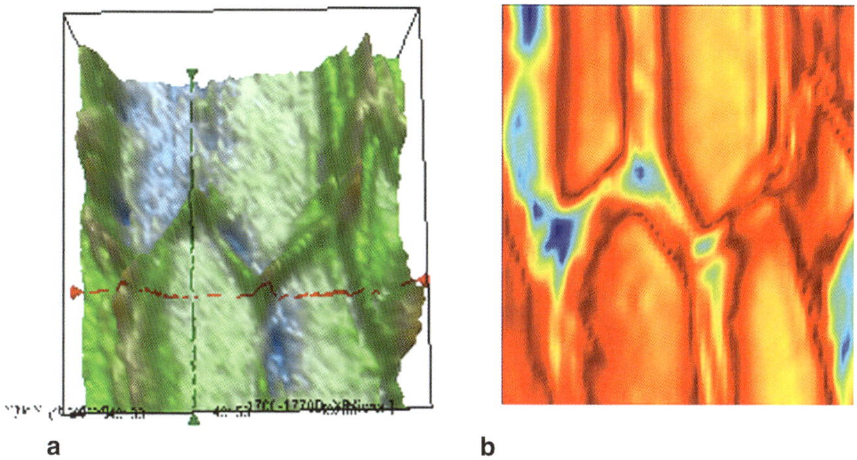

Fig. 3 **a** Surface plot of the cold acclimated *A. fistulosum* sample's FPA data shown in Fig. 2 integrated over the 1,700–1,778/cm wavenumber range, one of the published *Allium cepa* pectin peaks (Wilson et al. 2000), **b** Two-dimensional Principal Component Analysis of the original FPA data shown on Fig. 3a

Confocal Cryomicroscopic Analysis and Endoplasmic Reticulum Cryodynamics

Some studies have reported that membranes of plant cells are dynamically regulated even under freezing (Fujikawa and Takabe 1996; Yamazaki et al. 2008a, b). For example, when mulberry cortical parenchyma cells acquire extreme freezing tolerance in winter and are subsequently frozen, for example, at −5 °C, multiplex lamellae (MPL) form by the fusion of endoplasmic reticulum (ER) vesicles (Fujikawa and Takabe 1996). MPL completely cover the area beneath the plasma membrane and are composed of a parallel array of sheet-like ER cisternae. This cryodynamic process is completed within 10 min of freezing at −5 °C and is quickly reversed upon thawing. While the freeze-induced formation of MPL has been hypothesized to play a role in avoiding the close apposition of membranes, including the plasma membrane (Fujikawa and Takabe 1996), the actual role is unknown. Confocal microscopic technique is used as a powerful tool to reveal the dynamics of membranes and proteins, when the target is stained by fluorescent dye or tagged by the fluorescent protein. Thus, to know how plant cells live under freezing, live cell imaging during freezing and thawing with a confocal microscopic system is an essential approach.

Confocal laser microscopy combined with a cryostage allowed more detailed observation of membranes, for example, ER cryodynamics when using the ER selective dye of ER-Tracker (Invitrogen, Burlington, ON, Canada), which has been often used in *Arabidopsis* as a reliable ER marker at room temperatures (Cui et al. 2012; Furch et al. 2009). In our study, the confocal microscopic technique with ER-

Table 2 Cell survival and ER morphological analyses in nonacclimated and cold acclimated cells on the basis of confocal microscopic results[a]

	Nonacclimated sample	Cold acclimated sample
Survival rate at − 15 °C in the microscopic system[b] (% of control)	48.2±2.5 (n=67)	99.2±0.4 (n=63)
ER volume at 2 °C[c] (% of one cell volume)	21.5±2.7 (n=6)	34.0±3.1 (n=6)
Vesicle size of FIERVs at −15 °C[d] (μm^2)	13.7±7.3 (n=165)	19.2±12.1 (n=165)

[a] Confocal images at 2 and − 15 °C were captured in both cells nonacclimated and cold acclimated for 4 weeks

[b] In the cryomicroscopic system, the freezing tolerances at −15 °C were measured in nonacclimated cells and cold acclimated cells. Both values were significantly different (P<0.01)

[c] ER volume ratios at 2 °C per one cell were quantified using three-dimensional images of nonacclimated cells and cold acclimated cells. Both values were significantly different (P<0.01)

[d] Because the FIERV size is very small to measure its volume, the ostensible area of FIERV on picture was measured as a representative of the vesicle size in cold acclimated cells and nonacclimated cells. Both values were significantly different (P<0.01)

Tracker was performed at subzero temperatures, and the large cells of the freezing tolerant *A. fistulosum* enable this response to be more readily measured.

First, in this cryomicroscopic system, cell survival at − 15 °C was measured using a lipophilic fluorescence dye of FM1–43 (Invitrogen, Burlington, ON, Canada), which can strongly stain all endomembranes only when the plasma membrane is irreversibly broken. After freezing, samples were thawed at a rate of 1 °C/min and living cells were counted. Survival of nonacclimated cells and cold acclimated cells were 48.2 and 99.2 %, respectively (Table 2). Therefore, ER morphologies of nonacclimated and cold acclimated cells were compared at −15 °C. Next, the ER morphology in cold acclimated cells was also compared with that in nonacclimated cells. Before freezing, the ER membranes in cold acclimated cells were more abundant than those in nonacclimated cells (Fig. 4 and Table 2). Although the ER networks were observed in both treatments, those in cold acclimated cells were more developed. ER volume ratios per cell in nonacclimated and cold acclimated cells were quantified using three-dimensional images, and were 21.5 and 34.0 %, respectively (Table 2). In both cell types at −15 °C, there were no ER networks, and many freeze-induced ER vesicles (FIERVs) were observed (Fig. 4). However, the size of FIERVs in cold acclimated cells seemed to be larger than that in nonacclimated cells. Because the FIERV size is very small, to measure its volume, the ostensible area of FIERV on the picture was measured as a representative of the vesicle size. The mean area of FIERVs in cold acclimated cells was 19.2 μm^2 and was significantly larger than that of nonacclimated cells which was 13.7 μm^2 (Table 2). In *Arabidopsis* root cells, we also observed freeze-induced ER vesiculation (data not shown).

Interestingly, FIERVs which we observed using confocal laser microscopy are apparently different from MPLs which have been observed in woody plants using freeze-fracture electron microscopic system (Fujikawa and Takabe 1996). At present, we speculate some possibilities. First, the freeze-induced ER vesiculation

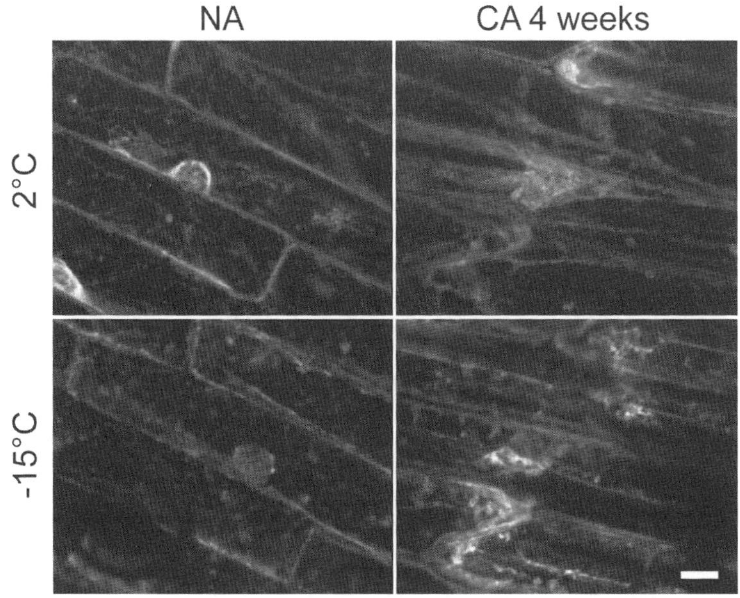

Fig. 4 Comparison of the ER morphology in cold acclimated *A. fistulosum* cells with that in nonacclimated cells. Confocal images at 2 and − 15 °C were captured in both cells nonacclimated and cold acclimated for 4 weeks. Confocal microscopic observations were performed on the cryostage (THMS600, Linkam) using a microscope with the confocal scanning unit of CSU 10 (Yokogawa), CCD camera, and 488 nm laser. The cryostage temperature is controlled through a silver block cooled with liquid nitrogen. The onion epidermal cell layer was sandwiched between two glass coverslips and placed onto the cryostage. At temperatures above zero, only the specimen corner on a metal peg cooled with liquid nitrogen was frozen. The temperature was then lowered at a rate of 0.08 °C/min with nucleation of the sample occurring around − 0.5 °C. In both nonacclimated and cold acclimated cells at − 15 °C, many freeze-induced ER vesicles (FIERVs) were observed. Bar indicates 50 μm

is specific in herbaceous plants. Second, after vesiculation, multiplex lamellar structures are reconstructed by a more severe freezing. However, to address these possibilities, it is necessary to perform more experiments with some plants including woody plants. The ER volume per cell before freezing increases during cold acclimation. The other possibility is that different results are due to different observational systems. While in freeze-fracture electron microscopic studies, the cortical ER near the plasma membrane has been observed at high-magnification, in confocal microscopic studies, the cytoplasmic ER was observed at low-magnification. The resolution of confocal microscopic system is lower than that of the electron microscopic system. In addition, when using the cryostage, high-magnification lenses cannot be used. To resolve this problem, high-resolution images must be obtained in the confocal cryomicroscopic system in future.

Changes in the interaction between ER and the cytoskeleton may be important for the cryodynamics of ER. In *A. cepa* onion cells, the chilling treatment brings the networks and strands of ER into the vesicle-like structures, and by pharmacologic studies, these changes are estimated to result in low temperature influence on the interaction of myosins and actin filaments (Quader et al. 1989). In their presented model, dynamic interactions among ER, actin filaments and myosins determined the morphology and dynamics of ER and caused the cytoplasmic streaming (Ueda et al. 2010). The inhibition of cytoplasmic streaming resulted from an increase in cytoplasmic calcium concentration, the mechanism of which is thought to involve the plasma membrane deformation and mechano-sensitive calcium ion channels (Hayashi and Takagi 2003). In protoplasts isolated from cold acclimated *Arabidopsis* leaves, the unique cryodynamics of the plasma membrane resulted from the freeze-induced mechanical stress and not dehydration (Yamazaki et al. 2008b). Taken together, it is possible that the cryodynamics of ER are caused by the changes of interactions between ER, actin filaments, and myosins which the mechanical deforming of the plasma membrane brings as a first signal.

Concluding Remarks

SR-FTIR and confocal cryomicroscopy are techniques which, when combined with a system such as *A. fistulosum*, can provide insight into the mechanism of cold acclimation and freezing stress resistance. The cell wall and ER cryodynamics appear to be centrally involved in freezing stress resistance. SR-FTIR requires further validation of responses and a means to combine the integration of peaks for a given organic molecule to enable identification of biological responses.

Acknowledgments The research performed in Iwate University was supported by Grant-in-Aid for Scientific Research on Innovative Areas (#22120003 to M. U.), Grant-in-Aid for Young Scientists (B) (#22780288 to Y. K.), Grant-in-Aid for Scientific Research (B) (#24370018 to M. U.), and "Live imaging of plant cells under freezing," Adaptable and Seamless Technology Transfer Program through Target-driven R&D, JST (to Y. K.). The research described in this chapter was performed at the Mid-IR beamline of the Canadian Light Source, which is supported by NSERC, NRC, CIHR, and the University of Saskatchewan. Tim May and Xia Liu are gratefully acknowledged for their assistance.

References

Byler DM, Susi H (1986) Examination of the secondary structure of proteins by deconvolved FTIR spectra. Biopolymers 25:469–487

Carr GL (2001) Resolution limits for infrared microspectroscopy explored with synchrotron radiation. Rev Sci Instr 72:1613–1619

Carr GL, Hanfland M, Williams GP (1995) Mid infrared beamline at the National Synchrotron Light Source port U2B. Rev Sci Instr 66:1643–1645

Cui F, Liu L, Zhao Q, Zhang Z, Li Q, Lin B, Wu Y, Tang S, Xie Q (2012) *Arabidopsis* ubiquitin conjugase UBC32 is an ERAD component that functions in brassinosteroid-mediated salt stress tolerance. Plant Cell 24:233–244

Doiron KJ, Yu P, Christensen CR, Christensen DA, McKinnon JJ (2009) Detecting molecular changes in 'Vimy' flaxseed protein structure using synchrotron FTIRM and DRIFT spectroscopic techniques: structural and biochemical characterization. Spectroscopy 23:307–322

Dokken KM, Davis LC (2005) Use of infrared microspectroscopy in plant growth and development. App Spectrosc Rev 40:301–326

Dokken KM, Davis LC (2007) Infrared imaging of sunflower and maize root anatomy. J Agric Food Chem 55:10517–10530

Dokken KM, Davis LC, Marinkovic NS (2005a) Using SR-IMS to study the fate and transport of organic contaminants in plants. Spectroscopy 20:14

Dokken KM, Davis LC, Erickson LE, Castro-Diaz S, Marinkovic NS (2005b) Synchrotron Fourier transform infrared microspectroscopy: a new tool to monitor the fate of organic contaminants in plants. Microchem J 81:86–91

Dučić T, Thieme J, Polle A (2012) Phosphorus compartmentalization on the cellular level of Douglas Fir root as affected by Mn toxicity: a Synchrotron-Based FTIR Approach. Spectroscopy: An Int J 27:265–272

Dumas P, Jamin N, Teillaud JL, Miller LM, Beccard B (2004) Imaging capabilities of synchrotron infrared microspectroscopy. Faraday Discuss 126:289–302

Duygu D, Baykal T, AÇikgöz İ, Yildiz K (2009) Fourier transform infrared (FT-IR) spectroscopy for biological studies. Gazi Univ. J Sci 22(3):117–121

Estevez JM, Fernández PV, Kasulin L, Dupree P, Ciancia M (2009) Chemical and *in situ* characterization of macromolecular components of the cell walls from the green seaweed *Codium fragile*. GlycoBiology 19:212–228

Fujikawa S, Takabe K (1996) Formation of multiplex lamellae by equilibrium slow freezing of cortical parenchyma cells of mulberry and its possible relationship to freezing tolerance. Protoplasma 190:189–203

Furch AC, Bel AJ van, Fricker MD, Felle HH, Fuchs M, Hafke JB (2009) Sieve element Ca^{2+} channels as relay stations between remote stimuli and sieve tube occlusion in *Vicia faba*. Plant Cell 21:2118–2132

Goff KL, Quaroni L, Wilson KE (2009) Measurement of metabolite formation in single living cells of *Chlamydomonas reinhardtii* using synchrotron Fourier Transform Infrared spectromicroscopy. Analyst 134:2216–2219

Hayashi T, Takagi S (2003) Ca^{2+}-dependent cessation of cytoplasmic streaming induced by hypertonic treatment in *Vallisneria* mesophyll cells: possible role of cell wall-plasma membrane adhesion. Plant Cell Physiol 44:1027–1036

Heraud P, Wood BR, Tobin MJ, Beardall J, McNaughton D (2005) Mapping of nutrient-induced biochemical changes in living algal cells using synchrotron infrared microspectroscopy. FEMS Microbiol Lett 249:219–225

Heraud P, Caine S, Sanson G, Gleadow R, Wood BR, McNaughton D (2007) Focal plane array imaging: a new way to analyze leaf tissue. New Phytol 173:216–225

Himmelsbach DS, Khalili S, Akin DE (1998) FT-IR microspectroscopic imaging of flax (*Linum usitatissimum* L.) stems. Cell Mol Biol (Paris) 44:99–108

Hirschmugl CJ, Bayarri ZE, Bunta M, Holt JB, Giordano M (2006) Analysis of the nutritional status of algae by Fourier transform infrared chemical imaging. Infrared Physics Technol 49:57–63

Marinkovic NS, Huang R, Bromberg P, Sullivan M, Toomey J, Miller LM, Sperber E, Moshe S, Jones KW, Chouparova E, Lappi S, Franzen S, Chance MR (2002) Center for synchrotron biosciences' U2B beamline: an international resource for biological infrared spectroscopy. J Synchrotron Radiation 9:189–197

McNear DH Jr, Chaney RL, Sparks DL (2010) The hyperaccumulator *Alyssum murale* uses complexation with nitrogen and oxygen donor ligands for Ni transport and storage. Phytochemistry 71:188–200

Michael J, Mantsch HH (1996) Biomedical infrared spectroscopy. In: Mantsch HH, Chapman D (eds) Infrared spectroscopy of biomolecules. Wiley-Liss, New York

Miller LM (2000) The impact of infrared synchrotron radiation on biology: past, present, and future. Synchrotron Radiat News 13:31–37
Miller LM, Dumas P (2006) Chemical imaging of biological tissue with synchrotron infrared light. Biochim Biophys Acta 1758:846–857
Nanba T, Kimura S, Okamura H, Sakurai M, Matsunami M, Kimura H, Moriwaki T, Ikemoto Y, Hirono T, Takahashi T, Shinoda K, Fukui K, Teragami M, Kondo Y (2003) Spring-8 as an IR-light source. In: Spectroscopy & material characterization. Laboratoire National des Champs Magnétiques Pulsés (LNCMP) ISBN 2–87649–035–8
Nasse MJ, Walsh MJ, Mattson EC, Reininger R, Kajdacsy-Balla A, Macias V, Bhargava R, Hirschmugl CJ (2011) High-resolution Fourier-transform infrared chemical imaging with multiple synchrotron beams. Nat Methods 8(5):413–418
Pandey KK, Pitman AJ (2003) FTIR studies of the changes in wood chemistry following decay by brown-rot and white-rot fungi. Int Biodet Biodegr 52:151–160
Pietrzak LN, Miller SS (2005) Microchemical structure of soybean seeds revealed in situ by ultra-spatially resolved synchrotron Fourier Transformed Infrared Microspectroscopy. J Agric Food Chem 53:9304–9311
Quader H, Hofmann A, Schnepf E (1989) Reorganization of the endoplasmic reticulum in epidermal cells of onion bulb scales after cold stress: involvement of cytoskeletal elements. Planta 177:273–280
Quaroni L, Zlateva T (2011) Infrared spectromicroscopy of biochemistry in functional single cells. Analyst 136:3219–3232
Raab TK, Martin MC (2001) Visualizing rhizosphere chemistry of legumes with mid-infrared synchrotron radiation. Planta 213:881–887
Reffner JA, Martoglio PA, Williams GP (1995) Fourier transform infrared microscopical analysis with synchrotron radiation: the microscope optics and system performance. Rev Sci Instr 66:1298–1302
Sacksteder C, Barry BA (2001) Fourier Transform Infrared Spectroscopy: a molecular approach to an organismal question. J Phycol 37:197–199
Salman A, Tsror L, Pomerantz A, Moreh R, Mordechai S, Huleihel M (2010) FTIR spectroscopy for detection and identification of fungal phytopathogenes. Spectroscopy 24:261–267
Stuart BH (2012) Infrared spectroscopy of biological applications: an overview. Encyclopedia of Analytical Chemistry
Susi H, Byler DM (1983) Protein structure by Fourier transform infrared spectroscopy: second derivative spectra. Biochem Biophys Res Commun 115:391–397
Tanino K, Kobayashi S, Hyett C, Hamilton K, Liu J, Li B, Borondics F, Pedersen T, Tse J, Ellis T, Kawamura Y, Uemura M (2013) *Allium fistulosum* L. as a novel system to investigate mechanisms of freezing resistance. Physiol Plant 147:101–111
Ueda H, Yokota E, Kutsuna N, Shimada T, Tamura K, Shimmen T, Hasezawa S, Dolja VV, Hara-Nishimura I (2010) Myosin-dependent endoplasmic reticulum motility and F-actin organization in plant cells. Proc Natl Acad Sci U S A 107:6894–6899
Wetzel DL, Eilert AJ, Pietrzak LN, Miller SS, Sweat JA (1998) Ultraspatially resolved synchrotron infrared microspectroscopy of plant tissue in situ. Cell Mol Biol (Paris) 44:145–167
Wetzel DL, Srivarin P, Finney JR (2003) Revealing protein infrared spectral detail in a heterogeneous matrix dominated by starch. Vib Spectrosc 31:109–114
Wetzel DL, Shi YC, Schmidt U (2010) Confocal Raman and AFM imaging of individual granules of octenyl succinate modified and natural waxy maize starch. Vib Spectrosc 53:173–177
Williams GP (1989) The initial scientific program at the NSLS infrared beamline. Lawrence Berkeley Laboratory BNL–43335.
Wilson RH, Smith AC, Kacuráková M, Saunders PK, Wellner N, Waldron KW (2000) The mechanical properties and molecular dynamics of plant cell wall polysaccharides studied by Fourier-transform infrared spectroscopy. Plant Physiol 124:397–405
Yamazaki T, Kawamura Y, Minami A, Uemura M (2008a) Calcium-dependent freezing tolerance in Arabidopsis involves membrane resealing via synaptotagmin SYT1. Plant Cell 20:3389–3404

Yamazaki T, Kawamura Y, Uemura M (2008b) Cryobehavior of the plasma membrane in protoplasts isolated from cold acclimated *Arabidopsis* leaves is related to surface area regulation. Plant Cell Physiol 49:944–957

Yu P (2005a) Molecular chemistry imaging to reveal structural features of various plants feed tissues. J Struct Biol 150:81–89

Yu P (2005b) Protein secondary structures (α-helix and β-sheet) at a cellular level and protein fractions in relation to rumen degradation behaviours of protein: a new approach. Brit J Nutr 94:655–665

Yu P (2007) Molecular chemical structure of barley proteins revealed by ultra-spatially resolved synchrotron light sourced FTIR microspectroscopy: comparison of barley varieties. Biopolymers 85(4):308–317

Yu P (2011) Microprobing the molecular spatial distribution and structural architecture of feed-type sorghum seed tissue (*Sorghum Bicolor* L.) using the synchrotron radiation infrared microspectroscopy technique. J Synchrotron Radiat 18:790–801

Yu P, Mckinnon JJ, Christensen CR, Christensen DA (2004) Using synchrotron transmission FTIR microspectroscopy as a rapid, direct, and nondestructive analytical technique to reveal molecular microstructural-chemical features within tissue in grain barley. J Agric Food Chem 52:1484–1494

Yu P, Christensen CR, Christensen DA, Mckinnon JJ (2005a) Ultrastructural-chemical makeup of yellow-seeded (*Brassica rapa*) and brown-seeded (*Brassica napus*) canola within cellular dimensions, explored with synchrotron reflection FTIR microspectroscopy. Can J Plant Sci 85(3):533–541

Yu P, McKinnon JJ, Soita HW (2005b) Use of synchrotron-based FTIR microspectroscopy to determine protein secondary structures of raw and heat treated brown and golden flaxseeds: a novel approach. Can J Animal Sci 85:437–448

Yu P, Wang R, Bai Y (2005c) Reveal protein molecular structural-chemical differences between two types of winterfat (forage) seeds with physiological differences in low temperature tolerance using synchrotron-based Fourier transform infrared microspectroscopy. J Agric Food Chem 53(24):9297–9303

Yu P, Block H, Niu Z, Doiron K (2007) Rapid characterization of molecular chemistry, nutrient make-up and microlocation of internal seed tissue. J Synchrotron Radiat 14:382–90

Yu P, Doiron K, Liu D (2008) Shining light on the differences in molecular structural chemical makeup and the cause of distinct degradation behavior between malting- and feed-type barley using synchrotron FTIR microspectroscopy: a novel approach. J Agric Food Chem 56(9):3417–26

Yu P, Jonker A, Gruber M (2009) Molecular basis of protein structure in proanthocyanidin and anthocyanin-enhanced Lc-transgenic alfalfa in relation to nutritive value using synchrotron-radiation FTIR microspectroscopy: a novel approach. Spectrochimica Acta Part A 73:846–853

Supercooling-Facilitating Hydrolyzable Tannins Isolated from Xylem Tissues of *Cercidiphyllum japonicum*

Donghui Wang, Chikako Kuwabara, Keita Endoh, Yukiharu Fukushi, Seizo Fujikawa and Keita Arakawa

Introduction

Freezing resistance is an essential physiological function for overwintering woody plants, especially plants in regions with a cold climate (Sakai and Larcher 1987). It is generally known that freezing resistance is controlled developmentally and seasonally. Thus, the level and mechanism of freezing resistance are different between tissues and/or organs in perennial woody plants. In the woody trunk, cells of the bark and cambium respond to subzero temperatures by extracellular freezing, which is generally observed in herbaceous plants. On the other hand, parenchyma cells in xylem respond to subzero temperatures by deep supercooling.

The mechanism of deep supercooling of xylem parenchyma cells (XPCs) is distinct from that of extracellular freezing in bark and cambium cells, because XPCs are hardly dehydrated after ice-seeding of tissues and freezable water inside the cells is kept in a supercooling state at subzero temperatures to avoid intracellular freezing (Quamme et al. 1982; Sakai and Larcher 1987; Kuroda et al. 2003). In boreal woody plants grown in the field, XPCs can survive from cooling at subzero temperatures lower than −40 °C to around −60 °C at maximum for long periods (more than several weeks in a longer case) in mid-winter. Despite the reception of physical stimuli from circumstances in habitats, supercooling of XPCs in trees grown in the field is more stable than that transiently observed in our daily life. This unique mechanism to maintain supercooled water in a metastable state in XPCs is called deep supercooling to distinguish it from temporal supercooling. However, supercooling of cellular water has a physical limit. Thus, cooling in excess of the limit of deep supercooling capability causes the breakdown of supercooling of cellular water, resulting in lethal damage due to intracellular freezing of XPCs (Fujikawa and Kuroda 2000; Fujikawa et al. 2009).

K. Arakawa (✉) · D. Wang · C. Kuwabara · K. Endoh · Y. Fukushi · S. Fujikawa
Research Faculty and Graduate School of Agriculture,
Hokkaido University, N9-W9, Kita-ku, Sapporo 080-8589, Japan
e-mail: keita-ar@for.agr.hokudai.ac.jp

R. Imai et al. (eds.), *Plant and Microbe Adaptations to Cold in a Changing World*,
DOI 10.1007/978-1-4614-8253-6_15,

In woody trunks of adult trees, especially boreal trees in regions with a cold climate in winter, freezing resistance of XPCs is lower than that of bark parenchyma cells (Fujikawa and Kuroda 2000). Lethal damage of XPCs caused by intracellular freezing resulted in lethal damage to the entire tree (Quamme et al. 1982). It has also been shown that the deep supercooling capability of XPCs changed with latitudinal distribution of trees from tropical to cold regions (Daniell and Crosby 1968; Quamme et al. 1972; Fujikawa and Kuroda 2000). Thus, deep supercooling capability of XPCs may be an inherent trait for determining winter survival and distribution of trees, especially boreal trees. Additionally, freezing resistance by deep supercooling is increased by seasonal cold acclimation and decreased by deacclimation. This feature is similar to the case of freezing tolerance by extracellular freezing in herbaceous plants.

Although results of many studies on the mechanism of extracellular freezing have been reported, studies on the mechanism of deep supercooling have been limited because the mechanism of deep supercooling of XPCs has been only explained by cell wall features by which cellular water is isolated physically from extracellular ice, resulting in maintenance of the supercooling state of cellular water in XPCs (Ashworth and Abeles 1984). This hypothesis was based on the isolated water droplet theory that was dependent on the results of in vitro experiments on the supercooling of isolated small water droplets (Fletcher 1970; MacKenzie 1977).

Contribution of the Cell Wall to Deep Supercooling Capability of XPCs by Isolation of Intracellular Water from Extracellular Ice

In deep supercooling XPCs, cell walls do not allow freeze-induced dehydration from the cells or penetration of extracellular ice into the protoplasm even after ice-seeding. Therefore, freezable water remaining in the protoplasm of XPCs is kept in a supercooled state at subzero temperatures. Also, it has been speculated that there are no heterogenous ice nucleators, which promote freezing at warm subzero temperatures, inside XPCs (Quamme et al. 1982). Thus, cellular water of XPCs is isolated in a small volume by thick and rigid cell walls from extracellular ice and supercooled to about −40 °C, which is similar to the limit of homogeneous ice nucleation of pure water (George and Burke 1977; Quamme et al. 1982; Ashworth and Abeles 1984). As mentioned before, the cell wall architecture may function as a barrier against cell dehydration and ice penetration in deep supercooling XPCs.

In previous studies, Wisniewski and co-workers found that deep supercooling capability of XPCs, which is represented as the low temperature exotherm in differential thermal analysis, was lowered or disappeared with modification of cell walls in xylem tissues of peach twigs by enzymatic digestion or chemical treatment such as treatment with pectinase or EGTA (Wisniewski et al. 1991a, b). These results suggested that structures of the pit membrane and amorphous layer in cell walls (especially pectin-related structures regulating the porosity of cell walls) were important for isolation of the protoplasm from the effects of extracellular ice in deep supercooling XPCs (Wisniewski 1995).

On the other hand, it is known that breakdown of supercooling (occurrence of ice nucleation via development of ice clusters) may be dependent on experimental conditions including the cooling rate, temperature, and cooling time. Also, it is known that ice nucleation is influenced by various factors such as conditions of the sample solution (volume, solute concentration, and composition), surface structure of containers, and physical impact in an experimental system. Considering that boreal trees in the field are occasionally stimulated by field conditions (wind, snowfall, feeding by animals, etc.) in their habitats and are subjected to low subzero temperatures for a long period (up to several weeks or months) in mid-winter, occurrence of ice nucleation may be avoided in deep supercooling XPCs in nature, and it seems that the mechanism of deep supercooling in XPCs may be more complicated than ideas according to the isolated droplet theory by only cell walls.

Thus, various factors may contribute to the deep supercooling mechanism in XPCs, and intracellular components that may be related to supercooling capability were examined.

Effects of Intracellular Substances on Deep Supercooling Capability of XPCs

When soluble components were released by washing with water after the cell structure of XPCs had been disrupted by freeze-thawing two or three times, supercooling capability represented by the limit temperature of supercooled water in XPCs (latent heat release of intracellular freezing) was shifted from −38 °C to a warmer temperature around −20 °C in differential thermal analysis (Kasuga et al. 2006). This meant that water surrounded by the cell wall structure was able to be kept in a supercooling state at around −20 °C, and that the decrease in supercooling capability may be caused by release of intracellular components. These results suggested that both the cell wall and intracellular components might contribute to deep supercooling capability of XPCs.

Secondary Metabolites with Supercooling-Facilitating (SCF) Activities

In order to find intracellular substances with supercooling-facilitating (SCF) activities from the extract of xylem tissues of woody plants, the droplet freezing assay according to Vali (1971) was carried out.

In the droplet freezing assay, the sample was diluted in phosphate buffer containing ice-nucleating bacteria, *Erwinia ananas* or silver iodide as an ice nucleator.

Two microliter-droplets of the solution were placed onto a copper plate and the plate was cooled at a rate of 0.2 °C/min from 0 to −30 °C. The frozen droplets were counted during the cooling process. Then a freezing spectrum was obtained by cumulative curve of freezing of 100 droplets by repeating the assay five times. For assessment of SCF activity, INT_{50}, which is the temperature at which 50 % of the droplets are frozen, was measured. SCF activity is represented by the difference between INT_{50} of the sample and that of the control.

By using this method, recent studies have shown SCF activities in the crude xylem extract of several hardwood species, indicating the existence of compounds with SCF activities and the possibility of contribution to deep supercooling capability of XPCs (Kasuga et al. 2007b). Since SCF activity in the crude xylem extract of Katsura tree was the highest among these trees, we identified the SCF substances to clarify the mechanism of deep supercooling capability in XPCs.

SCF-Flavonol Glycosides

After xylem tissues had been extracted by methanol, the extract was desiccated and redissolved in water. This methanol fraction was further extracted by ethyl acetate. Since SCF activity was detected in the ethyl acetate fraction, this fraction was further separated by silica gel column chromatography. SCF activity was detected in almost all silica gel fractions, and two major peaks of SCF activities at the concentration of 1 mg/mL were detected in silica gel fractions, as compared with the control. From the highest peak fraction among the two major peaks, four peaks with SCF activities were finally isolated by an octadecylsilane-C18 column, and structural analyses revealed that all four peaks were flavonol glycosides, quercetin 3-*O*-β-glucoside (Q3Glc), kaempferol 7-*O*-β-glucoside (K7Glc), 8-methoxykaempferol 3-*O*-β-glucoside (8mK3Glc), and kaempferol 3-*O*-β-glucoside (K3Glc), as novel SCF substances (Kasuga et al. 2008). One of them, K7Glc, showed markedly high SCF activities in the solution containing *E. ananas* as an ice nucleator. INT_{50} of K7Glc was 9.0 °C at 1 mg/mL, and other three compounds, Q3Glc, 8mK3Glc, and K3Glc, also showed high SCF activities of 2.8, 3.4, and 4.0 °C at 1 mg/mL, respectively. In addition, histochemical study showed localization of flavonol glycosides in the cytoplasm in XPCs (Kasuga et al. 2008).

Recent studies on characterization of SCF-flavonol glycosides have shown SCF activities of K7Glc, Q3Glc, and K3Glc at 0.1 and 0.5 mg/mL in solutions containing different kinds of ice nucleators, silver iodide, *Xanthomonas campestris* (different type of ice-nucleating bacteria of *E. ananas*), and benzene-1,3,5-triol (phloroglucinol) at appropriate concentrations instead of *E. ananas*, but levels of SCF activities were dependent on kinds and concentrations of SCF substances and kinds of ice nucleators used in the experiments. However, it was revealed that these SCF-flavonol glycosides did not affect the homogenous ice nucleation temperature as shown by differential thermal analysis profiles of an emulsion freezing assay (Kuwabara et al. 2011).

Furthermore, SCF activities of many kinds of flavonoids (0.5 or 1.0 mg/mL), the structures of which are similar to those of SCF-flavonol glycosides, were examined in solutions containing different kinds of ice nucleators. All of them showed SCF activities with various responses to ice nucleators added to the sample solutions (Kuwabara et al. 2012). Interestingly, some of them promoted ice nucleation of solutions under limited conditions. In solutions without addition of ice nucleators but with unintentional contamination of airborne ice nucleators, most of the flavonoids we examined promoted ice nucleation of solutions, although all of them showed SCF activities in solutions containing ice nucleators already known. However, the mechanism of the complicated reaction has not been elucidated yet.

SCF-Hydrolyzable Tannins

As mentioned before, flavonol glycosides were distributed in the first major peak of SCF activity in silica gel fractions originating from xylem extracts of Katsura tree. Next, we identified the SCF substances from the second major peak of SCF activity in silica gel fractions.

From three fractions in the second major peak of SCF activities, four main peaks of SCF activities were detected by HPLC analysis. We then isolated each peak of SCF compounds by using Sephadex LH-20 column chromatography and measured their activities in the solution containing *E. ananas*. SCF activities of purified peaks 1, 2, 3, and 4 were 2.3, 1.5, 2.9, and 2.1 °C at 1 mg/mL, respectively, and structural analyses were performed by using mass spectrum (MS) and nuclear magnetic resonance (NMR) analyses. According to MS and NMR analyses, their chemical structures were determined. All four compounds are categorized into hydrolyzable tannins, 2,2′,5-tri-*O*-galloyl-α,β-D-hamamelose (trigalloyl Ham), 1,2,6-tri-*O*-galloyl-β-D-glucopyranose, (trigalloyl Glc) 1,2,3,6-tetra-*O*-galloyl-β-D-glucopyranose (tetragalloyl Glc), and 1,2,3,4,6-penta-*O*-galloyl-β-D-glucopyranose (pentagalloyl Glc). It is thought that all of them are novel SCF compounds. In addition, SCF activity is a novel function of these tannins (Wang et al. 2012) because they are generally related to the defense system of plant cells.

For characterization of hydrolyzable tannins, INT_{50} was measured and SCF activities in solutions containing *E. ananas* at different concentrations were calculated from freezing spectra. SCF activities of hydrolyzable tannins were represented by shift of the freezing spectrum to a lower temperature. SCF activities increased with increase in the concentration. At high concentrations of tannins (0.5–1.0 mg/mL), significant SCF activities were detected in all tannins. Among them, tetragalloyl Glc showed the highest SCF activity in solution containing *E. ananas* (Wang et al. 2012).

Next, SCF activity was measured in solutions containing different ice nucleators for further characterization (Wang et al. 2011). For ice nucleators, silver iodide, *X. campestris*, and phloroglucinol at appropriate concentrations instead of *E. ananas*, and SCF activity was measured by the droplet freezing assay. INT_{50} values of buffer controls in solutions containing ice nucleators were dependent on the ice nucleators added. Silver iodide was the most effective ice nucleator among them and phloroglucinol was less effective, suggesting different kinetics or mode of action for ice nucleation.

SCF activities of four hydrolyzable tannins at concentrations of 0.5 and 1.0 mg/mL were detected in almost all solutions containing various ice nucleators, but the freezing spectra of solutions containing the ice nucleators were slightly different. Among them, trigalloyl Ham and trigalloyl Glc were effective in solutions containing *E. ananas* and silver iodide but were less effective in solutions containing *X. campestris* and phloroglucinol. Tetragalloyl Glc and pentagalloyl Glc were effective as SCF substances in solutions containing all four ice nucleators. At 1.0 mg/mL of the SCF substance, tetragalloyl Glc was the most effective in solutions containing *E. ananas* and *X. campestris*, trigalloyl Ham was the most effective in solutions containing silver iodide, and pentagalloyl Glc was the most effective in solutions containing phloroglucinol. For all tannins, high SCF activities were detected in solutions containing silver iodide, while relatively low SCF activities were detected in solutions containing phloroglucinol (Wang et al. 2011).

Additionally, tetragalloyl Glc and pentagalloyl Glc at 0.05 and 0.1 mg/mL did not show SCF activities in solutions containing phloroglucinol but showed ice-nucleating activities under the same conditions. Trigalloyl Ham and trigalloyl Glc at 0.05 and 0.1 mg/mL showed low or no SCF activities and ice-nucleating activities in solutions containing phloroglucinol. Ice-nucleating activities of tetragalloyl Glc and pentagalloyl Glc were found under limited conditions in our recent studies (Wang et al. unpublished data). The results indicated that levels of SCF activities of these hydrolyzable tannins were dependent on types and concentrations of SCF substances themselves and on types of ice nucleators used in the experiments.

To test the effects of SCF-hydrolyzable tannins on homogenous ice nucleation, an emulsion freezing assay was carried out in a way similar to that of SCF-flavonol glycosides (Wang et al. unpublished data). In this experiment, each hydrolyzable tannin was added to a buffer solution at a concentration of 1 mg/mL and then emulsion solutions were prepared in silicone oil. Emulsified droplets (around <10 μm in diameter) were prepared by using an ultrasonic cell disruptor. Since it has been shown that water droplets are too small to contain airborne ice nucleators in this system, these emulsified droplets might contain no ice nucleators. Freezing temperatures of these emulsified samples and control droplets in microtubes were measured by detection of latent heat release at a cooling rate of 0.2 °C/min by using a thermocouple. In this experiment, all of the exothermal peaks due to freezing of the droplets with and without tannins were detected at around -38.0 °C. This result indicated that homogenous nucleation temperature of emulsified water droplets of the solution was not affected by addition of tannins. Therefore, SCF activities of hydrolyzable tannins may be only effective for heterogeneous ice nucleation.

In order to estimate the localization of SCF-hydrolyzable tannins in the xylem, histochemical analysis of tannins was carried out according to Ruzin (1999). Small blocks of fresh xylem were prepared from 4- to 6-year-old twigs of Katsura tree harvested in winter and stained with a solution containing $FeSO_4$. Since histochemical analysis specific for hydrolyzable tannins is limited, a general staining method for tannins was used as a preliminary experiment. In this experiment, the protoplasm of XPCs was strongly stained rather than the cell walls. Therefore, it is thought that tannins including hydrolyzable tannins are localized mainly inside cells (Wang et al. 2011). However, the precise subcellular localization of tannins was not clearly determined by this method.

Levels of trigalloyl Ham and tetragalloyl Glc were altered around 0.3 and 0.2 μmol/g dry weight, respectively, through the year and were much higher than those of trigalloyl Glc and pentagalloyl Glc in XPCs of Katsura tree in winter (Wang et al. unpublished data). Levels of these tannins were not significantly changed seasonally in xylem tissue in Katsura tree. Therefore, it seemed that seasonal changes in levels of hydrolyzable tannins might be poorly correlated with seasonal changes in deep supercooling capability of xylem tissue, since it is thought that physiological roles of these tannins may not be only for winter hardiness but also for the defense system in plant. Nevertheless, SCF capability of these hydrolyzable tannins accumulated in XPCs can support maintenance of supercooled cellular water for a long period in winter in concert with SCF-flavonol glycosides.

Discussion

It is thought that SCF capability of an aqueous solution evaluated by in vitro experiments such as the droplet freezing assay is influenced by conditions of the sample solution (concentration, composition, volume), cooling process (cooling rate, given temperature, time), and conditions of the container (type, material, surface structure). According to this idea, the deep supercooling mechanism of XPCs might be influenced by conditions of the protoplasmic solution (solute concentration and compositions, cell volumes, cellular water contents), cooling process (seasonal and diurnal drops in ambient temperatures, depression of minimum air temperature, duration of mid-winter at subzero temperatures), and cell wall and cell membrane structures as the containers in vivo. Since the protoplasm of XPCs and the thick and rigid cell wall structures are regarded as tiny water droplets and cell container, respectively, the conditions of the protoplasmic solution of XPCs may be important for deep supercooling as well as the cell wall structure for isolation of the protoplasm from extracellular ice (isolated droplet theory). For maintenance of a supercooling state of XPCs at subzero temperatures for a long period in winter under the conditions of continual physical impact in nature, solute concentrations and compositions in the protoplasm might be important factors for deep supercooling XPCs in boreal woody plants to stably maintain a supercooling state. Actually, in our recent studies, we identified and characterized four flavonol glycosides and four hydrolyzable tannins, all of which are novel SCF substances, from the xylem extract of Katsura tree. SCF capabilities of these compounds are effective at low concentrations of several millimolars in a noncolligative manner. Furthermore, it is possible that other types of SCF substances exist in XPCs because many but not all structural analogs and derivatives of these secondary metabolites showed SCF activities.

In addition to these SCF substances, accumulation of compatible solutes such as soluble sugars at high concentrations in the cells would depress the melting point of cellular water in proportion to the concentration because accumulation of soluble sugars was induced during seasonal cold acclimation in XPCs in boreal hardwood species (Sauter et al. 1996; Sauter and Wellenkamp 1998, Kasuga et al. 2007a). This may depress the ice nucleation temperature because it has been shown that increase in the concentrations of various solutes added to an aqueous solution depresses homogeneous supercooling points of emulsified tiny droplets of the solutions about twice (from 1.3 to 2.5) as much as melting point depression corresponding to the solute concentrations (MacKenzie 1977, Wilson et al. 2003, Zachariassen and Kristiansen 2000). Therefore, solute accumulation in XPCs, including soluble sugars and other compatible solutes, might contribute to the increase in supercooling capability of XPCs as a function of the intracellular concentration.

It is also known that accumulation of soluble proteins is induced during seasonal cold acclimation in woody plants. In our preliminary experiment, the crude soluble protein fraction extracted from Katsura tree showed less supercooling-facilitating activity at a concentration of 0.1 % (unpublished data). There have been a few reports on cold-inducible proteins that possess SCF or thermal hysteresis activities in woody plants, although antifreeze proteins found in plants showed activity that inhibits ice crystal growth within a narrow temperature range or low thermal hysteresis activity (Griffith and Yaish 2004). A dehydrin, PCA60, accumulated in bark and xylem of peach during seasonal cold acclimation exhibited antifreeze activity but low thermal hysteresis activity (0.06 °C) (Wisniewski et al. 1999). PCA60 also showed cryoprotective activities against freeze-thawing of cold-labile proteins such as lactate dehydrogenase. Although there are some reports suggesting induction of other dehydrins and dehydrin-like proteins during cold acclimation in trees (Sarnighausen et al. 2002, Takata et al. 2007), their function for the mechanism of deep supercooling has not been clarified yet. Generally, dehydrins are highly soluble proteins with much hydrated water attached. Thus, it is thought that highly soluble proteins induced during cold acclimation may have protective effects on macromolecules or a limiting effect on free water movement at subzero temperatures in XPCs; however, further experiments are necessary to clarify the functions of these proteins for the mechanism of deep supercooling in XPCs. As mentioned before, it is thought that not only the cell wall but also intracellular components may function in the development of SCF capability in deep supercooling XPCs in boreal woody plants.

It is speculated that no substances with ice-nucleating activities at warm subzero temperatures exist in XPCs (Quamme 1982). However, we do not know whether or not substances exist in XPCs with ice-nucleating activities at low subzero temperatures because we have not examined the possibility of formation of ice-nucleating factors at low subzero temperatures due to, for example, structural, conformational, or interactional changes in cellular components. Interaction between SCF substances and possible intracellular ice-nucleating factors at low subzero temperature may be an interesting problem. Also, factors that regulate seasonal changes in deep supercooling capability of XPCs should be examined, although some candidates are known. Therefore, further studies are necessary to understand the molecular mechanism of deep supercooling capability of XPCs.

Acknowledgments The authors wish to thank Dr E. Fukushi and Mr. K. Watanabe (GC-MS and NMR Laboratory, Graduate School of Agriculture, Hokkaido University) for MS and NMR measurements. This work was partly supported by JSPS KAKENHI Grant Numbers 20380099 (to S.F.) and 2058036 (to K.A.).

References

Ashworth EN, Abeles FB (1984) Freezing behavior of water in small pores and the possible role in the freezing of plant tissues. Plant Physiol 76:201–204

Daniell JW, Crosby FL (1968) Occlusion of xylem elements in peach trees resulting from cold injury. Proc Am Soc Hortic Sci 93:128–134

Fletcher NH (1970) The chemical physics of ice. Cambridge University Press, London, pp 73–103

Fujikawa S, Kuroda K (2000) Cryo-scanning electron microscopic study on freezing behavior of xylem ray parenchyma cells in hardwood species. Micron 31:669–686

Fujikawa S, Kasuga J, Takata N, Arakawa K (2009) Factors related to change of deep supercooling capability in xylem parenchyma cells of trees. In: Gusta LV, Wisniewski ME, Tanino KK (eds) Plant Cold Hardiness: From the Laboratory to the Field. CAB International, London, pp 29–42

George MF, Burke MJ (1977) Cold hardiness and deep supercooling in xylem of shagbark hickory. Plant Physiol 59:319–325

Griffith M, Yaish MWF (2004) Antifreeze proteins in overwintering plants: a tale of two activities. Trends Plant Sci 9:399–405

Kasuga J, Mizuno K, Miyaji N, Arakawa K, Fujikawa S (2006) Role of intracellular contents to facilitate supercooling capability in beech (*Fagus crenata*) xylem parenchyma cells. Cryo-Letters 27:305–310

Kasuga J, Arakawa K, Fujikawa S (2007a) High accumulation of soluble sugars in deep supercooling Japanese white birch xylem parenchyma cells. New Phytol 174:569–579

Kasuga J, Mizuno K, Arakawa K, Fujikawa S (2007b) Anti-ice nucleation activity in xylem extracts from trees that contain deep supercooling xylem parenchyma cells. Cryobiology 55:305–314

Kasuga J, Hashidoko Y, Nishioka A, Yoshiba M, Arakawa K, Fujikawa S (2008) Deep supercooling xylem parenchyma cells of katsura tree (*Cercidiphyllum japonicum*) contain flavonol glycosides exhibiting high anti-ice nucleation activity. Plant Cell Environ 31:1335–1348

Kuroda K, Kasuga J, Arakawa K, Fujikawa S (2003) Xylem ray parenchyma cells in boreal hardwood species respond to subfreezing temperatures by deep supercooling that is accompanied by incomplete desiccation. Plant Physiol 131:736–744

Kuwabara C, Kasuga J, Wang D, Fukushi Y, Arakawa K, Koyama T, Inada T, Fujikawa S (2011) Change of supercooling capability in solutions containing different kinds of ice nucleators by flavonol glycosides from deep supercooling xylem parenchyma cells in trees. Cryobiology 63:157–163

Kuwabara C, Wang D, Kasuga J, Fukushi Y, Arakawa K, Koyama T, Inada T, Fujikawa S (2012) Freezing activities of flavonoids in solutions containing different ice nucleators. Cryobiology 64:279–285

MacKenzie AP (1977) Non-equilibrium freezing behavior of aqueous systems. Philos Trans R Soc Lond B 278:167–189

Quamme HA, Stushnoff C, Weiser CJ (1972) The relationship of exotherms to cold injury in apple stem tissues. J Am Soc Hortic Sci 97:608–613

Quamme HA, Chen PM, Gusta LV (1982) Relationship of deep supercooling and dehydration resistance to freezing injury in dormant stem tissues of 'Starkrimson Delicious' apple and 'Siberian C' peach. J Am Soc Hortic Sci 107:299–304

Ruzin ST (1999) Plant Microtechnique and Microscopy. Oxford University Press, New York

Sarnighausen E, Karlson D, Ashworth E (2002) Seasonal regulation of a 24-kDa protein from red-osier dogwood (*Cornus sericea*) xylem. Tree Physiol 22:423–430

Sakai A, Larcher W (1987) Frost Survival of Plants: Responses and Adaptation to Freezing Stress. Springer-Verlag, Berlin

Sauter JJ, Wellenkamp S (1998) Seasonal changes in content of starch, protein and sugars in the twig wood of *Salix caprea* L. Holzforschung 52:255–262

Sauter JJ, Wisniewski M, Witt W (1996) Interrelationships between ultrastructure, sugar levels, and frost hardiness of ray parenchyma cells during frost acclimation and deacclimation in poplar (*Populus × canadensis* Moench" robusta") wood. J Plant Physiol 149:451–461

Takata N, Kasuga J, Takezawa D, Arakawa K, Fujikawa S (2007) Gene expression associated with increased supercooling capability in xylem parenchyma cells of larch (*Larix kaempferi*). J Exp Bot 58:3731–3742

Vali G (1971) Quantitative evaluation of experimental results on the heterogeneous freezing nucleation of supercooled liquids. J Atmos Sci 28:402–409

Wang D, Kasuga J, Kuwabara C, Fukushi Y, Fujikawa S, Arakawa K (2011) Characterization of supercooling-facilitating (anti-ice nucleation) hydrolyzable gallotannins in supercooling xylem parenchyma cells in katsura tree (*Cercidiphyllum japonicum*). Cryobiol Cryotechnol 57:147–151

Wang D, Kasuga J, Kuwabara C, Fukushi Y, Fujikawa S, Arakawa K (2012) Presence of supercooling-facilitating (anti-ice nucleation) hydrolyzable tannins in deep supercooling xylem parenchyma cells in *Cercidiphyllum japonicum*. Planta 235:747–759

Wilson PW, Heneghan AF, Haymet ADJ (2003) Ice nucleation in nature: supercooling point (SCP) measurements and the role of heterogeneous nucleation. Cryobiology 46:88–98

Wisniewski M (1995) Deep supercooling and the role of cell wall structure. In: Lee RL Jr, Warren GJ, Gusta LV (eds) Biological Ice Nucleation and its Applications. APS Press, St Paul, pp. 163–182

Wisniewski M, Davis G, Arora R (1991a) Effect of macerase, oxalic acid, and EGTA on deep supercooling and pit membrane structure of xylem parenchyma of peach. Plant Physiol 96:1354–1359

Wisniewski M, Davis G, Scaffer K (1991b) Mediation of deep supercooling of peach and dogwood by enzymatic modifications in cell-wall structure. Planta 184:254–260

Wisniewski M, Webb R, Balsamo R, Close TJ, Yu X-M, Griffith M (1999) Purification, immunolocalization, cryoprotective, and antifreeze activity of PCA60: A dehydrin from peach (*Prunus persica*). Physiol Plant 105:600–608

Zachariassen KE, Kristiansen E (2000) Ice nucleation and antinucleation in nature. Cryobiology 41:257–279

ICE1, a Transcription Factor Involved in Cold Signaling and Tolerance

Kenji Miura

Introduction

Abiotic stress, including cold, drought, salinity, and heat, significantly influences the growth, productivity, and quality of crops and reduces the yields of major crops by more than 50 %. However, tolerant plants have evolved strategies to adapt to such conditions. Plants first need to sense environmental cues to be able to respond appropriately to stress, and because of the complex nature of stress factors, multiple sensors are probably responsible for the perception of stress stimuli. After the initial recognition of a stress stimulus, a signal transduction cascade is activated and transmitted to modulate the expression of stress-regulated genes and thus elicit stress responses.

Low-temperature stress can be classified as chilling (a low, nonfreezing temperature) and freezing stresses. Tropical or subtropical plants, such as rice, maize, soybean, tomato, and cucumber, generally exhibit symptoms of injury when exposed to chilling stress such as reduced leaf expansion, wilting, chlorosis, and in certain cases flower sterility. In contrast, temperate plants have evolved adaptive mechanisms that enable them to cold acclimate and acquire freezing tolerance when exposed to suboptimal nonfreezing temperatures. The molecular basis of cold acclimation and freezing tolerance has been extensively studied, mainly in *Arabidopsis* and cereals. A transcriptome analysis revealed that approximately 10 % of the genes in *Arabidopsis* and wheat were altered by cold treatment (Lee et al. 2005; Laudencia-Chingcuanco et al. 2011). Molecular analysis showed that the promoter region of many cold-regulated genes contain motifs known as C-repeat (CRT)/dehydration-responsive element (DRE), sequences that bind the transcription factors CRT-binding factor (CBF)/DRE-binding protein1 (DREB1). The expression of CBF/DREB1 is further regulated by the MYC-type transcription factor inducer of *CBF* expression1 (ICE1). This chapter briefly discusses the ICE1-dependent cold stress signaling and its functional conservation in crop plants.

K. Miura (✉)
Faculty of Life and Environmental Sciences, University of Tsukuba, Tsukuba, Japan
e-mail: kmiura@gene.tsukuba.ac.jp

R. Imai et al. (eds.), *Plant and Microbe Adaptations to Cold in a Changing World*,
DOI 10.1007/978-1-4614-8253-6_16,

Cold Stress Sensing

The thermosensor that perceives temperature remains unidentified to date. However, low temperature induces phase transitions of the plasma membrane, from a liquid crystalline to a rigid gel phase. Furthermore, Ca^{2+} is increased by cold shock in the cytosol and this induction may be mediated through membrane rigidification-activated mechanosensitive or ligand-activated Ca^{2+} channels, as shown in alfalfa and *Brassica napus* (Orvar et al. 2000; Sangwan et al. 2001). The application of a membrane rigidifier, dimethyl sulfoxide (DMSO), was shown to induce the *COR* genes at 25 °C, whereas the expression of the *COR* genes was suppressed by application of a membrane fluidizer, benzyl alcohol, even at 0 °C. These results suggest that membrane rigidification may trigger the first event that activates cold signaling.

ICE1 is a MYC-type Transcription Factor Involved in Cold Signaling

ICE1 is a MYC-type basic helix-loop-helix (bHLH) transcription factor that binds to the *cis*-elements in the promoter of C-repeat-binding factor3/dehydration-responsive element-binding protein1A (*CBF3/DREB1A*) and positively regulates expression of *CBF3/DREB1A* and cold signaling (Chinnusamy et al. 2003). The *ice1* mutation downregulated the expression of *CBF3/DREB1A* and other *COR* genes and decreased both chilling tolerance and cold acclimation, whereas transgenic *ICE1*-overexpressing plants exhibited higher freezing tolerance (Chinnusamy et al. 2003). Transcriptome analysis revealed that the *ice1* mutation downregulated about 40 % of *COR* gene expression. Among the cold-regulated transcription factors, 46 % of them were regulated by *ICE1* (Lee et al. 2005). Thus, ICE1 is thought to be a positive regulator that controls *CBF/DREB1* and other transcription factors, in addition to many cold-regulated regulons. ICE2 is another MYC-type bHLH transcription factor with a high similarity to ICE1 and activates the expression of *CBF1/DREB1B* and promotes freezing tolerance (Fursova et al. 2009).

The *CBF/DREB1*-dependent cold signaling pathway is the best characterized regulatory pathway for cold acclimation (Lissarre et al. 2010; Chinnusamy et al. 2010; Zhou et al. 2011). The CBF/DREB1 protein binds to *CRT/DRE cis*-elements (Liu et al. 1998; Stockinger et al. 1997), motifs found in the promoters of several *COR* genes. Recently, it was reported that histone modification and chromatin conformational changes are involved in the regulation of cold-induced *ZmDREB1* expression in maize (Hu et al. 2011). After cold treatment, the hyperacetylation of histones H3K9 and H4 and DNA methylation occurs in the ICE1-binding region of the *ZmDREB1* promoter, an event that is accompanied by an increase in chromatin accessibility. These results suggest that cold treatment increases the accessibility of ICE1 to the specific region of the *ZmDREB1* promoter (Hu et al. 2011).

ICE1 is functionally conserved in higher plants. In tomato (*Solanum lycopersicum*), the ICE1 homolog *SlICE1* activates *SlCBF1* and the overexpression of *SlICE1* results in increased chilling tolerance (Miura et al. 2012a). Furthermore, the overexpression of *SlICE1* enhanced the accumulation of antioxidants, several amino acids, amines, and sugars and increased the antioxidant activity in red tomato fruits (Miura et al. 2012b). *TaICE141* and *TaICE187* also regulated the expression of the wheat *CBF* group IV genes and the overexpression of *TaICE141* and *TaICE187* in *Arabidopsis* enhanced *CBF/DREB1*-dependent cold-responsive gene expression and cold acclimation (Badawi et al. 2008). In rice, cold stresses induced the expression of the OsICE1 and OsICE2 proteins and sequentially upregulated the expression of *OsDREB1B*, *OsHsfA3,* and *OsTPP1* suggesting that OsICE1 and OsICE2 function in the transcriptional regulation of cold response (Nakamura et al. 2011). In the case of the banana plant, *Musa acuminata*, MaICE1 interacted with MaMYC2, which is a key regulator in the activation of the jasmonate (JA) response pathway (Zhao et al. 2012). Furthermore, the expression of the cold response pathway genes was also significantly induced by methyl JA, suggesting that the JA-induced chilling tolerance of the banana plant is associated with induced expression of *MYC2*, which may act together with *ICE1* for the activation of CBF-dependent cold signaling (Zhao et al 2012). The overexpression of *Arabidopsis ICE1* also improved the chilling tolerance and enhanced the accumulation of soluble sugars and proline in cucumber demonstrating the function conservation of ICE among species (Liu et al. 2010).

Regulatory Mechanism of ICE1

ICE1 is regulated by posttranslational modifications. Small ubiquitin-related modifier (SUMO) conjugation, termed sumoylation, regulates cellular protein activity and signaling (Miura and Hasegawa 2010) and is mediated by the SUMO E3 ligase SIZ1 (Miura et al. 2005). ICE1 is regulated by sumoylation at lysine 393 and *in vitro* analysis demonstrated that the sumoylation of ICE1 blocks the ubiqutylation of ICE1 (Miura et al. 2007). Indeed, the sumoylation of ICE1 is likely to function in controlling the expression of *CBF3/DREB1A* and cold-responsive genes and in regulating cold tolerance (Miura et al. 2007).

Ubiquitylation is also involved in the regulation of ICE1. The RING-type ubiquitin E3 ligase high expression of osmotically responsive gene1 (HOS1) mediates the polyubiquitylation of ICE1 for proteasome-dependent degradation (Dong et al. 2006). The substitution of serine 403 to alanine blocked the polyubiquitylation of ICE1 and enhanced both the stability of ICE1 and cold tolerance (Miura et al. 2011). Furthermore, the expression of cold-regulated genes was upregulated in *ICE1* (*S403A*)-overexpressing plants (Miura et al. 2011). These results suggest that serine 403 is a key residue for the attenuation of the cold-induced HOS1-mediated degradation of ICE1; however, the molecular mechanism by which serine controls the polyubiquitylation remains elusive. Conversely, the substitution of S403 to alanine did not affect the sumoylation of ICE1 (Miura et al. 2011).

Cold Signaling and Salicylic Acid or Stomata Development

Salicylic acid (SA) is a well-known phytohormone involved in the responses to biotic stress and pathogenesis. Recent studies have demonstrated that SA is also involved in the regulation of abiotic stress responses such as drought, high and low temperature, and salinity (Hara et al. 2011). The application of a low concentration of SA improved chilling and freezing tolerance of various species, including potato, cucumber, rice, and wheat (Kang and Saltveit 2002; Tasgin et al. 2003; Mora-Herrera et al. 2005). SA treatment also enhances the accumulation of H_2O_2. Because a transient increase in the intracellular level of H_2O_2 triggers the transcriptional regulatory network for prolonged survival under chilling stress in rice (Yun et al. 2010), it is plausible that SA-induced transient H_2O_2 accumulation may enhance chilling tolerance. Conversely, a high concentration and the continual application of SA cause severe damage and decrease cold tolerance, most likely due to the large accumulation of H_2O_2.

Endogenous free SA and glucosyl SA were found to accumulate during chilling in *Arabidopsis* shoots, wheat and grape berry (Scott et al. 2004; Janda et al. 2007; Wan et al. 2009; Kosová et al. 2012). The *Arabidopsis*-enhanced disease susceptibility (*eds5*) mutant or transgenic plants expressing the salicylate hydroxylase gene (*nahG*), which are impaired in SA accumulation (Nawrath and Métraux 1999), exhibited better growth at 5 °C compared to the wild type, whereas the SA-accumulation mutant *cpr1* (constitutive expressor of *PR* genes) exhibited a strong reduction in growth under conditions of low temperature (Scott et al. 2004). The other SA-accumulating *Arabidopsis* mutants, *siz1* and (accelerated cell death (*acd6*), and *DEAR1* (DREB and EAR motif protein)-overexpressing plants also exhibited freezing sensitivity. The *nahG siz1* and *nahG acd6* mutants, in which the accumulation of SA was suppressed, have similar survival rates to those of the wild type (Tsutsui et al. 2009; Miura and Ohta 2010). Similarly, the overexpression of OsWRKY13 in rice enhanced disease resistance but suppressed cold tolerance (Qiu et al. 2008). The biochemical properties of *Arabidopsis* ICS1, which synthesizes isochorismate, a precursor of SA, indicates that the activity of ICS1 was, remarkably, still active at 4 °C (Strawn et al. 2007); thus, SA can be induced by cold treatment. ICE1 is also involved in the regulation of SA accumulation, as is SIZ1 in *Arabidopsis*. SA-inducible genes, such as *PR1* or *PR2*, were upregulated in the *ice1* mutant (Miura and Ohta 2010). ICE1 is also associated with the promoter of BON1-associated protein (*BAP1*), which negatively regulates defense responses (Zhu et al. 2011). The *ice1* mutant exhibits an enhanced resistance to bacterial pathogenesis, as does the *bap1* mutant (Zhu et al. 2010). CAMTA3, a member of the calmodulin-binding transcription activator family, is identical to AtSR1. The CAMTA3/AtSR1 protein recognizes the *CBF2/DREB1C* promoter to regulate cold signaling (Doherty et al. 2009) and also recognizes the *EDS1* promoter to repress its expression and disease resistance (Du et al. 2009). These data suggest that cold signaling and SA signaling are likely to be interrelated.

ICE1/SCRM was suggested to be involved in stomatal differentiation (Kanaoka et al. 2008) and *ICE1/SCRM* and *ICE2/SCRM2* are paralogous genes. The *ice1-2 scrm-2* double mutant exhibited lack of stomata differentiation from epidermal cells, whereas *ice1-1*, a dominant negative mutant, displayed many stomata (Kanaoka et al. 2008). Furthermore, ICE1 interacted with SPCH, MUTE, and FAMA, which are bHLH proteins. Because stomata are formed through a series of differentiation events that are mediated by these bHLH proteins, it is possible that ICE1 may be involved in integration of cold signaling and stomatal development. In general, the MYC-type transcription factors form dimers for transactivation or transrepression: the mammalian MYC-type transcription factors of the Myc/Mad/Max family form dimers in multiple combinations, generating several distinct biological functions (Davis and Halazonetis 1993; Ferre-D'Amare et al. 1993). Thus, it is possible that ICE1 functions in the regulation of stomatal development when it interacts with SPCH, MUTE, and FAMA, whereas the ICE1 protein acts as a positive regulator when it interacts with other MYC-type transcription factors. The identification of ICE1-interacting proteins may provide more detail about the molecular mechanism of cold adaptation.

Concluding Remarks

The ICE1-CBF/DREB1-dependent cold signaling cascade plays a crucial role in cold acclimation in diverse plant species. The genetic engineering of the ICE1-dependent cold signaling pathway can improve cold tolerance in several crops. Thus, the pathway seems to be conserved. Several molecular mechanisms for the regulation of ICE1-dependent cold signaling have been demonstrated: posttranslational modifications and the relationship between cold signaling and the accumulation of SA. Although significant progress has been made to unravel the molecular mechanism of cold acclimation, the cold sensor(s) has not yet been identified. Thus, further studies for the identification of the cold sensor(s) will be necessary to understand how plants sense temperature.

Acknowledgments The laboratory is supported by a Grant-in-Aid for Scientific Research on Innovative Areas from Japanese Ministry of Education, Culture, Sports, Science and Technology on "Environmental Sensing of Plants: Signal Perception, Processing Cellular Responses", Grant-in-Aid for Challenging Exploratory Research, Research Grant from Inamori Foundation, and Corporative Research Grant of the Gene Research Center, the University of Tsukuba.

References

Badawi M, Reddy YV, Agharbaoui Z, Tominaga Y, Danyluk J, Sarhan F, Houde M (2008) Structure and functional analysis of wheat ICE (inducer of CBF expression) genes. Plant Cell Physiol 49:1237–1249

Chinnusamy V, Ohta M, Kanrar S, Lee BH, Hong X, Agarwal M, Zhu JK (2003) ICE1: a regulator of cold-induced transcriptome and freezing tolerance in *Arabidopsis*. Gene Dev 17:1043–1054

Chinnusamy V, Zhu JK, Sunkar R (2010) Gene regulation during cold stress acclimation in plants. Methods Mol Biol 639:39–55

Davis LJ, Halazonetis TD (1993) Both the helix-loop-helix and the leucine zipper motifs of c-Myc contribute to its dimerization specificity with Max. Oncogene 8:125–132

Doherty CJ, Buskirk HA van, Myers SJ, Thomashow MF (2009) Roles for *Arabidopsis* CAMTA transcription factors in cold-regulated gene expression and freezing tolerance. Plant Cell 21:972–984

Dong CH, Agarwal M, Zhang Y, Xie Q, Zhu JK (2006) The negative regulator of plant cold responses, HOS1, is a RING E3 ligase that mediates the ubiquitination and degradation of ICE1. Proc Natl Acad Sci U S A 103:8281–8286

Du L, Ali GS, Simons KA, Hou J, Yang T, Reddy ASN, Poovaiah BW (2009) Ca^{2+}/calmodulin regulates salicylic acid-mediated plant immunity. Nature 457:1154–1158

Ferré-D'Amaré AR, Prendergast GC, Ziff EB, Burley SK (1993) Recognition by max of its cognate DNA through a dimeric b/HLH/Z domain. Nature 363:38–45

Fursova OV, Pogorelko GV, Tarasov VA (2009) Identification of *ICE2*, a gene involved in cold acclimation which determines freezing tolerance in *Arabidopsis thaliana*. Gene 429:98–103

Hara M, Furukawa J, Sato A, Mizoguchi T, Miura K (2011) Abiotic stress and role of salicylic acid in plants. In: Ahmad P, Prasad MNV (eds) Abiotic stress responses in plants: metabolism, productivity, and sustainability. Springer, New York, pp 235–251

Hu Y, Zhang L, Zhao L, Li J, He S, Zhou K, Yang F, Huang M, Jiang L, Li L (2011) Trichostatin A selectively suppresses the cold-induced transcription of the *ZmDREB1* gene in maize. PLoS One:e22132

Janda T, Szalai G, Leskó K, Yordanova R, Apostol S, Popova LP (2007) Factors contributing to enhanced freezing tolerance in wheat during frost hardening in the light. *Phytochemistry* 68:1674–1682

Kanaoka MM, Pillittere LJ, Fujii H, Yoshida Y, Bogenschutz NL, Takabayashi J, Zhu JK, Torii KU (2008) *SCREAM/ICE1* and *SCREAM2* specify three cell-state transitional steps leading to *Arabidopsis* stomatal differentiation. Plant Cell 20:1775–1785

Kang HM, Saltveit ME (2002) Chilling tolerance of maize, cucumber and rice seedling leaves and roots are differentially affected by salicylic acid. Physiol Plantarum 115:571–576

Kosová K, Prášil IT, Vítámvás P, Dobrev P, Motyka V, Floková K, Novák O, Turečková V, Rolčík J, Pešek B, Trávničková A, Gaudinová A, Galiba G, Janda T, Vlasáková E, Prášilová P, Vanková R (2012) Complex phytohormone responses during the cold acclimation of two wheat cultivars differing in cold tolerance, winter Samanta and spring Sandra. J Plant Physiol 169:567–576

Laudencia-Chngcuanco D, Ganeshan S, You F, Fowler B, Chibbar R, Anderson O (2011) Genome-wide gene expression analysis supports a developmental model of low temperature tolerance gene regulation in wheat (*Triticum aestivum L.*). BMC Genomics 12:299

Lee BH, Henderson DA, Zhu JK (2005) The Arabidopsis cold-responsive transcriptome and its regulation by ICE1. Plant Cell 17:3155–3175

Lissarre M, Ohta M, Sato A, Miura K (2010) Cold-responsive gene regulation during cold acclimation in plants. Plant Signal Behav 5:948–952

Liu L, Duan L, Zhang J, Zhang Z, Mi G, Ren H (2010) Cucumber (*Cucumis sativus* L.) over-expressing cold-induced Transcriptome regulator *ICE1* exhibits changed morphological characters and enhances chilling tolerance. Sci Hortic 124:29–33

Liu Q, Kasuga M, Sakuma Y, Abe H, Miura S, Yamaguchi-Shinozaki K, Shinozaki K (1998)Two transcription factors, DREB1 and DREB2, with an EREBP/AP2 DNA binding domain separate two cellular signal transduction pathways in drought- and low-temperature-responsive gene expression, respectively, in Arabidopsis. Plant Cell 10:1391–1406

Miura K, Rus A, Sharkhuu A, Yokoi S, Karthikeyan AS, Raghothama KG, Baek D, Koo YD, Jin JB, Bressan RA, Yun DJ, Hasegawa PM (2005) The *Arabidopsis* SUMO E3 ligase SIZ1 controls phosphate deficiency responses. Proc Natl Acad Sci U S A 102:7760–7765

Miura K, Jin JB, Lee J, Yoo CY, Stirm V, Miura T, Ashworth EN, Bressan RA, Yun DJ, Hasegawa PM (2007) SIZ1-mediated sumoylation of ICE1 controls *CBF3/DREB1A* expression and freezing tolerance in *Arabidopsis*. Plant Cell 19:1403–1414

Miura K, Hasegawa PM (2010) Sumoylation and other ubiquitin-like post-translational modifications in plants. Trends Cell Biol 20:223–232

Miura K, Ohta M (2010) SIZ1, a small ubiquitin-related modifier ligase, controls cold signaling through regulation of salicylic acid accumulation. J Plant Physiol 167:555–560

Miura K, Ohta M, Nakazawa M, Ono M, Hasegawa PM (2011) ICE1 Ser403 is necessary for protein stabilization and regulation of cold signaling and tolerance. Plant J 67:269–279

Miura K, Shiba H, Ohta M, Kang SW, Sato A, Yuasa T, Iwaya-Inoue M, Kamada H, Ezura H (2012a) *SlICE1* encoding a MYC-type transcription factor controls cold tolerance in tomato, *Solanum lycopersicum*. Plant Biotechnol 29:253–260

Miura K, Sato A, Shiba H, Kang SW, Kamada H, Ezura H (2012b) Accumulation of antioxidants and antioxidant activity in tomato, *Solanum lycopersicum*, are enhanced by the transcription factor *SlICE1*. Plant Biotechnol 29:261–269

Mora-Herrera ME, Lopez-Delgado H, Castillo-Morales A, Foyer CH (2005) Salicylic acid and H_2O_2 function by independent pathways in the induction of freezing tolerance in potato. Physiol Plantanum 125:430–440

Nakamura J, Yuasa T, Huong TT, Harano K, Tanaka S, Iwata T, Phan T, Iwaya-Inoue M (2011) Rice homologs of inducer of *CBF* expression (OsICE) are involved in cold acclimation. Plant Biotechnol 28:303–309

Nawrath C, Métraux JP (1999) Salicylic acid induction-deficient mutants of Arabidopsis express PR-2 and PR-5 and accumulate high levels of camalexin after pathogene inoculation. Plant Cell 11:1393–1404

Orvar BL, Sangwan V, Omann F, Dhindsa R (2000) Early steps in cold sensing by plant cells: the role of actin cytoskeleton and membrane fluidity. Plant J 23:785–794

Qiu D, Xiao J, Xie W, Liu H, Li X, Xiong L, Wang S (2008) Rice gene network inferred from expression profiling of plants overexpressing OsWRKY13, a positive regulator of disease resistance. Mol Plant 3:538–551

Scott IM, Clark SM, Wood JE, Mur LAJ (2004) Salicyliate accumulation inhibits growth at chilling temperature in Arabidopsis. Plant Physiol 135:1040–1049

Sangwan V, Foulds I, Singh J, Dhindsa RJ (2001) Cold activation of *Brassica napus* BN115 promoter is mediated by structural changes in membranes and cytoskeleton, and requires Ca^{2+} influx. Plant J 27:1–12

Stockinger EJ, Gilmour SJ, Thomashow MF (1997) *Arabidopsis thaliana CBF1* encodes an AP2 domain-containing transcriptional activator that binds to the C-repeat/DRE, a cis-acting DNA regulatory element that stimulates transcription in response to low temperature and water deficit. 94:1035–1040

Strawn MA, Marr SK, Inoue K, Inada N, Zubieta C, Wildermuth MC (2007) *Arabidopsis* isochorismate synthase functional in pathogen-induced salicylate biosynthesis exhibits properties consistent with a role in diverse stress responses. J Biol Chem 282:5919–5933

Tasgin E, Atici O, Nalbantoglu B (2003) Effects of salicylic acid and cold on freezing tolerance in winter wheat leaves. Plant Growth Regul 41:231–236

Tsutsui T, Kato W, Asada Y, Sako K, Sato T, Sonoda Y, Kidokoro S, Yamaguchi-Shinozaki K, Tamaoki M, Arakawa K, Ichikawa T, Nakazawa M, Seki M, Shinozaki K, Matsui M, Ikeda A, Yamaguchi J (2009) DEAR1, a transcriptional repressor of DREB protein that mediates plant defense and freezing stress responses in *Arabidopsis*. J Plant Res 122:633–643

Wan SB, Tian L, Tian RR, Pan QH, Zhan JC, Wen PF, Chen JY, Zhang P, Wang W, Huang WD (2009) Involvement of phospholipase D in the low temperature acclimation-induced thermotolerance in grape berry. Plant Physiol Biochem 47:504–510

Yun KY, Park MR, Mohanty B, Herath V, Xu F, Mauleon R, Wijaya E, Bajic VB, Bruskiewich R, los Reyes BG de (2010) Transcriptional regulatory network triggered by oxidative signals configures the early response mechanisms of japonica rice to chilling stress. BMC Plant Biol 10:16

Zhou MQ, Shen C, Wu LH, Tang KX, Lin J (2011) CBF-dependent signaling pathway: a key responder to low temperature stress in plants. Crit Rev Biotechnol 31:186–192

Zhao ML, Wang JN, Shan W, Fan JG, Kuang JF, Wu KQ, Li XP, Chen WX, He FY, Chen JY, Lu WJ (2012) Induction of jasmonate signalling regulators MaMYC2s and their physical interactions with MaICE1 in methyl jasmonate-induced chilling tolerance in banana fruit. Plant Cell Environ 36(1):30–51. doi:10.1111/j.1365–3040.2012.02551.x

Zhu Y, Yang H, Mang HG, Hua J (2011) Induction of *BAP1* by a moderate decrease in temperature is mediated by *ICE1* in Arabidopsis. Plant Physiol 155:1580–1588

Breeding for Improved Winter Survival in Forage Grasses

Odd Arne Rognli

Introduction

Winter survival of forage grasses is a very complex trait determined by components of abiotic stresses like low temperature/freezing, desiccation, water logging, ice encasement, anoxia, and snow cover, and in addition, biotic stresses caused by low-temperature fungi (Larsen 1994; Alm et al. 2011). Abiotic and biotic stresses may occur individually or more often in combination, which makes selection and breeding for improved winter survival (WS) a daunting task. Perennial plants like forage grasses have the ability to adapt physiologically to winter climates by cold acclimation (CA). CA is the exposure to low, nonfreezing temperatures, typically 0–10 °C and shortened photoperiod, which induce expression of a cascade of cold-responsive genes resulting in major changes in the content and composition of many types of metabolites like carbohydrates, proteins, nucleic acids, amino acids, growth regulators, phospholipids, and fatty acids. CA alters the physiological state of the plant and proper CA improves freezing tolerance (FT) and WS, and also improves tolerance towards low-temperature fungi (Sjøseth 1964; Årsvoll 1977; Larsen 1978; Tronsmo 1985). A prerequisite for proper induction of CA and attainment of FT is adapted plant material. Unadapted plant material is not able to respond properly to environmental cues, i.e., lower temperatures and/or shorter days, which induce the whole-plant physiological switch necessary for surviving winter (reduced leaf extension, accumulation of high-molecular weight sugars, etc.).

Although WS is a complex trait, and no component trait is able to explain variation in WS completely, experience and research have shown that FT is the single component that explains most of the variation in WS (Pulli et al. 1996). In this chapter, the current knowledge about the genetic and environmental control of FT, and how knowledge from modern genomic research can be utilized in breeding for improved WS, will be reviewed.

O. A. Rognli (✉)
Department of Plant and Environmental Sciences, CIGENE,
Norwegian University of Life Sciences, P.O. Box 5003, 1432 Ås, Norway
e-mail: odd-arne.rognli@umb.no

R. Imai et al. (eds.), *Plant and Microbe Adaptations to Cold in a Changing World*,
DOI 10.1007/978-1-4614-8253-6_17,

Methods for Quantifying FT and the Correlation with WS in the Field

Broadly, the methods of breeding for improved FT and WS are of three types: (1) adaptation by natural selection using multisite-year field trials, (2) direct selection using freezing tests of whole plants or plant parts like crowns and stolons in controlled climates (artificial freezing tests), and (3) indirect selection using physiological or biochemical properties, e.g., electrical conductivity/ion leakage (membrane damage) and chlorophyll fluorescence (photosynthetic acclimation).

Field Testing

Evaluation of winter hardiness in the field is efficient since a large number of genotypes can be tested at a relatively low cost. However, the success of field testing depends on whether the winter conditions differentiate genotypes for WS. The winter climate is characterized by stochastic seasonal weather events, e.g., fluctuations from year to year in the autumn temperature and onset of winter, influencing the potential for CA, and predictable events like the recurring seasonal changes in photoperiod. The variable spatial and temporal distribution of snow and ice cover exerts additional unpredictable factors to the ability to use field testing efficiently in breeding. Thus, a good year for field testing might occur rarely even in regions where high levels of cold tolerance is required. Ideally, genotypes should be tested over many years at many sites, which are not possible with the small amounts of seed available for the early generations in a breeding program. Despite all these shortcomings, field testing is still the most preferred method of testing WS among forage grass breeders.

Artificial Freezing Tests

Precise and efficient freezing tests would circumvent the problems associated with field testing and might be applied at all stages of the breeding scheme (screening genetic diversity, identifying elite genotypes, progeny testing in recurrent mass selection or other breeding systems, evaluating progress, and genetic studies).

Classical artificial freezing tests using whole plants or plant parts were first conducted by breeders of fruit trees and winter cereals early in the twentieth century (for reference, see Larsen 1978). The first accounts of the use of artificial freezing tests in breeding of forage grasses are from the 1950s (Wit 1952; Sjøseth 1957, 1964). A range of different methods is being used in artificial freezing tests, and in this chapter, I will only give a brief overview of the most important issues related to breeding. Detailed descriptions of artificial freezing tests for winter wheat have been published by Fowler et al. (1981) and reviewed by Săulescu and Braun

(2001); for forage grasses by Larsen (1978; 1979) and Limin and Fowler (1987). The methods vary in the way plants are established (seed plants, clones, or plants collected from the field) and prepared for testing, in the procedures for hardening and freezing, and in the way freezing injury is assessed. Tests are being conducted both on plant materials naturally hardened and collected from the field, and on artificially hardened materials raised in the green house. Testing of forage grasses can be based on families (half-sibs (HS) or full-sibs (FS)) or clones, depending on the purpose. Most often, clonal materials are tested. Clonal ramets are made in sufficient numbers and of similar size and developmental age, and either tested as whole plants (planted and allowed to establish in trays with standardized soil) or as crowns (trimmed by removing roots and leaves, except the lower leaf sheaths surrounding the apex, and placed in trays on moist sand). Probably, the most important step of the freezing test is the hardening conditions. Artificial hardening tries to imitate the natural conditions during the autumn, i.e., short days (10–12 h) and low, non-freezing temperatures, often a period (2–4 weeks) of prehardening at moderately low temperatures (8–10 °C) is included before the hardening period (2–3 weeks) at temperatures of 1–2 °C and relatively high light intensities (100–150 µmol/m/s). It has been shown that high light intensities during hardening improve the FT (Larsen 1978). The plants or crowns are then subjected to freezing, using programmable freezers or growth rooms by lowering the temperature stepwise (1–2 °C/h) to a series (2 °C intervals) of predetermined test temperatures for LT_{50} (50 % plant kill) determination or to a predetermined temperature in case of scoring of relative regrowth capacity. The plants are kept at the freezing temperatures for a relatively short time, e.g., 24 h, thawed at a rate of 1 °C/h and subsequently planted in soil (crowns) or placed (trays) in the green house under normal growing conditions for testing of regrowth capacity (cf. Pulli et al. 1996). FT is determined once or twice after a period of growth in the greenhouse, either as LT_{50} or as visual rating of relative survival and regrowth (Pulli et al. 1996; Alm et al. 2011). Many researchers prefer natural hardening in the field since it tends to give better FT, however, the climate conditions will vary more than in a standardized growth room test system and therefore will be more difficult to implement in a breeding scheme.

Very high correlations ($r=-0.95$) have been found between LT_{50} and field survival in winter wheat (Fowler et al. 1981) and between relative regrowth scores and field survival ($r=0.7$–0.8) in forage grasses (Larsen 1989). Fowler et al. (1981) found that a 1 °C difference in LT_{50} value was equal to about 30 % difference in field survival of winter wheat; therefore the LT_{50} method is not able to discriminate between small differences between genotypes (Gusta et al. 1982). The short period of freezing employed in LT_{50} tests does not reflect the stress imposed by the long periods of low temperature during the winter. Thus, a prolonged freezing test, which tests the lethal duration time (LD_{50}), i.e., the length of time at a fixed temperature representing an average midwinter temperature, which kills 50 % of the plants, was proposed by Gusta et al. (1997). Significant differences in ranking of genotypes judged by LT_{50} as compared with LD_{50} values have been found among winter wheat and winter oilseed rape cultivars (Gusta et al. 1997; Waalen et al. 2011). Gudleifsson et al. (1986) found that cold hardiness and ice encasement tolerance were weakly correlated ($r=0.36$), indicating that if ice cover is the main component of winter stress, we cannot expect selections based on freezing tests to be informative.

Although there has been considerable use of freezing tests for basic studies of the physiological and genetical bases of cold tolerance and to evaluate breeding methodology for winter hardiness, there is little evidence of effective application of freezing tests in breeding systems. There are several reasons for this: (1) high G × E interaction associated with direct freezing of plants in pots or trays, (2) screening of large numbers of genotypes is not efficient enough, (3) individual plant selection not precise enough, and (4) poor control over several distinct forms of possible freezing stress.

Indirect Physiological Tests

Measurements of electrical conductivity or ion leakage (LT_{50}) has been used for a long time in studies of freezing injuries of plant tissues, mostly in more fundamental plant physiology research (Dexter et al. 1932; Whitlow et al. 1992). These indirect methods have also been used for measuring FT in forage grasses. Yamada et al. (2004) mapped a Quantitative Trait Loci (QTL) for electrical conductivity on top of linkage group (LG) 4 in perennial ryegrass but failed to detect significant QTLs for WS in the field. Xiong et al. (2007) mapped 6 QTLs for FT measured by electrical conductivity or ion leakage, and 20 QTLs for WS in the field in an annual × perennial ryegrass mapping population in two consecutive years. They found a broad-sense heritability of 0.85 for WS in both years, while it was lowest (0.51) for FT measured by electrical conductivity. The correlations between FT and WS were 0.15 and 0.22, which is far too low for electrical conductivity to be useful in routine selection for WS, at least in forage grasses.

The photosynthetic apparatus is an environmental sensor of temperature and light intensity changes during CA, which influence the redox state of photosystem II (PSII; Ensminger et al. 2006). Acclimation to low temperature seems to be a complex interaction between photosynthetic redox state, expression of cold-regulated genes, and sugar signaling. During prolonged exposure to cold, the photosynthetic apparatus acclimate to avoid photoinhibition caused by high PSII excitation pressures by increasing light energy consumption (increased photochemical quenching—qP) or improved light energy dissipation (increased nonphotochemical quenching—NPQ) of chlorophyll fluorescence (Huner et al. 1993). As the changes in PSII excitation pressure influence improve both FT and acclimation of photosynthesis to avoid low-temperature photoinhibition, measurements of apparent quantum yield of PS II (F_v/F_m) as well as qP and NPQ coefficients of chlorophyll fluorescence quenching during exposure to low temperature can be used to quantify the degree of CA and FT.

How well do measurements of F_v/F_m correlate with direct freezing tests and WS in the field? Measuring F_v/F_m is attractive compared to direct tests since it is rapid, more objective, and nondestructive. It should therefore be particularly suitable to rank a large number of genotypes required in a breeding program. Rizza et al. (2001)

found very high correlations (0.60–0.87) between F_v/F_m measured after freezing (−10 to −16 °C) and field survival in a set of spring and winter oats. Measurements taken after 1 or 2 days of recovery were especially useful because larger variation in response were evident among genotypes. Rizza et al. (2011) studied the diversity in response to low temperature in a set of 54 European barley genotypes. They found similar correlations with winter hardiness both for F_v/F_m values of plants at the first-leaf stage hardened for 4 weeks at 3/1 °C day/night temperature and frozen at −14 °C, and LT_{50} of plants at the three-leaf stage hardened for 3 weeks at 3 °C. Large genotype × year interactions for winter hardiness in the field was observed making it very difficult to use indirect screening methods. It seems like F_v/F_m values are better correlated with FT estimated under controlled conditions (LT_{50}) than WS in the field as demonstrated both for winter barley (Rapacz et al. 2008) and triticale (Rapacz et al. 2011). One possible explanation for this is that the hardening conditions in the field varies much more with seasons and years than the standardized hardening conditions used in controlled growth rooms. Thus, the usefulness of this indirect test might depend strongly on the type and stability of the winter climate in the target environments for the breeding.

What is the current status for using chlorophyll fluorescence screening in perennial forage grasses? Acclimation of the photosynthetic apparatus to high light-low temperature conditions is essential for perennial forage grasses to develop winter hardiness and FT (see review by Sandve et al. 2011). Changes of the redox state of PSII during hardening and prehardening are related to the level of low temperature tolerance and changes in the expression of selected genes involved in CA (Ndong et al. 2001; Rudi et al. 2011). Based on several years of field trials with *Festulolium*, it was shown that avoidance of photoinhibition (NPQ) was associated with winter hardiness (Rapacz et al. 2004). Furthermore, genotypes with impaired photosynthetic acclimation were unable to undergo proper CA and increase their FT. However, the correlation between winter hardiness (or FT) and the ability of photosynthetic acclimation was ~0.7, indicating that FT and tolerance to cold-induced photoinhibition are at least partially independent mechanisms (Rapacz et al. 2004). In a backcross population derived from an androgenic dihaploid *Festulolium* (*L. multiflorum* × *F. pratensis* amphiploid), a good relationship was found between increased NPQ and FT following CA (Humphreys et al. 2007). The authors argued that the intensity of NPQ observed after CA could serve as a physiological marker of FT in *Lolium* × *Festuca* hybrids. We are currently comparing populations selected by direct freezing tests (whole plants) or chlorophyll fluorescence for two generations in high and low direction for FT from a *Festulolium* population (FuRs0357) in field experiments at three locations in Norway. The idea is to test how the two methods of screening for FT correlate with the WS in the field. Preliminary results indicate that selection for higher frost tolerance using the freezing test has led to earlier autumn growth cessation than selection based on F_v/F_m. This is interesting since a major reason for inferior winter hardiness in *L. perenne* and *Festulolium*, at least at higher latitudes in the Nordic countries, is the late growth cessation. These species are not properly adapted and do not sense the shorter photoperiods and lower temperatures, which are environmental cues indicating onset of winters.

Genetics of Winter Hardiness/FT

Genetic Variation and Heritability

The first study of the genetic control of winter hardiness was conducted in winter wheat by Nilsson-Ehle in Sweden in 1913 (Nilsson-Ehle 1913). He observed transgressive segregation in crosses between two winter wheat varieties, which were intermediate in winter hardiness and concluded that winter hardiness was a quantitative trait controlled by many genes. Quite a few studies of the genetic control winter hardiness and FT have been conducted in winter cereals; much less effort has been devoted to forage grasses. The results of FT studied in diallel crosses of winter wheat were summarized by Larsen (1994) as follows: both additive and nonadditive gene action present; nonadditive gene action only due to incomplete dominance, detected both in high and low direction; additive gene action much more important than dominance; and both general combining ability (GCA) and specific combining ability (SCA) significant. The heritabilities in the studies of FT in winter wheat were very high, in the range of 0.7–0.9. Freezing tests and selection for FT in forage grasses were initiated in Norway in the 1960s (Sjøseth 1964). He crossed timothy (*Phleum pratense*) genotypes of North- and South-Norwegian origin and observed large variation and transgressive segregation among F_1 families and also performed mass selections in meadow fescue (*Festuca pratensis*) and perennial ryegrass for ice cover resistance and FT, respectively, which gave positive results. Larsen (1979) performed an extensive series of selections for high and low FT based on artificial freezing tests in orchardgrass (*Dactylis glomerata*), meadow fescue (2x and 4x), and perennial ryegrass. Selection for one generation gave significant responses both in high and low direction. Broad sense heritabilities for FT were relatively high (0.55–0.74) while the realized heritabilities after one generation of selection were 0.3–0.5 for selection in high direction and in general lower (0.13–0.42) in low direction. Similar results were obtained for divergent selection for FT within an F_1 mapping family (B14/16 × HF2/7) of meadow fescue (Rudi et al. 2011). Realized heritabilities were very high (0.89) for selection in high direction and much lower (0.31) for selection in low direction. Tronsmo (1993) studied FT and resistance against pink (*Microdochium nivale*) and gray (*Typhula ishikariensis*) snow mould in HS families of orchardgrass. Broad sense heritabilities for resistance to snow mould were in the range 0.4–0.5 and 0.5–0.7 for FT, while the genotypic correlations between FT and snow mould resistance were very high indicating pleiotropic effects of the same genes. More genetic variation and higher broad-sense heritabilities for resistance to low-temperature fungi in perennial ryegrass were observed in diploid (local Norwegian population "Kleppe") than in tetraploid material generated from "Kleppe" (Larsen and Tronsmo 1991).

The relatively high realized heritabilities obtained in the selection experiments with forage grass species indicate that, similar to the results obtained with winter wheat, FT is a typical quantitative trait governed mainly by additive gene action. In theory, good progress by selection should therefore be possible.

Breeding by Phenotypic Selection

The range of variation will always be important for discrimination between a large numbers of genotypes in a breeding program. Limin and Fowler (1987) tested 35 western Canadian forage grass cultivars from 18 species sown in the spring and in the fall, and compared FT (artificial freezing tests, LT_{50}) with autumn sown most hardy winter wheat "Norstar" and "Puma" rye. They found that early established forage grasses had LT_{50} values from −22 to about –39 °C, while late-sown (seedlings) were less hardy (−17 to −30 °C). Autumn-sown "Norstar" and "Puma" reached LT_{50} values of −24 and –31 °C, respectively. The actual level of FT (LT_{50}) depends on the hardening; if the conditions during hardening are not optimal and the duration not sufficiently long, the LT_{50} values will be much higher than those mentioned before. In general, hardening under natural conditions will give a lower LT_{50} value then under controlled conditions. This is most probably due to the joint effect of several factors like average temperature and day/night fluctuations, duration of the hardening, irradiance, light quality, etc. The question is: what are the optimal conditions for discriminating between genotypes in a breeding program? This depends on the genetic variation for FT in the plant materials in question. For adapted materials, the differences in LT_{50} values will be small; thus, the experimental error needs to be very low to optimize heritability. This is a challenge with freezing tests, especially with heterozygous and segregating plant materials typical of forage grasses. Freezing tests with clonal materials lower the experimental error and increase the heritability, but they are only realistic for research purposes and not for practical breeding since the number of genotypes that can be screened are far too small. Brule-Babel and Fowler (1988) only found transgressive segregants in crosses between relatively nonhardy genotypes of winter wheat and lowest heritability for crosses between cultivars with high FT. For exotic crosses of genotypes with larger differences in FT, the variation will be large and often transgressive, as demonstrated in the F_1 mapping family (B14/16 × HF2/7) of meadow fescue by Alm et al. (2011).

QTL for WS Traits and Prospects of Marker-Assisted Selection (MAS)

Few QTLs for frost tolerance and WS have been mapped in forage grasses. Alm et al. (2011) mapped frost (FT) and drought tolerance (DT) QTLs using regrowth tests, and QTLs for WS based on field survival, in the "B14/16 × HF2/7" FS family of meadow fescue (*Festuca pratensis*). QTLs for several stress tolerance traits mapped to the same regions on *Festuca* chromosomes 1F, 4F, and 5F. Frost and drought both induce cell dehydration, which induces many stress-responsive genes via the DREB1A/CBF transcription factors. Thus, one interpretation of these results is that WS QTLs colocated with DT QTLs indicate dehydration as the main winter stress, while FT and WS QTL colocation indicates freezing stress as most important.

Some WS QTLs did not colocate with any component stress factors and were most probably caused by genes affecting seasonal adaptation, e.g., photoperiodic sensitivity, since the mapping family was an exotic cross between parents from Norway and the former Yugoslavia (Alm et al. 2011). Two FT/WS QTLs on chromosome 5F most likely correspond to major frost tolerance genes/QTLs Fr-A1 (Galiba et al. 1995) and Fr-A2 (Vágújfalvi et al. 2003) on wheat homoeologous group 5A, while a small QTL for FT on chromosome 4F colocated with the FpVRN1 gene, an ortholog of the wheat VRN1 gene (Ergon et al. 2006), most likely is a pleiotropic effect of vernalization and/or photoperiodic genes as in wheat (Limin and Fowler 2002; Dhillon et al. 2010). Multiple CBF genes are located on Triticeae group 5 chromosomes in the same region as the Fr-A2/H2 frost resistance locus (Francia et al. 2007; Skinner et al. 2006; Miller et al. 2006) and in a syntenic region on perennial ryegrass LG5 (Tamura and Yamada 2007). In meadow fescue, the CBF gene Fp-CBF6, which is induced and has peak expression 2 h after start of CA, is colocated with the major FT/WS QTL, which most likely is orthologous to the Fr-A2/H2 frost resistance QTLs in wheat and barley (Alm et al. 2011). In perennial ryegrass, a QTL for electrical conductivity was detected on LG4, which might correspond to the vernalization/FT (VRN1/Fr1) region on homoeologous chromosomes 5 of wheat. Xiong et al. (2007) mapped QTLs for FT, WS, and fall growth (FG) in an annual (*L. multiflorum*) × perennial (*L. perenne*) ryegrass interspecific hybrid population and found common QTLs between FG, FT, and WS. WS QTLs on LG 4 and LG 5, and a QTL for FT on LG 5, were consistently identified over years and based on the map locations they are probably corresponding to the same FT/WS QTLs as present on Triticeae group 5 chromosomes, like the situation in meadow fescue.

The coincident location of several of the QTLs in *Festuca* and *Lolium* with QTLs and genes in Triticeae species indicates the action of structural or regulatory genes that are conserved across evolutionarily distant species. In this respect, CBF-transcription factors and dehydrin genes regulating the expression of cold- and drought-regulated genes, and the vernalization response genes appear to play decisive roles. Whether the conserved FT/WS QTLs on chromosome 5 of cereals and forage grasses are useful in MAS is questionable. Firstly, the QTLs need to be confirmed in unrelated plant material; so far they have been detected only in mapping populations. Secondly, such major conserved QTLs will probably be fixed through adaptation by natural selection rather quickly and thus show very little variation in adapted material. Thirdly, much more detailed information about the genes residing within the QTLs and their function is needed in order to develop MAS strategies. Also, the effect of VRN1 and other vernalization genes on initiation of hardening and development of FT need to be better understood. This is an active field of research in winter cereals (e.g., Dhillon et al. 2010) and it is expected that knowledge obtained in winter cereals will be transferable to forage grasses and help develop molecular breeding tools.

Studies of transcriptional and metabolomics differential changes during CA have identified a number of potential candidate genes for FT, such as ice recrystallization inhibition protein (IRIP), fructan, tumor suppressor (QM), and triose phosphate/phosphate translocator (TPT) genes (see Rudi et al. 2011; Sandve et al. 2011).

The likelihood that allelic variation in a few candidate genes will be able to explain WS in the field well enough to be used as functional markers in selection is very low. A complex and costly phenotype trait like WS might be a good candidate for application of genomic selection (GS) in the future. GS is being developed for several crop species; also forage grasses (Hayes et al. in press). A major challenge in forage grasses is the extreme outcrossing habit, which leads to very low linkage disequilibrium (LD) and the need for very high marker density to obtain sufficient resolution. Hopefully, the very rapid development of high-throughput, low-cost sequencing technology will make it possible to implement GS in forage grass breeding schemes.

Breeding for WS Under Climate Change Conditions

At higher latitudes, the predicted effects of the current climate changes are increased temperature, longer growth seasons, higher precipitation especially in the autumn, and more irregular winter climate with periods of thawing and freezing not only in coastal but also in more continental regions. These rapid changes are challenging for perennial plants like forage grasses. We need to breed new cultivar types and adapt new species, which combine utilization of the longer growth season with sufficient winter hardiness. This combination is currently not present in the cultivars adapted at higher latitudes; they are sensitive to photoperiod and cease growth early in the autumn to prepare for winter. Thus, it is necessary to increase the genetic variation by introgression of exotic materials into adapted species like timothy, and improve adaptation in species like perennial ryegrass and *Festulolium*.

The rate and extent of dehardening is a critical factor in WS. Dehardening resistance and rehardening capacity after warm spells during winter are thus important breeding objectives especially under climate change conditions. Dehardening is a very complex process influenced by changes in content and composition of carbohydrates, membrane lipids, proteins, and antioxidants, and dependent on photosynthesis and gene expression patterns (Kalberer et al. 2006). Dehardening may be reversible or irreversible; if elongation growth has started as a result of high temperature, long days, or completed vernalization, the process is irreversible. Increase in temperature is an absolute requirement for full dehardening (Gay and Eagles 1991) and day temperatures, which promote growth, might trigger irreversible dehardening. Cultivar-dependent photoperiodic effects on deacclimation have been demonstrated in timothy (Eagles 1994). Vernalized plants are particularly sensitive to deacclimation, which may trigger loss of frost resistance even under cold-acclimating conditions (Fowler et al. 1996). However, deacclimation under short days prevents transition to reproductive phase (Mahfoozi et al. 2001). Increased ability to reharden after mild spells may be a supplementary or alternative plant strategy to avoid frost injury under fluctuating temperature conditions (Kalberer et al. 2007).

Whether genetic variation in dehardening resistance and rehardening capacity is present within our forage grass germplasms is currently unknown. A recent study found that very winter hardy cultivars of timothy and perennial ryegrass were not more resistant against dehardening than less winter hardy cultivars (Jørgensen et al. 2010). However, much more research is needed on diverse plant materials to conclude on these matters.

References

Alm V, Busso C, Ergon A, Rudi H, Larsen A, Humphreys M, Rognli O (2011) QTL analyses and comparative genetic mapping of frost tolerance, winter survival and drought tolerance in meadow fescue (*Festuca pratensis* Huds.). Theor Appl Genet 123:369–382

Årsvoll K (1977) Effects of hardening, plant age, and development in *Phleum pratense* and *Festuca pratensis* on resistance to snow mould fungi. Meld Norg Landbruks 56:1–14

Brule-Babel AL, Fowler DB (1988) Genetic control of cold hardiness and vernalization requirement in winter wheat. Crop Sci 28:879–884

Dexter ST, Tottingham WE, Graber LF (1932) Investigations of the hardiness of plants by measurement of electrical conductivity. Plant Physiol 7:63–78

Dhillon T, Pearce S, Stockinger E, Distelfeld A, Li C, Knox A, Vashegyi I, Vagujfalvi A, Galiba G, Dubcovsky J (2010) Regulation of freezing tolerance and flowering in temperate cereals: the VRN-1 connection. Plant Physiol 153:1846–1858

Eagles CF (1994) Temperature, photoperiod and dehardening of forage grasses and legumes. In: Dorffling K, Brettschneider B, Tantau H, Pithan K (eds) Crop adaptation to cool climates. Report of Cost 814 European Commission workshop, Hamburg 12th–14th Oct 1994

Ensminger I, Busch F, Huner NPA (2006) Photostasis and cold acclimation: sensing low temperature through photosynthesis. Physiol Plant 126:28–44

Ergon Å, Fang C, Jørgensen Ø, Aamlid TS, Rognli OA (2006) Quantitative trait loci controlling vernalisation requirement, heading time, and number of panicles in meadow fescue (*Festuca pratensis* Huds.). Theor Appl Genet 112:232–242

Fowler DB, Gusta LV, Tyler NJ (1981) Selection for winter hardiness in wheat. III. Screening methods. Crop Sci 21:896–901

Fowler DB, Limin AE, Wang SY, Ward RW (1996) Relationship between low-temperature tolerance and vernalization response in wheat and rye. Can J Plant Sci 76:37–42

Francia E, Barabaschi D, Tondelli A, Laido G, Rizza F, Stanca A, Busconi M, Fogher C, Stockinger E, Pecchioni N (2007) Fine mapping of a HvCBF gene cluster at the frost resistance locus Fr-H2 in barley. Theor Appl Genet 115:1083–1091

Galiba G, Quarrie S, Sutka J, Morgounov A, Snape J (1995) RFLP mapping of the vernalization (Vrn1) and frost resistance (Fr1) genes on chromosome 5A of wheat. Theor Appl Genet 90:1174–1179

Gay AP, Eagles CF (1991) Quantitative analysis of cold hardening and dehardening in *Lolium*. Ann Bot (Lond) 67:339–345

Gudleifsson BE, Andrews CJ, Bjørnsson H (1986) Cold hardiness and ice tolerance o pasture grasses grown and tested in controlled environments. Can J Plant Sci 66:601–608

Gusta LV, Fowler DB, Tyler NJ (1982) Factors influencing hardening and survival in winter wheat. In: Sakai A, Li PH (eds) Plant cold hardiness and freezing stress mechanisms and crop implications. Academic Press, New York

Gusta LV, O'Connor BJ, MacHutcheon MG (1997) The selection of superior winter-hardy genotypes using a prolonged freeze test. Can J Plant Sci 77:15–21

Humphreys MW, Gasior D, Lesniewska-Bocianowska A, Zwierzykowski Z, Rapacz M (2007) Androgenesis as a means of dissecting complex genetic and physiological controls: selecting useful gene combinations for breeding freezing tolerant grasses. Euphytica 158:337–345

Huner NPA, Öquist G, Hurry VM, Krol M, Falk S, Griffith M (1993) Photosynthesis, photoinhibition and low temperature acclimation in cold tolerant plants. Photosynth Res 37:19–39

Jørgensen M, Østrem L, Höglind M (2010) De-hardening in contrasting cultivars of timothy and perennial ryegrass during winter and spring. Grass Forage Sci 65:38–48
Kalberer SR, Arora R, Leyva-Estrada N, Krebs SL (2007) Cold hardiness of floral buds of deciduous azaleas: dehardening, rehardening, and endodormancy in late winter. J Am Soc Hortic Sci 132:73–79
Kalberer SR, Wisniewski M, Arora R (2006) Deacclimation and reacclimation of cold-hardy plants: current understanding and emerging concepts. Plant Sci 171:3–16
Larsen A (1978) Freezing tolerance in grasses: methods for testing in controlled environments. Meld Norg Landbruks 57(23):1–56
Larsen A (1979) Freezing tolerance in grasses: variation within populations and response to selection. Meld Norg Landbruks 58(42):1–28
Larsen A (1989) Foredling for overvintringsevne hos engvekster [Breeding for winter survival in forage crops]. Norsk Landbruksforskning [Norwegian Agricultural Research], Supplement no. 5, 1989, pp 75–80
Larsen A, Tronsmo AM (1991) Selection for resistance against low-temperature fungi in perennial ryegrass (*Lolium perenne* L.) (in Norwegian). Nordisk Jordbruksforskning 73:516
Larsen A (1994) Breeding winter hardy grasses. Euphytica 77:231–237
Limin AE, Fowler DB (1987) Cold hardiness of forage grasses grown on the Canadian prairies. Can J Plant Sci 67:1111–1115
Limin AE, Fowler DB (2002) Developmental traits affecting low-temperature tolerance response in near-isogenic lines for the vernalization locus Vrn-A1 in wheat (*Triticum aestivum* l. em Thell). Ann Bot (Lond) 89:579–585
Mahfoozi S, Limin AE, Fowler DB (2001) Influence of vernalization and photoperiod responses on cold hardiness in winter cereals. Crop Sci 41:1006–1011
Miller A, Galiba G, Dubcovsky J (2006) A cluster of 11 CBF transcription factors is located at the frost tolerance locus Fr-Am2 in *Triticum monococcum*. Mol Genet Genomics 275:193–203
Ndong C, Danyluk J, Huner NPA, Sarhan F (2001) Survey of gene expression in winter rye during changes in growth temperature, irradiance or excitation pressure. Plant Mol Biol 45:691–703
Nilsson-Ehle H (1913) Zur kenntnis der erblichkeitsverhältnisse der eigenschaft winterfestigkeit beim weizen. Z Pflanzenzucht 1:3–12
Pulli S, Hjortsholm K, Larsen A, Gudleifsson B, Larsson S, Kristiansson B, Hömmö L, Tronsmo AM, Ruuth P, Kristensson C (1996) Development and evaluation of laboratory testing methods for winter hardiness breeding. Nordic Gene Bank 32:1–68
Rapacz M, Gasior D, Zwierzykowski Z, Lesniewska-Bocianowska A, Humphreys MW, Gay AP (2004) Changes in cold tolerance and the mechanisms of acclimation of photosystem II to cold hardening generated by another culture of *Festuca pratensis* × *Lolium multiflorum* cultivars. New Phytol 162:105–114
Rapacz M, Sasal M, Gut M (2011) Chlorophyll fluorescence-based studies of frost damage and the tolerance for cold-induced photoinhibition in freezing tolerance analysis of Triticale (× *Triticosecale* Wittmack). J Agron Crop Sci 197:378–389
Rapacz M, Tyrka M, Kaczmarek W, Gut M, Wolanin B, Mikulski W (2008) Photosynthetic acclimation to cold as a potential physiological marker of winter barley freezing tolerance assessed under variable winter environment. J Agron Crop Sci 194:61–71
Rizza F, Pagani D, Gut M, Prášil IT, Lago C, Tondelli A, Orrù L, Mazzucotelli E, Francia E, Badeck F-W, Crosatti C, Terzi V, Cattivelli L, Stanca AM (2011) Diversity in the response to low temperature in representative barley genotypes cultivated in Europe. Crop Sci 51:2759–2779
Rizza F, Pagani D, Stanca AM, Cattivelli L (2001) Use of chlorophyll fluorescence to evaluate the cold acclimation and freezing tolerance of winter and spring oats. Plant Breeding 120:389–396
Rudi H, Sandve SR, Opseth LM, Larsen A, Rognli OA (2011) Identification of candidate genes important for frost tolerance in *Festuca pratensis* Huds. by transcriptional profiling. Plant Sci 180:78–85
Sandve SR, Kosmala A, Rudi H, Fjellheim S, Rapacz M, Yamada T, Rognli OA (2011) Molecular mechanisms underlying frost tolerance in perennial grasses adapted to cold climates. Plant Sci 180:69–77

Săulescu NN, Braun H-J (2001) Cold tolerance. In: Reynolds MP, Ortiz-Monasterio JI, McNab A (eds) Application of physiology in wheat breeding. CIMMYT, Mexico

Sjøseth H (1957) Undersøkelser over frostherdighet hos engvekster. Forskn Fors Landbr 8:77–98 [In Norwegian]

Sjøseth H (1964) Studies on frost hardening in plants. Acta Agric Scand 14:178–192

Skinner J, Szucs P, Zitzewitz J von, Marquez-Cedillo L, Filichkin T, Stockinger E, Thomashow M, Chen T, Hayes P (2006) Mapping of barley homologs to genes that regulate low temperature tolerance in Arabidopsis. Theor Appl Genet 112:832–842

Tamura K, Yamada T (2007) A perennial ryegrass CBF gene cluster is located in a region predicted by conserved synteny between Poaceae species. Theor Appl Genet 114:273–283

Tronsmo AM (1985) Induced resistance to biotic stress factors in grasses by frost hardening. In: Kaurin Å, Junttila O, Nilsen J (eds) Plant production in the north. Universitetsforlaget, Oslo, pp 127–133

Tronsmo AM (1993) Resistance to winter stress factors in half-sib families of *Dactylis glomerata*, tested in a controlled environment. Acta Agric Scand Sect B Soil Plant Sci 43:89–96

Vagujfalvi A, Galiba G, Cattivelli L, Dubcovsky J (2003) The cold-regulated transcriptional activator Cbf3 is linked to the frost-tolerance locus Fr-A2 on wheat chromosome 5A. Mol Genet Genomics 269:60–67

Waalen W, Tanino KK, Olsen JE, Eltun R, Rognli OA, Gusta LV (2011) Freezing tolerance of winter canola cultivars is best revealed by a prolonged freeze test. Crop Sci 51:1988–1996

Wit F (1952) Techniques of breeding cold-resistant grasses and clovers. Proc 6th Int Grassld Congr, USA, pp 1607–1612

Whitlow TH, Bassuk NL, Ranney TG, Reichert DL (1992) An improved method for using electrolyte leakage to assess membrane competence in plant tissues. Plant Physiol 98:198–205

Xiong YW, Fei SZ, Arora R, Brummer EC, Barker RE, Jung GW, Warnke SE (2007) Identification of quantitative trait loci controlling winter hardiness in an annual x perennial ryegrass interspecific hybrid population. Mol Breeding 19:125–136

Yamada T, Jones ES, Cogan NOI, Vecchies AC, Nomura T, Hisano H, Shimamoto Y, Smith KF, Hayward MD, Forster JW (2004) QTL analysis of morphological, developmental, and winter hardiness-associated traits in perennial ryegrass. Crop Sci 44:925–935

Molecular Changes in Recurrently Selected Populations of Forage Legumes

Annick Bertrand and Yves Castonguay

Introduction

The perennial growth habit of forage legumes combined with their capacity to fix atmospheric N_2 symbiotically markedly contributes to long-term sustainability of agriculture through reduced reliance on fossil fuel consumption, improved soil fertility and structure, and by providing a renewable source of dietary protein for ruminants. Alfalfa (*Medicago sativa* spp. *sativa* L.) is the most widely grown temperate forage legume with over 32 million hectares worldwide (Russelle 2001). Lack of winter hardiness constitutes a major limitation to alfalfa persistence in areas experiencing harsh winter conditions. Red clover (*Trifolium pratense* L) is another leading forage legume in Canada, USA, and northern and eastern Europe. It is distinguished by a rapid spring establishment and it performs better than alfalfa on acid and wet soils (Bosworth and Stringer 2012). Because of its poor winter hardiness, red clover is, however, considered as a short-lived perennial. Even though alfalfa possesses a higher inherent level of freezing tolerance than red clover as shown by a lower level of winterkill for alfalfa than for red clover in the field following a prolonged exposure to subfreezing temperature (Suzuki 1972), both these perennial species lack sufficient winter hardiness.

The nutrient composition of alfalfa and red clover including crude protein (16–22 %), acid detergent fiber (25–35 %), neutral detergent fiber (35–45 %), and mineral concentrations are similar (Hoffman and Broderick 2001). Nevertheless, differences in nutritional value have been documented for harvested forage from both species. The high content of polyphenol oxidases in red clover reduces proteolysis in silage and in the rumen due to complexing of o-quinones with dietary protein (Eickler et al. 2011). This results in a higher proportion of rumen escape proteins and more efficient N utilization by the animal. In addition, red clover fiber may be more digestible providing more energy to lactating dairy cows than alfalfa

A. Bertrand (✉) · Y. Castonguay
Soils and Crops Research and Development Centre, Agriculture and Agri-Food Canada, 2560 Hochelaga Boulevard, Québec City, QC G1V 2J3, Canada
e-mail: annick.bertrand@agr.gc.ca

R. Imai et al. (eds.), *Plant and Microbe Adaptations to Cold in a Changing World*,
DOI 10.1007/978-1-4614-8253-6_18,

(Hoffman and Broderick 2001). Feeding a combination of red clover and alfalfa that were grown and ensiled together reduced nonprotein nitrogen in the silage and had positive effects on feed intake and milk production of dairy cows. Thus, the two species possess different attributes that make them desirable for feeding ruminants and the choice will be guided by considerations such as soil properties, management, and particularly their adaptation to the environment.

Recurrent Selection for the Improvement of Freezing Tolerance in Forage Legumes

Freezing tolerance is the single most important factor that determines the survival of alfalfa and red clover to severe winter conditions. Field selection for freezing tolerance is a long and costly process due to the unpredictability of test winters. In that context, we applied a recurrent selection scheme entirely performed indoor in growth chambers and walk-in freezers to identify genotypes with superior freezing tolerance. In each cycle of selection, three successive freezing stresses were applied to progressively eliminate 90 % of the population and to retain only the 10 % more cold-tolerant genotypes (Fig. 1). Assessment of populations putatively more tolerant to freezing (TF) obtained by intercrossing selected genotypes revealed a marked increase in freezing tolerance of alfalfa and red clover populations obtained after repeated cycles of selection (Fig. 2a–c). Plants of alfalfa acclimated to natural hardening conditions in a unheated greenhouse and sampled midwinter showed a 4.0–5.0 °C increase in freezing tolerance after four to six cycles of selection (Fig. 2b). This remarkable increase in freezing tolerance translated into improvement in persistence and yield of alfalfa exposed to harsh winter conditions in the field (Castonguay et al. 2009). Preliminary analyses indicated a 2–3 °C increase in freezing tolerance in red clover after only four cycles of recurrent selection (Bertrand et al. in preparation). Field trials are currently underway to confirm the impact on crop productivity and persistence.

Biochemical Changes Linked to Freezing Tolerance in Alfalfa

The characterization of cold-induced molecular changes in overwintering crowns showed that improvement in freezing tolerance in alfalfa TF populations was associated with differences in traits typically associated with the cold acclimation process (Castonguay et al. 2011). Interestingly, some of these responses were consistently observed in selections performed in two unrelated cultivars: Apica and Evolution. For instance, starch concentration decreased during cold acclimation and throughout the overwintering period from around 200 mg/g in nonacclimated to 50 mg/g in fully acclimated plants. Concomitant to starch hydrolysis, there was a

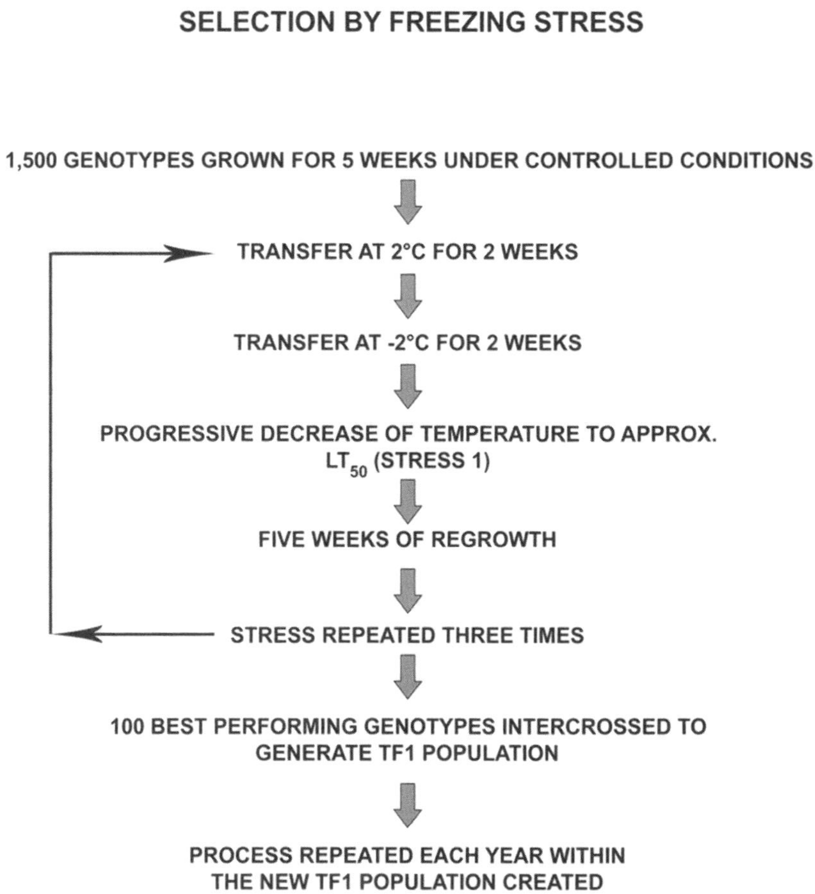

Fig. 1 Schematic illustration of the procedure used for the selection of forage legumes with improved freezing tolerance. The recurrent selection by three successive freezing stresses is conducted entirely under environmentally controlled conditions. The target temperature of each freezing stress is the lethal temperature for 50% of the plants. The first cycle reduces the number of plants from 1,500 to around 750; the second cycle reduces this number to around 375 and, from the last cycle, the 100 best performing genotypes are selected to be intercrossed and generate a TF population. (Source: Castonguay et al. 2009, Reprinted by Permission, ASA, CSSA, SSSA)

large accumulation of soluble sugars in crowns of alfalfa (Castonguay et al. 2011). The concentrations of cryoprotective sugars (sucrose, stachyose, and raffinose) were generally higher in fully acclimated plants of advanced cycles of selection (TF5 and TF6 for Apica and TF3 and TF4 for Evolution) as compared to the initial TF0 populations (Fig. 3a, b). Soluble sugars and other compatible metabolites increase cytoplasmic viscosity at very high concentrations in freeze-desiccated cells, allowing molecules to be locked in an amorphous glassy matrix. This process slows down further dehydration and reduces deleterious interactions between reactive metabolites and ultrastructural cell constituents like proteins and membranes. Sucrose

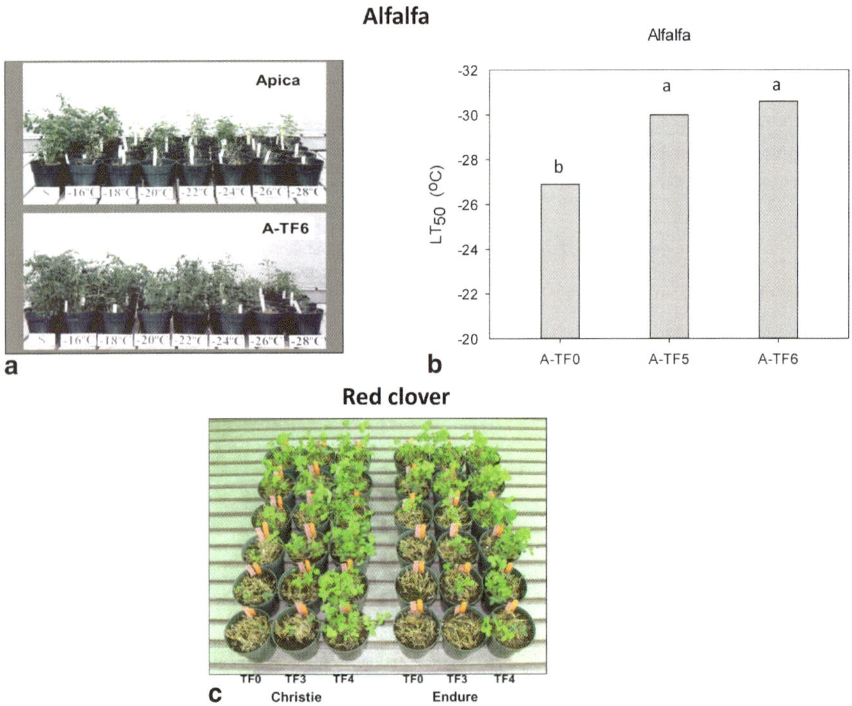

Fig. 2 **a** Visual assessment of the survival of alfalfa exposed to −2 °C (S) or to freezing temperature decreasing from −16 to −28 °C in the original genetic background (cv. Apica) and in population A-TF6 obtained after six cycles of selection. **b** Freezing tolerance assessed as lethal temperature for 50 % of the plants (LT_{50}) for alfalfa in the original genetic background (cv. Apica, A-TF0) and in populations A-TF5 and A-TF6 obtained after five and six cycles of selection, respectively. Freezing tolerance was measured in January after cold acclimation under natural conditions in an unheated greenhouse (Source: Castonguay et al. 2011, Reprinted by Permission, ASA, CSSA, SSSA). **c** Visual assessment of survival after exposure to −18 °C for populations TF3 and TF4 respectively obtained after three and four cycles of selection within two initial backgrounds (TF0) of red clover (cv. Christie and Endure)

and the oligosaccharides raffinose and stachyose are thought to be major promoters of cytoplasmic vitrification (Strimbeck and Schaberg 2009) and their higher concentration in advanced cycles of selection could partly contribute to their superior freezing tolerance.

We also observed an increase in the concentration of free amino acids during cold acclimation of alfalfa TF populations. Total amino acid concentration reached a maximum in January when plants were fully acclimated and was greater in recurrently selected populations than in the initial backgrounds (Fig. 3c, d). These results support the determinant role of N reserves in winter stress tolerance and in the vigor of spring regrowth of alfalfa that has been previously highlighted by the work of Volenec et al. (1996) and Dhont et al. (2003). Concentrations of proline, asparagine, and arginine were significantly higher in recurrently selected populations than in the initial backgrounds (Castonguay et al. 2011). Proline is a major solute which accumulates in plants in response to environmental stress and its superior accumulation

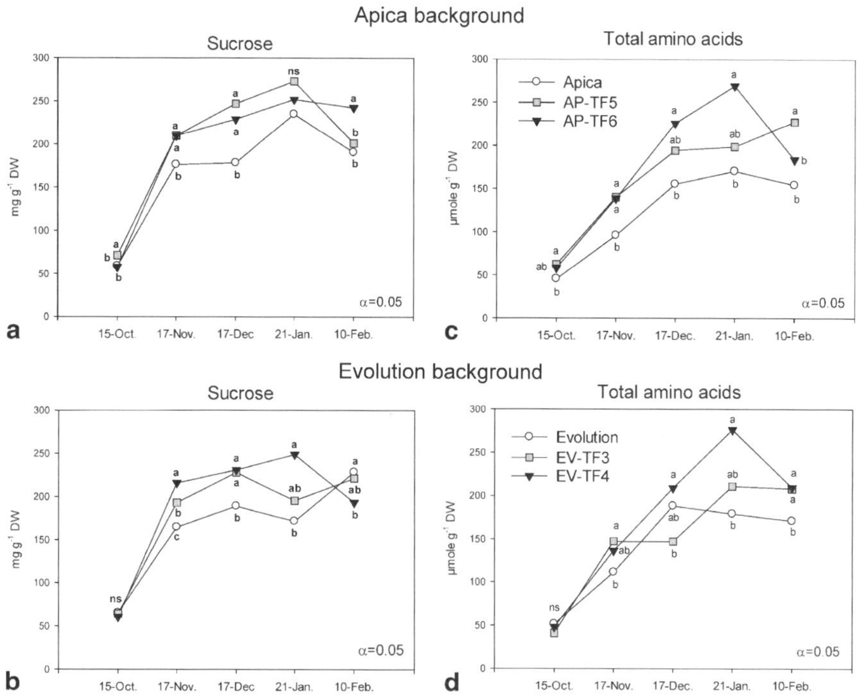

Fig. 3 Sucrose (**a**, **b**) and total amino acid (**c**, **d**) concentrations in crowns of alfalfa plants from two cultivars (Apica and Evolution) and from two advanced cycles of recurrent selection for improved tolerance to freezing performed within these two cultivars (AP-TF5, AP-TF6 and EV-TF3 and EV-TF4). Plants were hardened under natural variations of light and temperature in an unheated greenhouse during fall and winter. Significance was declared at $p<0.05$, $n=5$. (Source: Castonguay et al. 2011, Reprinted by Permission, ASA, CSSA, SSSA)

in TF populations further supports the adaptive value of its stress-induced accumulation. In addition, higher levels of proline likely contributed to the reservoir of organic nitrogen and the vigor of regrowth once plants are exposed to warmer temperatures.

Taken together, our assessment of cold-induced biochemical changes in alfalfa TF populations provides compelling evidence that the pressure of selection by freezing stress improved adaptation to cold by modifying the underlying molecular traits typically linked with the acquisition of freezing tolerance.

Changes in Gene Expression in Alfalfa

The observed enhancement of biochemical traits associated with the acquisition of freezing tolerance led us to explore the impact of recurrent selection on the expression of cold-regulated (COR) genes. Transcripts levels of *GaS*, which codes for the enzyme galactinol synthase that catalyzes the first step in the biosynthesis of the cryoprotective oligosaccharides of the raffinose family and of *cas15*, encoding a nuclear-targeted protein (Monroy et al. 1993) were strongly upregulated during acclimation to declining temperatures in the fall (Fig. 4). Interestingly, the marked induction of the expression of these two COR genes was significantly increased in response to selection in Apica. The significant impact of recurrent selection for freezing tolerance on the accumulation of transcripts of the two COR genes could be indicative that adaptive alleles at multiple loci across the genome were subject to selection. Alternatively, it is possible that these two genes belong to a regulon under the control of a common signal transduction pathway that responded to selection.

Transcripts levels were generally higher in populations obtained after five or six cycles of selection as compared to the initial genetic background (TF0) within the cultivar Apica. In the cultivar Evolution, transcript levels did not generally differ between advanced cycles of selection (TF3 and TF4) and the initial (TF0) population. The more mitigated response of the two COR genes in the recurrent selections derived from Evolution than in those from the cultivar Apica could be attributable to the lower number of selection cycles that were performed within this cultivar. It could also be reflective of the fact that enrichment or depletion of some alleles or paralogs through successive rounds of selection may be specific to a given genetic background.

Whatever the underlying causes, this observation underscores the relevance of performing several cycles of selection within multiple genetic backgrounds. Comparative analysis of advanced cycles of selection from distinct genetic backgrounds will help identify common and specific adaptive gene complements. This information will help guide breeding programs in their efforts to create new genotype combinations having a synergistic impact on the freezing tolerance phenotype.

Variation in Alfalfa Genome Associated with Superior Freezing Tolerance

The identification of markers targeting regions of the genome that affect freezing tolerance would greatly assist breeding efforts towards the improvement of this quantitatively inherited trait. In that perspective, analysis of gene diversity among alfalfa and red clover TF populations could facilitate the identification of DNA variations that correlate with the freezing tolerance phenotype.

Random PCR amplification of pooled DNA samples from Apica-TF populations using the sequence-related amplified polymorphism (SRAP) technique identified several polymorphic fragments that varied in intensity in response to selection (Castonguay et al. 2010). Detection of these fragments in individual genotypes confirmed that polymorphisms detected with pools reflect the frequency of their occurrence within populations. The anonymous nature of the SRAP polymorphisms along with the absence of a genome of reference constitutes an enormous challenge for the development of robust markers. The identification of functional variants of candidate genes known to be associated with freezing tolerance would facilitate the development of markers more broadly applicable to diverse genetic backgrounds. Dehydrins belong to a complex family of intrinsically disordered proteins that has been linked to adaptation to low temperature in many

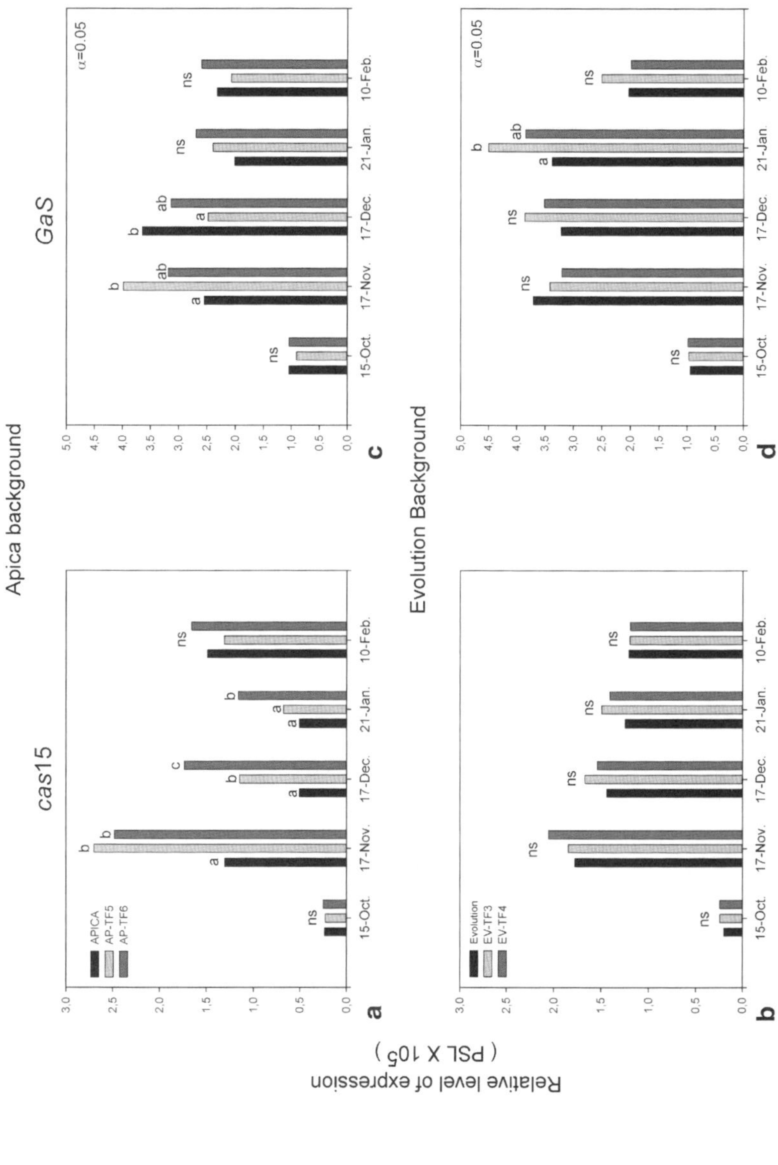

Fig. 4 Relative level of expression of genes *cas*15 (**a**, **b**) and *GaS* (**c**, **d**) in crowns of alfalfa before (15 October) and during fall hardening and overwintering (17 November, 17 December, 21 January, and 10 February). Measurements were made in two cultivars (Apica and Evolution) and in two advanced cycles of recurrent selection for improved tolerance to freezing performed within these two cultivars (AP-TF5, AP-TF6 and EV-TF3 and EV-TF4). Each mean is the average photostimulated luminescence (PSL) of five pots containing 15 plants per pot. (Source: Castonguay et al. 2011, Reprinted by Permission, ASA, CSSA, SSSA)

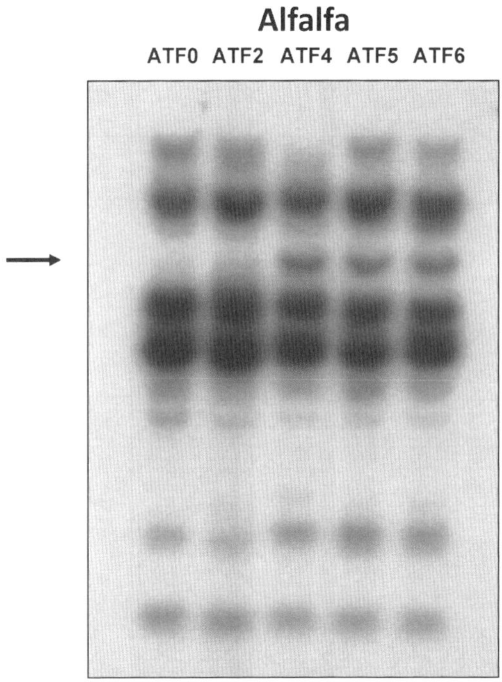

Fig. 5 Southern hybridization of the cold-inducible dehydrin homolog *msa*CIG with DraI-digested pool of DNA from Apica (TF0) and from populations derived from Apica obtained after two (TF2), four (TF4), five (TF5), and six (TF6) cycles of recurrent selection for improved freezing tolerance (45 genotypes for each TF population). The *arrow* indicates polymorphism that progressively intensified with the number of selection cycles. (Source: Remus-Borel et al. 2010, Reprinted with kind permission from Springer Science and Business Media)

plant species (Rorat 2006). Rémus-Borel et al. (2010) uncovered a Y_2K_4 dehydrin restriction fragment length polymorphism (RFLP) that progressively intensifies with the number of selection cycles for freezing tolerance in alfalfa (Fig. 5). Assessment of freezing tolerance of populations derived from crosses between genotypes with (D+) and without (D−) the polymorphic dehydrin confirmed a significant impact on freezing tolerance. This result clearly illustrates the value of recurrent selections for the identification of functional variants of candidate genes affecting traits of interest. In-depth characterization of the specific dehydrin sequence associated with the RFLP led to the identification of a paralog with a characteristic short intron (Castonguay et al. 2012). Increased copy number of the short-intron variant was raised as a possibility for strong amplifications noted in genotypes that were scored positive for the dehydrin RFLP with a demonstrated impact on freezing tolerance.

Conclusion

A recurrent selection protocol entirely performed indoor significantly increased the freezing tolerance of two forage legumes. Extensive analyses of recurrent selections performed within heterogeneous populations of allogamous alfalfa and red clover provided cumulative evidence that superior freezing tolerance is associated with marked changes in the molecular and genetic composition of these species. Our results illustrate that recurrent selections within heterogeneous populations of open-pollinated species are powerful resources for the identification of DNA variants associated with the improvement of quantitative traits. Further research is being pursued to identify and to validate functional polymorphisms in other candidate genes using the alfalfa and red clover populations that were selected for superior tolerance to freezing.

References

Bosworh SC, Stringer WC (2012) Cutting Management of Alfalfa, Red Clover, and Birdsfoot Trefoil. http://pubs.cas.psu.edu/freepubs/pdfs/uc044.pdf. Accessed on July 3, 2013

Castonguay Y, Michaud R, Nadeau P, Bertrand A (2009) An indoor screening method for improvement of freezing tolerance in alfalfa. Crop Sci 49:809–818

Castonguay Y, Cloutier J, Bertrand A, Michaud R, Laberge S (2010) SRAP polymorphisms associated with superior freezing tolerance in alfalfa (*Medicago sativa* spp. *sativa*). Theor Appl Genet 120:1611–1619

Castonguay Y, Bertrand A, Michaud R, Laberge S (2011) Cold-induced biochemical and molecular changes in alfalfa populations selectively improved for freezing tolerance. Crop Sci 51:2132–2144

Castonguay Y, Dubé M-P, Cloutier J, Michaud R, Bertrand A, Laberge S (2012) Intron-length polymorphism identifies a Y_2K_4 dehydrin variant linked to superior freezing tolerance in alfalfa. Theor Appl Genet 124:809–819

Dhont C, Castonguay Y, Nadeau P, Bélanger G, Chalifour F-P (2003) Alfalfa root nitrogen reserves and regrowth potential in response to fall harvest. Crop Sci 43:181–194

Eickler B, Gierus M, Kleen J, Taube F (2011) Specific polyphenol oxidase activity of red clover (*Trifolium pratense*) and its relation with forage quality in field experiments. Acta Agric Scand Sect B Soil Plant Sci 61:39–49

Hoffman PC, Broderick GA (2001) Red clover forages for lactating dairy cows. Focus on forage. 3:1–2. http://www.uwex.edu/ces/crops/uwforage/RedCloverCows.html. Accessed on July 3, 2013.

Monroy AF, Castonguay Y, Laberge S, Sarhan F, Vezina L-P, Dhindsa RS (1993) A new cold-induced alfalfa gene is associated with enhanced hardening at subzero temperature. Plant Physiol 102:873–879

Rémus-Borel W, Castonguay Y, Cloutier J, Michaud R, Bertrand A, Desgagnés R, Laberge S (2010) Dehydrin variants associated with superior freezing tolerance in alfalfa (*Medicago sativa* L.). Theor Appl Genet 120:1163–1174

Rorat T (2006) Plant dehydrins—tissue location, structure and function. Cell Mol Biol Lett 11:536–556

Russelle MP (2001) Alfalfa. American Scientist 89:252–261

Strimbeck GR, Schaberg PG (2009) Going to extremes: low-temperature tolerance and acclimation in temperate and boreal conifers. In: Gusta LV, Wisniewski ME, Tanino KK (eds) Plant cold hardiness. From the laboratory to the field. CABI, London, pp 226–239

Suzuki M (1972) Winterkill patterns of forage crops and winter wheat in P.E.I in 1972. Canadian Plant Disease Survey 52:156–159

Volenec JJ, Ourry A, Joern BC (1996) A role for nitrogen reserves in forage regrowth and stress tolerance. Physiol Plant 97:185–193

Abiotic Stress Signal Network with Expression QTLs for Cold-Responsive Genes in Common Wheat

Julio C. M. Iehisa, Yoichi Motomura, Fuminori Kobayashi and Shigeo Takumi

Introduction

Low temperature may cause serious damage and losses in agricultural productivity of crops. Common wheat (*Triticum aestivium* L.) and its related species are able to grow under wide range of environmental conditions. Wheat and its relatives show broad genetic diversity in the ability to tolerate various abiotic stresses such as low temperature (Fowler and Gusta 1979). Cold acclimation, or hardening is an adaptive process in which the short-term exposure of plants to low, nonfreezing temperatures leads to an increase in freezing stress tolerance. Freezing-tolerant cultivars show greater ability for cold acclimation than susceptible varieties (Sakai and Larcher 1987). Another response to low temperature is vernalization, where long-term exposure leads to acceleration of competence to flowering (Sung and Amasino 2005). Winter wheat cultivars are freezing-tolerant and tend to exhibit requirement for vernalization, whereas spring cultivars are generally freezing-sensitive and do not require vernalization for flowering.

In common wheat, major loci controlling freezing tolerance (*Fr-1*) and vernalization requirement (*Vrn-1*) have been assigned to the long arm of chromosomes 5A (*Fr-A1*/*Vrn-A1*), 5B (*Fr-B1*/*Vrn-B1*), and 5D (*Fr-D1*/*Vrn-D1*; Galiba et al. 1995; Snape et al. 1997; Tóth et al. 2003). Previous studies showed that the *Fr-A1* locus was located 2 cm proximal to the *Vrn-A1* (Galiba et al. 1995), *Fr-B1* and *Vrn-B1* interval was 40 cm (Tóth et al. 2003), and that of *Fr-D1* and *Vrn-D1* was 10 cm (Snape et al. 1997). Physical mapping study using Chinese Spring (CS) deletion lines indicated that *Fr-A1* and *Vrn-A1* loci are closely linked, but physically separated (Sutka et al. 1999). However, a recent study suggests that freezing tolerance is

S. Takumi (✉) · J. C. M. Iehisa · Y. Motomura
Graduate School of Agricultural Science, Kobe University,
1-1 Rokkodai, Nada, Kobe, Hyogo 657-8501, Japan
e-mail: takumi@kobe-u.ac.jp

F. Kobayashi
Plant Genome Research Unit, National Institute of Agrobiological Sciences,
2-1-2 Kannondai,Tsukuba, Ibaraki 305-8602, Japan

R. Imai et al. (eds.), *Plant and Microbe Adaptations to Cold in a Changing World*,
DOI 10.1007/978-1-4614-8253-6_19,

a pleiotropic effect of *Vrn-1* rather than the effect of *Fr-1* (Dhillon et al. 2010). *Fr-1* is either tightly linked to *Vrn-1* or is tightly associated with the pleiotropic effects of *Vrn-1*; chromosomal relationship between *Fr-1* and *Vrn-1* is unknown in wheat and barley (Galiba et al. 1995, 2009; Snape et al. 1997; Casao et al. 2011). On the other hand, a second locus for freezing tolerance, *Fr-A2*, has been identified on a distal region of the long arm of chromosome 5A in einkorn and common wheat (Vágújfalvi et al. 2003, 2005, Båga et al. 2007). The recent advances clarified that the *Fr-B1* locus was orthologous to *Fr-2* rather than *Fr-1* according to the marker position and the location of *Fr-B1* QTL (Catalogue of gene symbols, http://www.shigen.nig.ac.jp/wheat/komugi/). *Fr-2* is coincident with a cluster of genes encoding C-repeat (CRT)-binding factors (CBFs; Miller et al. 2006), which are transcription factors that induce downstream genes during cold acclimation (Thomashow et al. 2001). However, relationship between the *Fr* loci and cold-responsive gene expression patterns still remains unclear in common wheat.

Here, we compare cold stress signaling pathways of *Arabidopsis* and wheat, and discuss the usefulness of analysis at transcript levels, particularly eQTL, in dissection of complex traits such as cold-stress responses in common wheat.

Cold Acclimation in *Arabidopsis thaliana* and Wheat

Cold acclimation is a very complex process and involves drastic physiological, biochemical, and metabolic changes. Most of these alterations are regulated through changes in gene expression. One of the mechanisms behind the development of freezing tolerance is induction of the cold-responsive (*Cor*)/late-embryogenesis-abundant (*Lea*) gene family; COR/LEA proteins promote the development of freezing tolerance by protecting cellular components from stress (Thomashow 1999). Most *Cor/Lea* genes contain CRT/dehydration-responsive element (DRE) and ABA-responsive element (ABRE) motifs in their promoters and these *cis*-acting elements are considered to function independently. In wheat, CRT/DRE, ABRE, and other cold-responsive *cis*-acting elements have been identified in the promoter region of many *Cor/Lea* genes such as *Em*, *Wcor15*, *Wdhn13*, *Wrab17*, *Wrab18*, and *Wrab19* (Marcotte et al. 1989; Takumi et al. 2003; Kobayashi et al. 2008a). These *Cor/Lea* genes are responsive at the transcription level to low temperature, drought, and exogenous ABA (Morris et al. 1990; Ohno et al. 2003; Takumi et al. 2003; Kobayashi et al. 2004, 2006; Egawa et al. 2006).

Cold Stress-Signaling Pathway

Arabidopsis CBF/DREB1 (*DRE-binding factor 1*) genes, which contain a highly conserved AP2 (APETALA2)/ERF (Ethylene-Responsive Factor) domain, *CBF1/DREB1B*, *CBF2/DREB1C*, and *CBF3/DREB1A* are rapidly induced by low-temperature treatment (Mizoi et al. 2012). These three genes are organized in tandem on chromosome IV of *Arabidopsis* (Gilmour et al. 1998; Medina et al. 1999). Natural variation in freezing tolerance between ecotypes is correlated with the expression levels of *CBF* genes, indicating that regulation of their expression levels is a major regulatory mechanism for the CBF activity (Mizoi et al. 2012). Inducer of *CBF* Expression 1 (ICE1), an MYC-like basic helix-loop-helix (bHLH) transcription factor, binds to the MYC recognition site in the *CBF3* promoter and activates its expression (Chinnusamy et al. 2003). On the other hand, ICE2, an ICE1-like protein, has been shown to activate *CBF1* expression (Fursova et al. 2009). *CBF2* promoter contains two *cis*-acting elements Induction of *CBF* Expression region 1 and 2 (ICEr1 and ICEr2) that are sufficient to impart cold-responsive gene expression (Zarka et al. 2003). It was found that the binding site of a calmodulin-binding transcription activator (CAMTA) protein, CAMTA3, overlaps with ICEr2. The *camta3* mutation reduces cold-induced transcript accumulation of *CBF1*, but not *CBF3* (Doherty et al. 2009). A negative regulator of *CBFs* was also identified. MYB15, an R2R3-MYB transcription factor, binds to the MYB recognition sites in the promoters of *CBF1*, *CBF2*, and *CBF3*. MYB15 is upregulated by cold and physically interacts with ICE1. The expression of *MYB15* is downregulated by ICE1 (Agarwal et al. 2006).

As in *Arabidopsis*, a cluster of *CBF* genes has been mapped on chromosome 5A, at the *Fr-A^{m}2* and *Fr-A2* loci in einkorn and common wheat, respectively (Miller et al. 2006; Badawi et al. 2007). There are up to 25 CBF genes in the wheat genome, and at least 11 *CBF* copies are located as a gene cluster on the *Fr-A^{m}2* region (Miller et al. 2006; Badawi et al. 2007). Phylogenetic analysis revealed that the *Poaceae CBFs* can be classified into ten groups, of which six groups are found only in the Pooideae (a subfamily containing the temperate cereals as wheat, barley, and rye) suggesting that they represent the CBF machinery that evolved recently during colonization of temperate habitats (Badawi et al. 2007). Expression analysis revealed that five of the Pooideae-specific groups display higher constitutive and low-temperature-inducible expression in the winter cultivars, and may contribute to the superior low-temperature tolerance of these cultivars (Badawi et al. 2007). In common wheat, the cultivar difference in freezing tolerance has also been associated with the difference in *WCBF2* expression pattern (Kume et al. 2005). WCBF2 directly activates downstream *Cor*/*Lea* genes through binding the *Cor*/*Lea* gene promoters (Fig. 1, Takumi et al. 2008). Two inducers of CBF expression (*ICE1*-like genes), *TaICE87* and *TaICE41*, were also isolated in wheat, and belonged, to ICE1 and ICE2 clades, respectively (Badawi et al. 2008). Both ICE transcription factors bind to different MYC elements in the *TaCBFIVd-B9* promoter and activate its transcription (Fig. 1, Badawi et al. 2008). Therefore, the *ICE-CBF-Cor*/*Lea* signal pathway is evolutionally and functionally conserved between *Arabidopsis* and wheat.

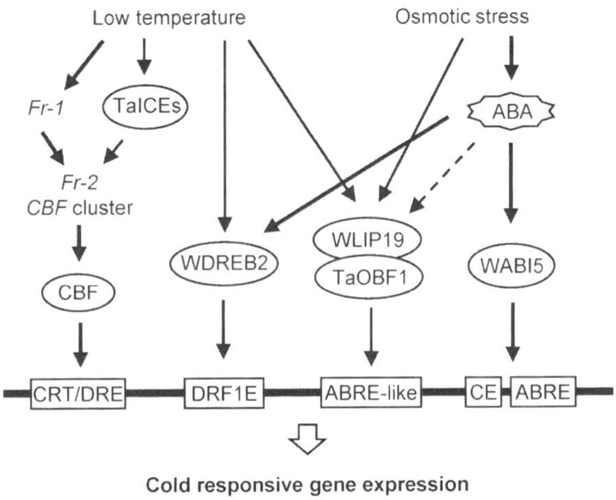

Fig. 1 Cold stress-signaling pathways in common wheat. Low temperature leads to accumulation of transcription factors (indicated by *circles*) through ABA-dependent and -independent pathways. Specific binding of each transcription factor to the *cis*-acting elements (indicated by *boxes*) activates the cold-responsive gene expression. *Dashed lines* indicate putative interaction

Osmotic Stress Signaling

Cold can also cause osmotic stress in addition to their direct effect on metabolism. Therefore, osmotic stress appears to be common consequence of exposure to drought, salinity, and cold (Huang et al. 2012). Two major regulatory systems for stress-responsive gene expression are well known in *Arabidopsis*: ABA-dependent and -independent pathways. In the ABA-independent gene expression, AP2/ERF transcription factors, DREB2A and DREB2B, induce abiotic stress-responsive genes through binding to DRE/CRT elements (Nakashima et al. 2000).

Unlike in *Arabidopsis*, wheat *DREB2* homologs, *TaDREB1* and *WDREB2*, not only are induced by low temperature, salinity, and drought, but also by ABA (Shen et al. 2003; Egawa et al. 2006). *WDREB2* represents significant homology to barley *HvDRF1*. HvDRF1 shows little activity for binding to the DRE element, while preferably binds to a CT-rich element (TWACCGCCTT) in which the core sequence is called as a DRF1E motif (ACCGCC; Fig. 1, Xue and Loveridge 2004). WDREB2 directly binds to *Cor*/*Lea* promoters for upregulation of the *Cor*/*Lea* genes (Kobayashi et al. 2008a). Another significant difference between *Arabidopsis* and Triticeae is in regulation of the *DREB2* expression. *AtDREB2s* are regulated by posttranslational mechanisms such as protein phosphorylation, whereas alternative splicing posttranscriptionally controls the *WDREB2* and *HvDRF1* transcripts (Xue and Loveridge 2004; Egawa et al. 2006).

Although ABA-independent pathways exist, ABA-dependent pathways play essential roles during osmotic stress (Huang et al. 2012). ABRE is a major *cis*-acting element for the ABA-responsive gene expression. However, a single copy of ABRE is not sufficient for efficient ABA-responsive transcription, requiring additional copies of ABRE sequences or coupling elements (Fujita et al. 2011). Transcription factors that interact with ABREs are basic-domain leucine zipper (bZIP) proteins named ABRE-binding factors (ABFs) or ABRE-binding proteins (AREBs), of which AREB1/ABF2, AREB2/ABF4, and ABF3 are induced both by ABA and by osmotic stress in *Arabidopsis* (Fujita et al. 2011). Other transcription factors such as MYC-like bHLH, MYB, NAC, and WRKY are also involved in the ABA-dependent pathways (Fujita et al. 2011).

The bZIP transcription factors were also identified in common wheat. *WABI5* expression is activated by low temperature, drought, and exogenous ABA and its expression level corresponded to cultivar difference in freezing tolerance and ABA sensitivity (Kobayashi et al. 2008b). *WLIP19* expression is also activated by low temperature and drought and the expression level is slightly increased by exogenous ABA (Kobayashi et al. 2008c). Although low-temperature-induced expression of *WLIP19* is not affected in ABA-less-sensitive mutant lines of common wheat, *WLIP19* functions at least in part in the ABA-signaling pathway under abiotic stress conditions (Kobayashi et al. 2006, 2008c). WLIP19 can form homodimer or interact with another bZIP transcription factor TaOBF1 to form heterodimer. *TaOBF1* positively responds to drought stress and is weakly induced by low temperature (Fig. 1, Kobayashi et al. 2008c). WABI5 and WLIP19 directly bind to *Cor*/*Lea* gene promoters and activate their downstream gene expression to develop abiotic stress tolerance (Kobayashi et al. 2008b, c). A number of NAC and MYB transcription factors that respond to drought, salinity, low temperature, and ABA have also been reported in common wheat (Xia et al. 2010; Cai et al. 2011; Mao et al. 2011, 2012).

Genetic Dissection of Complex Traits at Transcript Levels

Natural genetic variations are useful resource for crop improvement. Most of the natural variation is a result of polygenic variation, which may complicate the dissection of molecular causes (Alonso-Blanco et al. 2009). Diversity at the molecular level usually far exceeds phenotypic variation, described as phenotypic buffering or robustness (Fu et al. 2009; Delker and Quint 2011). Through QTL analysis, many of the large-effect phenotypes have been shown to be the result of alterations in amino acid sequences as well as alterations in noncoding genomics regions such as promoters and introns that affect transcript accumulation level (Alonso-Blanco et al. 2009). Expression-level polymorphisms, which are natural variations in gene expression between accessions, are considered more sensitive to molecular diversity than phenotypic differences. Most expression-level polymorphisms are regulated in *trans* or in *cis* by multiple eQTLs (Delker and Quint 2011). Mapping of eQTLs is an efficient approach to identify genetic loci controlling complex traits such as seed development and disease resistance in higher plants (Jordan et al. 2007; Chen et al. 2010a, b).

As described in the above sections, a correlation between both *CBF* and *Cor*/*Lea* gene expression patterns and freezing tolerance has been reported in various studies. In cultivated einkorn wheat, a diploid wheat species *Triticum monococcum*, eQTL for *Cor*/*Lea* gene *Cor14b* have been performed in an RIL population derived from winter and spring habit varieties (Vágújfalvi et al. 2003). The parental winter-type accession G3116 was significantly more freezing tolerant and accumulated more *Cor14b* transcripts at 15 °C than the parental spring-type accession DV92. The eQTL for *Cor14b* overlapped with the QTL for freezing tolerance on the long arm of chromosome $5A^m$, coinciding with freezing tolerance locus *Fr-A^{m}2*. The einkorn wheat ortholog of barley *CBF3* was also mapped at the chromosomal region of the *Cor14b* eQTL and *Fr-A^{m}2* QTL, suggesting that allelic variance at the *CBF* locus might result in differential regulation of *Cor14b* expression and freezing tolerance in einkorn wheat (Vágújfalvi et al. 2003).

In common wheat, two loci on the long arm of chromosome 5A have been suggested to control the threshold induction temperature of *Cor14b* using nine single-chromosome recombinant lines derived from a cross between common wheat accessions CS/*Triticum spelta* 5A and CS/Cheyenne (Cnn) 5A (Vágújfalvi et al. 2000). These two loci are corresponding to *Fr-A^{m}1* and *Fr-A^{m}2* of einkorn wheat (Vágújfalvi et al. 2000, 2005). In another study, it was also found that three of eight *CBF* copies (*CBF1A*, *CBF1C*, and *CBF7*), mapped at the putative *Fr-A2* region (originally named as *Rcg1*), showed significantly higher levels of the *Cor14b* transcript accumulation in line with putatively distinct alleles of *Fr-A2* (Vágújfalvi et al. 2005).

Mironovskaya 808 (M808) is well known as one of the most tolerant wheat cultivars to freezing stress (Veisz and Sutka 1990). eQTL analysis performed in an RIL population derived from the cross between freezing-sensitive cultivar CS and freezing-tolerant M808 found eQTLs for *WCBF2* and three *Cor/Lea* genes (*Wlt10*, *Wdhn13*, and *Wcor14*) on long arm of chromosome 5A and coincided with the *Fr-A2* locus (Fig. 2, Motomura et al. 2013). An eQTL for *TaCFB12* was located on chromosome 2A and additional eQTLs for *Wlt10* and *Wcor14* were detected on chromosomes 1D and 4B, respectively. The bioassay results for freezing tolerance validated that the 1D and 5A eQTLs play important roles in development of freezing tolerance, and that the M808 allele of the 5A eQTL contributes to the high freezing tolerance of M808. The chromosomal location of the 1D eQTL seems to coincide with a freezing tolerance QTL as reported by Båga et al. (2007), and this locus might act as a positive regulator of the *Cor/Lea* gene expression during cold acclimation in common wheat. The 5A eQTL regulated not only the *Cor/Lea* genes in *trans* but also the *CBF* copies in *cis*, suggesting that one or some of the *CBF* copies at the *Fr-2* region positively regulate other copies, which might amplify the positive effects of the *CBF* cluster on the downstream *Cor/Lea* gene activation. Allelic difference at *Fr-A2* might be a major cause for cultivar differences of freezing tolerance level in common wheat.

Previous studies using ABA-sensitive and less-sensitive mutants of common wheat suggested that both positive and negative regulation of ABA response are involved in the basic mechanisms for development of freezing tolerance (Kobayashi et al. 2006, 2008d). Cultivar difference in ABA responsiveness has also been associated with differential expression levels of ABA-inducible genes (Egawa et al. 2006; Kobayashi et al. 2004, 2006, 2008b, c, d). The common wheat cultivars Cnn and Hope have been reported as highly ABA-responsive varieties compared with CS (Iehisa et al. 2011). The substitution of CS chromosomes 3A or 5A, to that of Cnn

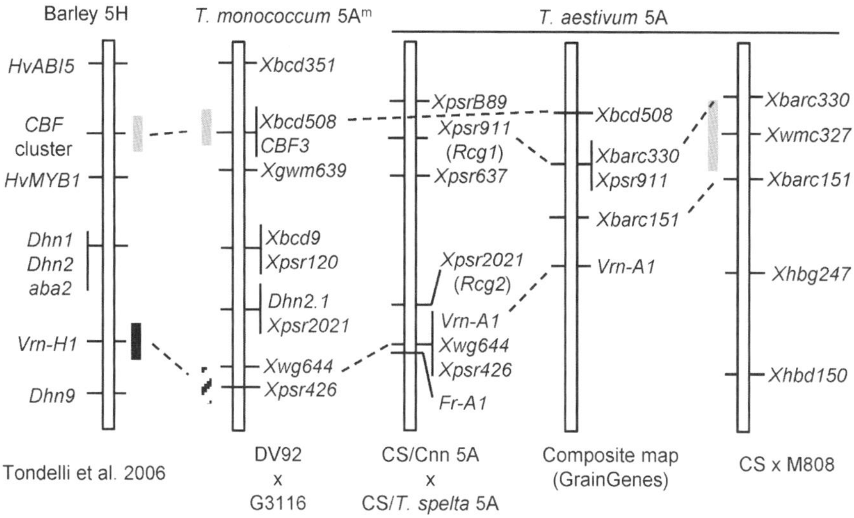

Fig. 2 Chromosomal synteny of homoeologous group 5 chromosomes with *Vrn-1* and frost resistance loci among barley, *T. monococcum* and common wheat. Chromosomal positions of QTLs for *Fr-1* (*black bar*), *Fr-2* (*gray bars*), and *Vrn-1* (*shaded bar*) are indicated. *Gray bar* indicates the position of eQTL for four cold-responsive genes in the genetic map of CS × M808 RILs. *Rcg1* and *Rcg2*, involved in the *Cor14b* regulation, were originally identified in Vágújfalvi et al. (2000)

or Hope, significantly increased the ABA responsiveness. Exogenous ABA-induced expression of *Wdhn13* and *Wrab17* was also higher in the 3A or 5A chromosomes substituted lines compared with CS, and they showed improved abiotic stress tolerance (Iehisa et al. 2011). Therefore, the induction levels of ABA-responsive genes are tightly associated with abiotic stress tolerance levels. On the other hand, QTL analysis for ABA responsiveness in an RIL population obtained from CS × M808 found major QTL on chromosome 6D and minor QTLs on chromosomes 1B, 2A, 3A, and 7B and they appeared to be involved in the regulatory system of *Cor/Lea* expression (Kobayashi et al. 2010). These results suggest that eQTL analysis of ABA-inducible genes might be useful in dissecting genetically not only ABA responsiveness, but also abiotic stress tolerance in common wheat.

Concluding Remarks

eQTL analysis is a powerful tool for genetic dissection of complex traits such as abiotic stress tolerance. Although eQTL analysis of five cold-responsive genes demonstrated to identify two major freezing tolerance loci, analysis of other cold-responsive genes such as WRKY transcription factors, ice recrystallization inhibitors, and ABA-responsive genes might lead to better understanding of regulatory networks involved in low temperature response of common wheat. eQTL analysis also suggests that cultivar differences in gene expression levels could be used as markers in molecular breeding program for abiotic stress tolerance. Wide variation has been found for ABA responsiveness and dehydration tolerance in accessions of *Aegilops tauschii*, a wild diploid progenitor of common wheat, and their derived synthetic hexaploids (Kurahashi et al. 2009; Sohail et al. 2011; Iehisa and Takumi 2012). Synthetic hexaploid wheat lines, obtained from interspecific crosses between tetraploid wheat and various *Ae. tauschii* accessions, can be considered as a potential source of new genetic variations for abiotic stress tolerance in improvement of common wheat cultivars.

References

Agarwal M, Hao Y, Kapoor A, Dong CH, Fujii H, Zheng X, Zhu JK (2006) A R2R3 type MYB transcription factor is involved in the cold regulation of CBF genes and in acquired freezing tolerance. J Biol Chem 281:37636–37645

Alonso-Blanco C, Aarts MGM, Bentsink L, Keurentjes JJB, Reymond M, Vreugdenhil D, Koornneef M (2009) What has natural variation taught us about plant development, physiology, and adaptation? Plant Cell 21:1877–1896

Badawi M, Danyluk J, Boucho B, Houde M, Sarhan F (2007) The *CBF* gene family in hexaploid wheat and its relationship to the phylogenetic complexity of cereal *CBFs*. Mol Genet Genomics 277:533–554

Badawi M, Reddy YV, Agharbaoui Z, Tominaga Y, Danyluk J, Sarhan F, Houde M (2008) Structure and functional analysis of wheat *ICE* (Inducer of CBF Expression) genes. Plant Cell Physiol 49:1237–1249

Båga M, Chodaparambil SV, Limin AE, Pecar M, Fowler DB, Chibbar RN (2007) Identification of quantitative trait loci and associated candidate genes for low-temperature tolerance in cold-hardy winter wheat. Funct Integr Genomics 7:53–68

Cai H, Tian S, Liu C, Dong H (2011) Identification of a MYB3R gene involved in drought, salt and cold stress in wheat (*Triticum aestivum* L.). Gene 485:146–152

Casao MC, Igartua E, Karsai I, Bhat PR, Cuadrado N, Gracia MP, Lasa JM, Casas AM (2011) Introgression of an intermediate *VRNH1* allele in barley (*Hordeum vulgare* L.) leads to reduced vernalization requirement without affecting freezing tolerance. Mol Breed 28:475–484

Chen X, Hackett CA, Niks RE, Hedley PE, Booth C, Druka A, Marcel TC, Vels A, Bayer M, Milne I et al (2010a) An eQTL analysis of partial resistance to *Puccinia hordei* in barley. PLoS ONE 5:e8598

Chen X, Niks RE, Hedley PE, Morris J, Druka A, Marcel TC, Vels A, Waugh R (2010b) Differential gene expression in nearly isogenic lines with QTL for partial resistance to *Puccinia hordei* in barley. BMC Genomics 11:629

Chinnusamy V, Ohta M, Kanrar S, Lee B, Hong X, Agarwal M, Zhu JK (2003) ICE1: a regulator of cold-induced transcriptome and freezing tolerance in *Arabidopsis*. Genes Dev 17:1043–1054

Delker C, Quint M (2011) Expression level polymorphisms: heritable traits shaping natural variation. Trend Plant Sci 16:481–488

Dhillon T, Pearce SP, Stockinger EJ, Distelfeld A, Li C, Knox AK, Vashegyi I, Vágújfalvi A, Galiba G, Dubcovsky J (2010) Regulation of freezing tolerance and flowering in temperate cereals: the *VRN-1* connection. Plant Physiol 153:1846–1858

Doherty CJ, Van Buskirk HA, Myers SJ, Thomashow MF (2009) Roles for *Arabidopsis* CAMTA transcription factors in cold-regulated gene expression and freezing tolerance. Plant Cell 21:972–984

Egawa C, Kobayashi F, Ishibashi M, Nakamura T, Nakamura C, Takumi S (2006) Differential regulation of transcript accumulation and alternative splicing of a *DREB2* homolog under abiotic stress conditions in common wheat. Genes Genet Syst 81:77–91

Fowler DB, Gusta LV (1979) Selection for winterhardiness in wheat. I. Identification of genotypic variability. Crop Sci 19:769–772

Fu J, Keurentjes JJB, Bouwmeester H, America T, Verstappen FWA, Ward JL, Beale MH, Vos RCH de, Dijkstra M, Scheltema RA et al (2009) System-wide molecular evidence for phenotypic buffering in Arabidopsis. Nat Genet 41:166–167

Fujita Y, Fujita M, Shinozaki K, Yamaguchi-Shinozaki K (2011) ABA-mediated transcriptional regulation in response to osmotic stress in plants. J Plant Res 124:509–525

Fursova OV, Pogorelko GV, Tarasov VA (2009) Identification of *ICE2*, a gene involved in cold acclimation which determines freezing tolerance in *Arabidopsis thaliana*. Gene 429:98–103

Galiba G, Quarrie SA, Sutka J, Morgounov A, Snape JW (1995) RFLP mapping of the vernalization (*Vrn1*) and frost resistance (*Fr1*) genes on chromosome 5 A of wheat. Theor Appl Genet 90:1174–1179

Galiba G, Vágújfalvi A, Li C, Soltész A, Dubcovsky J (2009) Regulatory genes involved in the determination of frost tolerance in temperate cereals. Plant Sci 176:12–19

Gilmour SJ, Zarka DG, Stockinger EJ, Salazar MP, Houghton JM, Thomashow MF (1998) Low temperature regulation of the *Arabidopsis* CBF family of AP2 transcriptional activators as an early step in cold-induced *COR* gene expression. Plant J 16:433–442

Huang GT, Ma SL, Bai LP, Zhang L, Ma H, Jia P, Liu J, Zhong M, Guo ZF (2012) Signal transduction during cold, salt, and drought stresses in plants. Mol Biol Rep 39:969–987

Iehisa JCM, Kurahashi Y, Takumi S (2011) Identification of chromosomes controlling abscisic acid responsiveness and transcript accumulation of *Cor–Lea* genes in common wheat seedlings. Funct Plant Biol 38:758–766

Iehisa JCM, Takumi S (2012) Variation in abscisic acid responsiveness of Ae. tauschii accessions and hexaploid wheat synthetics due to the D-genome diversity. Genes Genet Syst 87:9–18

Jordan MC, Somers DJ, Banks TW (2007) Identifying regions of the wheat genome controlling seed development by mapping expression quantitative trait loci. Plant Biotech J 5:442–453

Kobayashi F, Takumi S, Nakata M, Ohno R, Nakamura T, Nakamura C (2004) Comparative study of the expression profiles of the *Cor/Lea* gene family in two wheat cultivars with contrasting levels of freezing tolerance. Physiol Plant 120:585–594

Kobayashi F, Takumi S, Egawa C, Ishibashi M, Nakamura C (2006) Expression patterns of low temperature responsive genes in a dominant ABA-less-sensitive mutant line of common wheat. Physiol Plant 127:612–623

Kobayashi F, Ishibashi M, Takumi S (2008a) Transcriptional activation of *Cor/Lea* genes and increase in abiotic stress tolerance through expression of a wheat *DREB2* homolog in transgenic tobacco. Transgenic Res 17:755–767

Kobayashi F, Maeta E, Terashima A, Takumi S (2008b) Positive role of a wheat *HvABI5* ortholog in abiotic stress response of seedlings. Physiol Plant 134:74–86

Kobayashi F, Maeta E, Terashima A, Kawaura K, Ogihara Y, Takumi S (2008c) Development of abiotic stress tolerance via bZIP-type transcription factor LIP19 in common wheat. J Exp Bot 59:891–905

Kobayashi F, Takumi S, Nakamura C (2008d) Increased freezing tolerance in an ABA-hypersensitive mutant of common wheat. J Plant Physiol 165:224–232

Kobayashi F, Takumi S, Handa H (2010) Identification of quantitative trait loci for ABA responsiveness at the seedling stage associated with ABA-regulated gene expression in common wheat. Theor Appl Genet 121:629–641

Kume S, Kobayashi F, Ishibashi M, Ohno R, Nakamura C, Takumi S (2005) Differential and coordinated expression of *Cbf* and *Cor/Lea* genes during long-term cold acclimation in two wheat cultivars showing distinct levels of freezing tolerance. Genes Genet Syst 80:185–197

Kurahashi Y, Terashima A, Takumi S (2009) Variation in dehydration tolerance, ABA sensitivity and related gene expression patterns in D-genome progenitor and synthetic hexaploid wheat lines. Int J Mol Sci 10:2733–2751

Mao X, Jia D, Li A, Zhang H, Tian S, Zhang X, Jia J, Jing R (2011) Transgenic expression of *TaMYB2 A* confers enhanced tolerance to multiple abiotic stresses in *Arabidopsis*. Funct Integr Genomics 11:445–465

Mao X, Zhang H, Qian X, Li A, Zhao G, Jing R (2012) *TaNAC2*, a NAC-type wheat transcription factor conferring enhanced multiple abiotic stress tolerances in *Arabidopsis*. J Exp Bot 63:2933–2946

Marcotte WR Jr, Russell SH, Quatrano RS (1989) Abscisic acid-responsive sequences from the Em gene of wheat. Plant Cell 1:969–976

Medina J, Bargues M, Terol J, Pérez-Alonso M, Salinas J (1999) The Arabidopsis *CBF* gene family is composed of three genes encoding AP2 domain-containing proteins Whose expression is regulated by low temperature but not by abscisic acid or dehydration. Plant Physiol 119:463–469

Miller AK, Galiba G, Dubcovsky J (2006) A cluster of 11 *CBF* transcription factors is located at the frost tolerance locus *Fr-A^{m}2* in *Triticum monococcum*. Mol Genet Genomics 275:193–203

Mizoi J, Shinozaki K, Yamaguchi-Shinozaki K (2012) AP2/ERF family transcription factors in plant abiotic stress responses. Biochim Biophys Acta 1819:86–96

Morris PC, Kumar A, Bowles DJ, Cuming AC (1990) Osmotic stress and abscisic acid induce expression of the wheat Em genes. Eur J Biochem 190:625–630

Motomura Y, Kobayashi F, Iehisa JCM, Takumi S (2013) A major quantitative trait locus for cold-responsive gene expression is linked to frost-resistance gene *Fr-A2* in common wheat. Breed Sci 63:58–67

Nakashima K, Shinwari ZK, Sakuma Y, Seki M, Miura S, Shinozaki K, Yamaguchi-Shinozaki K (2000) Organization and expression of two *Arabidopsis DREB2* genes encoding DRE-binding proteins involved in dehydration- and high-salinity-responsive gene expression. Plant Mol Biol 42:657–665

Ohno R, Takumi S, Nakamura C (2003) Kinetics of transcript and protein accumulation of a low-molecular-weight wheat LEA D-11 dehydrin in response to low temperature. J Plant Physiol 160:193–200

Sakai A, Larcher W (1987) Frost survival of plants: responses and adaptation to freezing stress. Springer, Berlin

Shen YG, Zhang WK, He SJ, Zhang JS, Liu Q, Chen SY (2003) An EREBP/AP2-type protein in *Triticum aestivum* was a DRE-binding transcription factor induced by cold, dehydration and ABA stress. Theor Appl Genet 106:923–930

Snape JW, Semikhodskii A, Fish L, Sarma RN, Quarrie SA, Galiba G, Sutka J (1997) Mapping frost resistance loci in wheat and comparative mapping with other cereals. Acta Agron Hung 45:265–270

Sohail Q, Inoue T, Tanaka H, Eltayeb AE, Matsuoka Y, Tsujimoto H (2011) Applicability of *Aegilops tauschii* drought tolerance traits to breeding of hexaploid wheat. Breed Sci 61:347–357

Sung S, Amasino RM (2005) Remembering winter: toward a molecular understanding of vernalization. Ann Rev Plant Biol 56:491–508

Sutka J, Galiba G, Vagujfalvi A, Gill BS, Snape JW (1999) Physical mapping of the *Vrn-A1* and *Fr1* genes on chromosome 5A of wheat using deletion lines. Theor Appl Genet 99:199–202

Takumi S, Koike A, Nakata M, Kume S, Ohno R, Nakamura C (2003) Cold-specific and light-stimulated expression of a wheat (*Triticum aestivum* L.) *Cor* gene *Wcor15* encoding a chloroplast-targeted protein. J Exp Bot 54:2265–2274

Takumi S, Shimamura C, Kobayashi F (2008) Increased freezing tolerance through up-regulation of downstream genes via the wheat *CBF* gene in transgenic tobacco. Plant Physiol Biochem 46:205–211

Thomashow MF (1999) Plant cold acclimation: freezing tolerance genes and regulatory mechanisms. Annu Rev Plant Biol Plant Mol Biol 50:571–599

Thomashow MF, Gilmour SJ, Stockinger EJ, Jaglo-Ottosen KR, Zarka DG (2001) Role of the Arabidopsis CBF transcriptional activators in cold acclimation. Physiol Plant 112:171–175

Tóth B, Galiba G, Fehér E, Sutka J, Snape W (2003) Mapping genes affecting flowering time and frost resistance on chromosome 5B of wheat. Theor Appl Genet 107:509–514

Vágújfalvi A, Crosatti C, Galiba G, Dubcovsky J, Cattivelli L (2000) Two loci on wheat chromosome 5 A regulate the differential cold-dependent expression of the *cor14b* gene in frost-tolerant and frost-sensitive genotypes. Mol Genet Genomics 263:194–200

Vágújfalvi A, Galiba G, Cattivelli L, Dubcovsky J (2003) The cold-regulated transcriptional activator *Cbf3* is linked to the frost-tolerance locus *Fr-A2* on wheat chromosome 5 A. Mol Genet Genomics 269:60–67

Vágújfalvi A, Aprile A, Miller A, Dubcovsky J, Delugu G, Galiba G, Cattivelli L (2005) The expression of several *Cbf* genes at the *Fr-A2* locus is linked to frost resistance in wheat. Mol Genet Genomics 274:506–514

Veisz O, Sutka J (1990) Frost resistance studies with wheat in natural and artificial conditions. In: Panayotov I, Pavlova S (eds) Proceedings of the international symposium on cereal adaptation to low temperature stress. Albena, Bulgaria, pp 12–17

Xia N, Zhang G, Liu XY, Deng L, Cai GL, Zhang Y, Wang XJ, Zhao J, Huang LL, Kang ZS (2010) Characterization of a novel wheat NAC transcription factor gene involved in defense response against stripe rust pathogen infection and abiotic stresses. Mol Biol Rep 37:3703–3712

Xue GP, Loveridge CW (2004) *HvDRF1* is involved in abscisic acid-mediated gene regulation in barley and produces two forms of AP2 transcriptional activators, interacting preferably with a CT-rich element. Plant J 37:326–339

Zarka DG, Vogel JT, Cook D, Thomashow MF (2003) Cold induction of Arabidopsis *CBF* genes involves multiple ICE (Inducer of *CBF* Expression) promoter elements and a cold-regulatory circuit that is desensitized by low temperature. Plant Physiol 133:910–918

Molecular Analysis of Fructan Metabolism Associated with Freezing Tolerance and Snow Mold Resistance of Winter Wheat

Midori Yoshida and Akira Kawakami

Introduction

In northern regions, winter cereals are exposed to severe winter conditions such as low freezing temperatures, ice encasement, and long periods of persistent snow cover. Wheat (*Triticum aestivum* L.) is one of the major crops in Hokkaido, the northernmost island of Japan located at about 41° to 45° latitude. The climate in Hokkaido differs depending on the region. The eastern and northern regions of Hokkaido receive little snowfall and severe freezing temperature stresses prevail. Conversely, in central and southwest regions of Hokkaido, deep snow accumulation that persists for 4–5 months occurs in agricultural fields. This situation encourages the development of several different snow mold species and the distribution and severity of disease caused by individual species varies with the regional climate (Matsumoto 2009; Fig. 1a). Severe snow mold attacks occur once every several years and can cause fatal damage to wheat fields resulting in extensive loss of yield (Fig. 1b). Winter wheat in Hokkaido also suffers from sublethal temperatures before establishment of a snow cover (Yoshida et al. 1998). Therefore, enhancement of both freezing tolerance and snow mold resistance to endure winter survival is one of the main objectives for wheat breeding programs in Hokkaido (Iriki et al. 2001).

Fructans, which are fructose oligo- or polymers attached to sucrose, are water-soluble carbohydrates that accumulate in vacuoles and act as storage carbohydrates in a large number of plant species mainly belonging to Asteraceae, Liliaceae, and temperate Poaceae (Pollok and Cairns 1991; Hendry 1993; Ritsema and Smeekens 2003). Wheat also uses fructan as temporal photoassimilates instead of starch in plastids. In overwintering plants, accumulated fructans are the main carbohydrate

M. Yoshida (✉)
NARO Hokkaido Agricultural Research Center, Hitsujigaoka, Sapporo 062-8555, Japan
e-mail: midori@affrc.go.jp

A. Kawakami
NARO Kyushu Okinawa Agricultural Research Center, Suya, Koshi 861-1192, Japan

R. Imai et al. (eds.), *Plant and Microbe Adaptations to Cold in a Changing World*, DOI 10.1007/978-1-4614-8253-6_20, © Springer Science+Business Media New York 2013

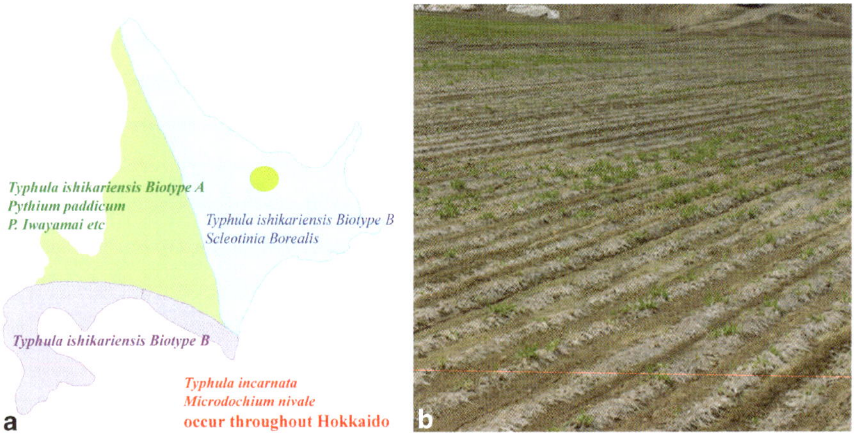

Fig. 1 Snow mold distribution and wheat damage in Hokkaido, Japan. **a** Distribution of the different snow mold species in Hokkaido. **b** Wheat field damaged by snow mold photographed at Oketo, Hokkaido in April 2004

source for survival under a persistent snow cover. In addition, fructans are highly hydrophilic and are thought to play an important role in protecting plants against environmental stresses such as freezing and drought (Pilon-Smits et al. 1995; Konstantinova et al. 2002; Livingston et al. 2009). Accumulation of fructan in winter cereals during cold hardening has been shown to be associated with winter survival (Olien 1984; Pontis 1989 Tognetti et al. 1990; Mohammad et al. 1997; Yoshida et al. 1998). In order to improve freezing tolerance and snow mold resistance of wheat, it is essential to understand the molecular mechanisms of fructan metabolism in wheat tissues.

Structure of Wheat Fructan

Annual and perennial cereals growing in the temperate and northern boreal ecosystems such as wheat, barley, rye, and temperate grasses, accumulate a levan type of fructan composed primarily of β(2→6)-linked fructosyl units (Gallagher et al. 2007; Yoshida et al. 2007; Yoshida and Tamura 2011), while plants in Asteraceae and Liliaceae such as chicory and onion accumulate an inulin type of fructan composed of β(2→1) fructosyl linkage (Van den Ende and Van Laere 2007). Wheat accumulates "graminan," a complex-branched levan type of fructan, in vegetative tissues such as leaves, stems, roots, and ears, but β(2→6)-linked linear levans accumulate in the stem before and after anthesis (Bancal et al. 1992, 1993; Yoshida et al. 2007). Graminans consist of both β(2→6)- and β(2→1)-linked fructosyl units. The main components are synthesized by sucrose:sucrose 1-fructosyltransferase (1-SST,

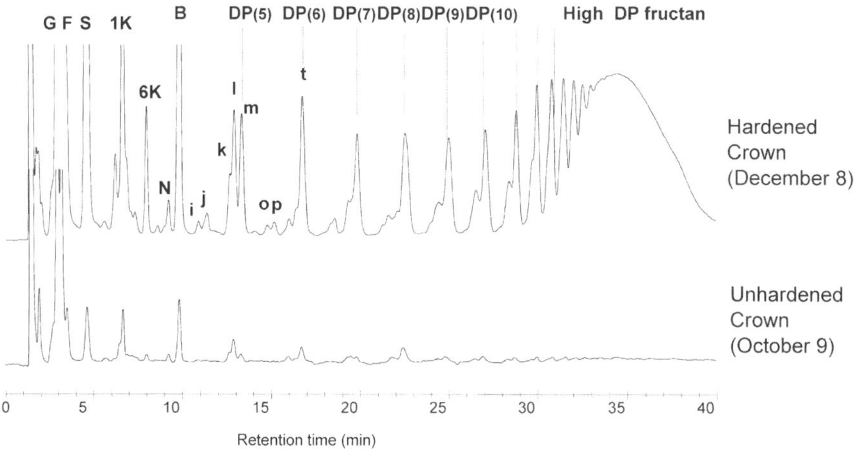

Fig. 2 Fructan profiling in winter wheat during hardening. Chromatograms of high-performance anion exchange separation (detected by pulse amperometric detector, DX500, Dionex) showing fructan compositions in crown tissues of wheat cv. PI 173438 (Kawakami and Yoshida 2005). *Dotted lines* indicate main graminans with β(2,6)-fructofuranosyl units linked to bifurcose. *G* glucose, *F* fructose, *S* sucrose, *1K* 1-kestose, *6K* 6-kestose, *N* 1,1-nystose, *B* bifurcose, *i* 6,1-nystose, *j* 6,6-nystose, *k* 1,1&6-kestopentaose, *l* 1&6,1-kestopentaose, *m* 1&6,6-kestopentaose, *o* 6,6,1-kestopentaose, *p* 6,6,6-kentopentaose, *t* 1&6,6,6-kestohexaose, *DP* degree of polymerization. Peaks were identified by authentic standards and estimated by the productions in rice transgenic plants expressing wheat fructosyltransferase genes (Kawakami et al. 2008; Yoshida et al. 2009)

EC 2.4.1.99), sucrose:fructan 6-fructosyltransferase (6-SFT), and fructan:fructan 1-fructosyltransferase (1-FFT, EC 2.4.1.100) and are degraded by several fructan exohydrolases (FEHs, EC 3.2.1.153,154; Yoshida et al. 2007). 1-SST synthesizes 1-kestose (1-kestotriose) by sucrose. 6-SFT is an enzyme responsible for the formation of bifurcose (1&6-kestotetraose) in cooperation with 1-SST and for elongation of β(2→6) fructosyl linkages using sucrose and fructan. Since there are few long β(2→1)-linked inulin in wheat, wheat 1-FFT may work to form branches of β(2→1)-linked fructosyl units attached to levans or graminans (Kawakami and Yoshida 2005).

The degree of polymerization of wheat fructans increases in tissues during cold hardening from autumn before snow cover in early winter (Fig. 2; Kawakami and Yoshida 2005). A few oligosaccharides are detected in wheat seedling early in the autumn before hardening whereas highly polymerized graminan are observed in fully hardened plants in early winter (Fig. 2). The degree of polymerization in wheat (generally DP3–20 or higher in some cultivars) is lower than that of the high-DP levan types of fructans as observed in some temperate grasses such as timothy (*Phleum pratense* L.) and orchardgrass (*Dactylis glomerata* L.; Suzuki 1989; Cairns et al. 1999; Yoshida and Tamura 2011). One of the main components of wheat fructan is bifurcose, a tetrasaccharide composed of both β(2→1)- &

β(2→6)-linked fructose. The major components (indicated by peaks with dotted lines in Fig. 2) are β(2→6)-linked fructosyl polymers attached to bifurcose.

Fructan Accumulation and Degradation Associated with Freezing Tolerance and Snow Mold Resistance of Wheat

The total nonstructural water-soluble carbohydrate content in wheat crown tissues has been shown to be associated with winter survival (Bruehl and Cunfer 1971; Kiyomoto and Bruehl 1977). Fructan is a dominant water-soluble carbohydrate that accumulates in wheat leaf and crown tissues in response to low temperature during hardening (Jeong and Housley 1990; Tognetti et al. 1990). In northern regions that receive deep snow, the fructan content in winter wheat cultivars increases from autumn to winter until snow cover and decreases throughout the winter until early spring under snow (Yoshida et al. 1998). Coincidently, freezing tolerance and snow mold resistance of field winter wheat also increase during hardening and decrease under snow cover (Nakajima and Abe 1994; Yoshida et al. 1998; Gaudet et al. 1999).

Clear differences exist between freezing-tolerant cultivars and snow mold-resistant cultivars in carbohydrate accumulation during hardening by the onset of snow cover and depletion under snow cover during the winter until early spring (Table 1; Yoshida et al. 1998). At hardening temperatures above 0 °C, levels of simple sugars (mono- and disaccharides) and fructan (oligo- and polysaccharides) increase in winter wheat. After the ambient minimum temperature drop to subzero temperatures, freezing-tolerant cultivars cease accumulation of fructan to increase mono- and disaccharides that are known as strong cryoprotectants. Increase in freezing tolerance was observed in freezing-tolerant cultivars but not in snow mold-resistant cultivars under the conditions with exposure to subzero temperatures (Yoshida et al. 1997, 1998). On the other hand, snow mold-resistant cultivars continued to accumulate more fructan until the onset of snow cover (Yoshida et al. 1998), which results in less content of mono- and disaccharides in the tissues related to their low level of freezing tolerance (Yoshida et al. 1998).

This low content of mono- and disaccharides may be a component of snow mold resistance since snow mold fungi have a decreased ability to grow on a medium using fructan as a sole carbon source compared to their ability to grow on a medium using glucose, fructose, and sucrose (Kawakami and Yoshida 2012). In addition, higher total sugar content in a snow mold-resistant cultivar would result in lower water potential of tissues (Kuwabara et al. 1997; Yoshida et al. 1997), which might influence the infection and spread of snow mold (Tronsmo 1986; Gaudet et al. 1999). Furthermore, the rate of catabolism of fructan in snow mold-resistant cultivars under snow is lower compared to that in susceptible cultivars (Yoshida et al. 1998). The sugar content remaining at the end of snow cover was related to wintering ability as well as snow mold resistance (Table 1; Yoshida et al. 1998). It is likely that high fructan levels in plants provide a sufficient energy supply for plant cells to survive a snow mold attack (Kawakami and Yoshida 2012). In snow mold-infected

Table 1 Characteristics of sugar accumulation in wheat varieties during hardening and consumption under snow

Before snow cover				
Wheat variety	Freezing tolerance (LT_{50}, minus °C)	Simple sugar content in crown (mg/g.fw.)	Fructan content in crown (mg/g.fw.)	Total sugar content in crown (mg/g.fw.)
Hardy winter cultivars	24–25	27–35	70–100	100–135
Snow mold-resistant cultivars	16–20	17–25	90–130	110–150
Japanese snow mold-resistant cultivars	20–23	28–30	90–110	120–135
Hokkaido leading winter cultivar (1990s)	ca. 20	26	ca. 65	ca. 90
Japanese northern winter cultivars	17–19	27–28	50–70	80–90
Spring wheat cultivars	<13	15–23	30–40	30–67
40 days before snow melt				
Survived wheat variety	Snow mold resistance	Consumption of fructan under snow	Fructan content in crown (mg/g.fw.)	Total sugar content in crown (mg/g.fw.)
Hardy winter cultivars	Moderately resistant	Very high and moderate	15–20	40–45
Snow mold-resistant cultivars	Highly resistant	Very low to high	55–80	55–105
Japanese snow mold-resistant cultivars	Resistant	Moderate	35–60	50–90
Hokkaido leading winter cultivar (1990s)	Moderately resistant	Moderate	ca. 10	ca. 28
Japanese northern winter cultivars	Susceptible	Low and moderate	5–20	25–30
Spring wheat cultivars	Susceptible	Low	<3	10–15

Data are shown as approximate values with reference to Yoshida et al. (1998). Field-grown wheat was sampled on December 1, before snow cover, and on February 21, about 40 days before snow melt. LT_{50} the temperature at which 50 % plants are killed by freezing wheat tissues, fructan levels declined more rapidly compared to those in noninfected tissues and significant sugar loss was observed in a snow mold-susceptible wheat cultivar (Kawakami and Yoshida 2012). The variation in fructan content associated with wintering ability of wheat might be caused by the difference in balance of fructan synthesis and degradation in response to low temperatures (Yukawa et al. 1995).

High freezing tolerance is also necessary for winter wheat to survive freezing stress under severe low temperatures in the northern region. Freezing tolerance of wheat cultivars excluding snow mold-resistant cultivars were correlated both with total contents of mono- and disaccharides and with fructan contents in crown tissues before snow cover (Yoshida et al. 1998). The higher level of fructan in snow mold-resistant cultivars is incompatible with high content of simple sugars. This problem has still remained in improving both freezing tolerance and snow mold resistance of winter wheat.

Molecular Characteristics of Genes for Fructan Synthesis in Winter Wheat

Cloning Genes Encoding Enzymes 1-SST, 6-SFT, and 1-FFT in Wheat

Fructosyltransferases (1-SST, 6-SFT, and 1-FFT) belong to the glycoside hydrolase family 32, GH32 (http://www.cazy.org) and are thought to have evolved from vacuolar invertases. Complementary DNAs (cDNAs) of 1-SST (*wft2*, AB029888), 6-SFT (*wft1*, AB029887; Kawakami and Yoshida 2002), and 1-FFT (*wft3* and *wft4*, AB088409 and AB088410; Kawakami and Yoshida 2005) were cloned from a cDNA library constructed by mRNA in hardened wheat crown tissues using sequences of the typical highly conserved consensus amino acids in β-fructosidase motif (NDPNG) and catalytic site (WECV/ID) of invertases, and we examined their catalysis using recombinant enzymes produced by *Pichia pastoris*. In addition, the production by Wft1, Wft2, and the combination of Wft1 and Wft2 were analyzed in vivo in transgenic rice plants expressing wheat fructosyltransferase (Kawakami et al. 2008; Yoshida et al. 2009). The results demonstrated the following: (1) wheat 1-SST, the first enzyme of fructan synthesis, is able to produce inulin oligomers from 1-kestose (DP3) to at least 1,1,1,1,1-kestoheptose (DP7) under the condition where excess sucrose is continuously supplied; (2) wheat 6-SFT synthesizes β(2→6) fructosyl-linked levan including dominant products of 6-kestose (6-kestotriose) by itself. It, however, preferentially produces bifurcose and elongates β(2→6)-linked fructosyl units attached to bifurcose rather than produce 6-kestose and levan in combination with 1-SST; and (3) wheat 1-FFT directs the formation of β(2→1) fructosyl-branched units attached to levans or graminans than β(2→1)-linked inulin formation as general catalysis.

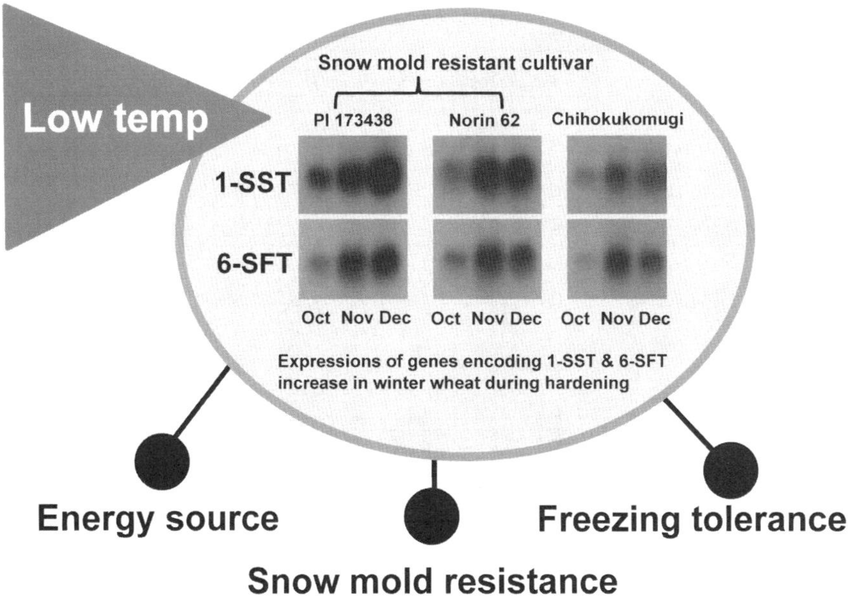

Fig. 3 Low-temperature-induced expressions of 1-SST and 6-SFT in leaves associated with wintering abilities of wheat varieties (Kawakami and Yoshida 2002)

Expression Analysis of Genes for Fructan Synthesis in Winter Wheat During Hardening

The expressions of *wft1* (6-SFT) and *wft2* (1-SST) in crown tissues of field-grown winter wheat during hardening were consistently higher than those in leaves in all wheat cultivars examined and transcript levels increased with the duration of hardening (Kawakami and Yoshida 2002). The transcript levels of these genes during hardening were higher in snow mold-resistant cultivars and levels of both genes, especially *wft2* (1-SST) in leaves increased markedly until the onset of snow cover. Conversely, in some other cultivars, transcript levels of *wft1* and *wft2* in leaves ceased to increase or decrease in December (freezing period). The transcript levels of both genes were correlated with fructan accumulation during hardening in the autumn and early winter and were associated with varietal differences in resistance to snow mold (Kawakami and Yoshida 2002). Thus, the regulation of gene expression of 1-SST or 6-SFT might affect fructan content in tissues, which in turn, are associated with freezing tolerance and snow mold resistance as well as energy storage in wheat (Fig. 3). It was proved that transgenic perennial ryegrass (*Lolium perenne* L.) expressing *wft1* or *wft2* had increased in fructan content and increased freezing tolerance of leaf tissues (Hisano et al. 2004). In wheat, fructan contents in crown tissues are highly correlated with freezing tolerances of varied cultivars from spring wheat to winter wheat as mentioned above (Yoshida et al. 1998). On the other hand, we

Table 2 Substrate specificities of recombinant enzymes of wheat FEHs produced by *Pichia pastoris*

	1-FEH w1[a]	6-KEH w1[b]	6&1-FEH w1[c]	6-FEH w1[d]	Wfh-sm3[e]
Sucrose (DP2)	×	×	×	×	±
1-Kestose (DP3)	+++	×	++	×	+
6-Kestose (DP3)	×	+++	++	++	+++
1,1-Nystose (DP4)	++	×	++	×	+
Bifurcose (DP4)	×	±	+++	–	++
Levan (high DP) [β(2,6)]	±	×	±	+++	±
Inulin (high DP) [β(2,1)]	+	×	×	×	±
Phlein (high DP)	–	±	–	+++	–
Wheat graminan (high DP)	+	×	±	–	+++

Symbols show the affinity levels of degradation: +++ most highly hydrolyzed, ++ highly hydrolyzed, + hydrolyzed, ± weakly hydrolyzed, × hardly hydrolyzed, – not examined
[a] Van den Ende et al. 2003
[b] Van den Ende et al. 2005
[c] Kawakami et al. 2005
[d] Van Riet et al. 2006
[e] Kawakami and Yoshida 2012

also demonstrated that snow mold-resistant cultivars did not conform to the correlation (Yoshida et al. 1998). The upregulation of expressions of fructan-synthesizing genes in highly snow mold-resistant cultivars likely leads to low content of simple sugar resulting in lower freezing tolerance.

Molecular Characteristics of Fructan Hydrolase Genes in Winter Wheat

Cloning of Genes Encoding FEH

Several clones of wheat fructan exohydrolases, 1-FEH (w1, FJ184989; w2, FJ184991; w3, FJ184990; Van den Ende et al. 2003; Van Riet et al. 2008), 6-FEH (w1, AM075205; Wfh-sm3, AB196524; Van Riet et al. 2006; Kawakami and Yoshida 2012), 6-KEH (w1, AB089270; w2, AB089371; Van den Ende et al. 2005), and 6&1-FEH (6&1-FEH w1, AB089269; Kawakami et al. 2005) have been cloned and functionally analyzed (Table 2). These studies revealed that there are multiple genes and homologs of wheat FEHs with each possessing different substrate specificities. Recombinant proteins produced by *P. pastoris* showed enzymatic activities with various substrate specificities. 6-KEH and 6&1-FEH are highly expressed in crown tissues of hardened wheat (Kawakami et al. 2005; Van den Ende et al. 2005) together with genes encoding *wft1*, *2*, *3*, and *4*. The preferred substrates for hydrolysis by these two enzymes are simple tri- and tetrasaccharides in wheat. 1-FEH can hydrolyze high-DP inulin but activities for 1-kestose (DP3) and 1-nystose (DP4) are much higher. This enzyme is considered a trimming enzyme because the inulin types of fructans excluding 1-kestose are not major components of wheat fructans (Van den Ende et al. 2003). 6-FEH can degrade higher-DP $\beta(2\rightarrow6)$ fructosyl-linked type of fructans. Wfh-sm3 can hydrolyze all forms of wheat graminan.

FEHs show high homology to cell wall type of invertases, while frucrosyltransferases show homology to vacuolar invertases (Ritsema and Smeekens 2003). FEHs are considered to play a role in the catabolism of fructans that accumulate in vacuoles in plant cells. On the other hand, it has been suggested that wheat 6-KEH is localized in the apoplast (Van den Ende et al. 2005). In winter oats, Livingston and Henson (1998) reported that fructans and FEH activity were detected in apoplastic fluid extracted from crown tissues. They demonstrated that both fructan content and FEH activity in the apoplastic fluid increased after the second phase cold hardening at −3 °C following the first-phase cold hardening at 2 °C. They proposed that FEHs might be involved in the degradation of apoplastic fructans, which may play a role in protection against stresses. The localization of FEH should be investigated in future studies.

Expression Analysis of Genes For Fructan Degradation In Winter Wheat During Hardening And Under Snow

There are many homologs of wheat FEHs and their expression varies in different tissues (Kawakami et al. 2005; Van den Ende et al. 2005; Kawakami and Yoshida 2012). FEHs may play an important role in physiological adaptation of winter wheat to different abiotic and biotic stresses by regulating fructan catabolism. Figure 4 shows results of northern hybridization for changes in transcript levels of cloned wheat FEH genes in leaves and crown tissues of field-grown winter wheat under snow. We demonstrated that 6&1-FEH and 6-KEH cloned from crown tissues are continuously expressed in crown tissues, but not in leaves, during hardening and under snow cover (Kawakami et al. 2005; Van den Ende et al. 2005). Among the FEH genes cloned to date, only Wfh-sm3 can degrade almost all components of wheat graminan (Table 2). The gene *wfh-sm3* was cloned from a cDNA library constructed by wheat leaves inoculated with snow mold, *Typhula ishikariensis* (Kawakami and Yoshida 2012). Interestingly, *Wfh-sm3* transcripts markedly increased in wheat leaf tissues inoculated with snow mold under a snow cover and its upregulation coincided with the rapid decline in fructan content in leaves and crown tissues. These results suggested that the expression of *wfh-sm3* was induced in response to snow mold infection and that Wfh-sm3 but not enzyme derived from fungi might be responsible for degradation of fructan in the same period (Kawakami and Yoshida 2012).

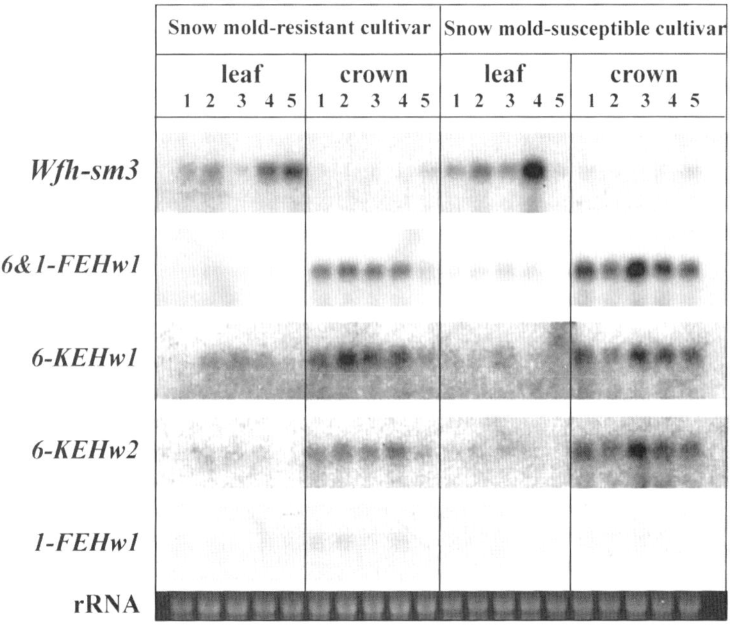

Fig. 4 Profiling of FEH genes in winter wheat in response to *T. ishikariensis* infection (Kawakami and Yoshida 2012). The genes examined are shown in Table 2. Wheat cultivars PI 173438 (snow mold-resistant) and Valuevskaya (snow mold-susceptible) were sown in a soil container and grown in a field. Crown tissues were sampled: (1) just before snow cover, (2) 10 days after snow cover without snow mold inoculation, (3) 20 days after snow cover without snow mold inoculation, (4) 10 days after snow cover with *T. ishikariensis* inoculation, and (5) 20 days after snow cover with *T. ishikariensis* inoculation

Other FEH genes remain to be cloned and investigation of the physiological function of each FEH is underway. Studies on the role and control mechanism of FEHs in fructan consumption and degradation against freezing stress and snow mold infection will be crucial for understanding the wintering abilities of wheat varieties.

Conclusion

To improve freezing tolerance and snow mold resistance of wheat, it will be necessary to understand gene expression system in fructan synthesis and degradation. Recent molecular studies on fructan metabolism have demonstrated the following: (1) due to the complex structure of fructan, wheat possesses several kinds of enzymes for both synthesis and degradation; (2) there are many tissue-specific FEH genes and homologs in wheat that regulate fructan degradation; and (3) the coordinate expressions of genes involved in both fructan synthesis and degradation endure an adequate energy supply to permit winter survival. The varietal differences in expressions of these genes likely play an important role in freezing tolerance and snow mold resistance in wheat.

References

Bancal P, Carpita NC, Gaudillére JP (1992) Difference in fructan accumulated in induced and field-grown wheat plants: an elongation-trimming pathway got their synthesis. New Phytol 120:313–321

Bancal P, Gibeaut DM, Carpita NC (1993) Analysis methods for the determination of fructan structure and biosynthesis. In: Suzuki M, Catterton NJ (eds) Science and technology of fructans. CDC Press, Boca Raton, ISBN:0–8493–5111–1, pp 83–118

Bruehl GW, Cunfer B (1971) Physiologic and environmental factors that affect the severity of snow mold of wheat. Phytopathology 61:792–799

Cairns AJ, Nash B, De Carvalho MM, Sims IM (1999) Characterization of the enzymatic polymerization of 2,6-linked fructan by leaf extracts from timothy grass (*Phleum pratense*). New Phytol 142:79–91

Gallagher JA, Cairns AJ, Turner LB (2007) Fructan in temperate forage grasses: agronomy, physiology (Chapter 10). In: Shiomi N, Benkeblia N, Onodera S (eds) Recent advances in fructooligosaccharides research. Research Signpost Publisher, Kerala, ISBN: 81–308–014609 pp 15–46

Gaudet DA, Laroche A, Yoshida M (1999) Low temperature-wheat-fungal interactions: a carbohydrate connection. Physiol Plant 106:437–444

Hendry GAF (1993) Evolutionary origins and natural functions of fructans—a climatological, biogeographic and mechanistic appraisal. New Phytol 123:3–14

Hisano H, Kanazawa A, Kawakami A, Yoshida M, Shimamoto Y, Yamada T (2004) Transgenic perennial ryegrass plants expressing wheat fructosyltransferase genes accumulate increased amounts of fructan and acquire increased tolerance on a cellular level to freezing. Plant Sci 167:861–868

Iriki N, Kawakami A, Takata K, Kuwabara T, Ban T (2001) Screening relatives of wheat for snow mold resistance and freezing tolerance. Euphytica 122:335–341

Jeong BR, Housley TL (1990) Fructan metabolism in wheat in alternating warm and cold temperatures. Plant Physiol 93:902–906

Kawakami K, Sato Y, Yoshida M (2008) Genetic engineering of rice capable of synthesizing fructans and enhancing chilling tolerance. J Exp Bot 59:793–802

Kawakami A, Yoshida M (2002) Molecular characterization of sucrose:sucrose 1-fructosyltransferase and sucrose:fructan 6-fructosyltransferase associated with fructan accumulation in winter wheat during cold hardening. Biosci Biotechnol Biochem 66:2297–2305

Kawakami A, Yoshida M (2005) Fructan:fructan 1-fructosyltransferase, a key enzyme for biosynthesis of graminan oligomers in hardened wheat. Planta 223:90–104

Kawakami A, Yoshida M, Van den Ende W (2005) Molecular cloning and functional analysis of a novel 6&1-FEH from wheat (*Triticum aestivum* L.) preferentially degrading small graminans like bifurcose. Gene 358:93–101

Kawakami A, Yoshida M (2012) Graminan breakdown by fructan exohydrolase induced in winter wheat inoculated with snow mold. J Plant Physiol 169:294–302

Kiyomoto RK, Bruehl GW (1977) Carbohydrate accumulation and depletion by winter cereals differing in resistance to *Typhula idahoensis*. Phytopathology 67:206–211

Konstantinova T, Parvanova D, Atanassov A, Djilianov D (2002) Freezing tolerant tobacco, transformed to accumulate osmoprotectants. Plant Sci 163:157–164

Kuwabara T, Abe J, Moriyama M, Iriki N, Yoshida M, Takata K (1997) Relationship between physiological and ecological trait and snow mold damage in winter wheat varieties (*Triticum aestivum* L.). Breed Sci 47:271–277. (In Japanese with English summary, figs and tables)

Livingston III DP, Henson CA (1998) Apoplastic sugars, fructans, fructan exohydrolase, and invertase in winter oat: responses to second-phase cold hardening. Plant Physiol 116:403–408
Livingston III DP, Hincha DK, Heyer AG (2009) Fructan and its relationship to abiotic stress tolerance in plants. Cell Mol Life Sci 66:2007–2023
Matsumoto N (2009) Snow molds: a group of fungi that prevail under snow. Microbes Environ 24:14–20
Mohammad F, Windes JM, Souza E (1997) Total non-structural carbohydrates in winter wheat populations segregating for snow mold tolerance. Crop Sci 37:108–112
Nakajima T, Abe J (1994) Development of resistance to *Microdocium nivale* in winter wheat during autumn and decline of the resistance under snow. Can J Bot 72:1211–1215
Olien CR (1984) An adaptive response of rye to freezing. Crop Sci 24:51–54
Pontis HG (1989) Fructans and cold stress. J Plant Physiol 134:148–150
Pilon-Smits EAH, Ebskamp MJM, Paul MJ, Jeuken MJW, Weisbeek PJ, Smeekens SCM (1995) Improved performance of transgenic fructan-accumulating tobacco under drought stress. Plant Physiol 107:125–130
Pollok CJ, Cairns AJ (1991) Fructan metabolism in grasses and cereals. Annu Rev Plant Physiol Plant Mol Biol 42:77–101
Ritsema T, Smeekens S (2003) Fructans: beneficial for plant and humans. Curr Opin Plant Biol 3:223–230
Suzuki M (1989) Fructan in forage grasses with varying degrees of cold hardiness. J Plant Physiol 134:224–231
Tognetti JA, Salerno CL, Crespi MD, Pontis HG (1990) Sucrose and fructan metabolism of different wheat cultivars at chilling temperatures. Physiol Plant 78:554–559
Tronsmo AM (1986) Host water potentials may restrict development of snow mold fungi in low temperature-hardened grasses. Physiol Plant 68:175–179
Van den Ende W, Clerens S, Vergauwen R, Van Riet L, Van Laere A, Yoshida M, Kawakami A (2003) Fructan 1-exohydrolases. β-(2,1)-trimmers during graminan biosynthesis in stems of wheat? Purification, characterization, mass mapping, and cloning of two fructan 1-exohydrolase isoforms. Plant Physiol 131:621–631
Van den Ende W, Van Laere A (2007) Fructans in dicotyledous plants: occurrence and metabolism (Chapter 10). In: Shiomi N, Benkeblia N, Onodera S (eds) Recent advances in fructooligosaccharides research. Research Signpost Publisher, Kerala, ISBN: 81–308–014609 pp 1–14
Van den Ende W, Yoshida M, Clerens S, Vergauwen R, Kawakami A (2005) Cloning, characterization and functional analysis of novel 6-kestose exohydrolases (6-KEHs) from wheat (*Triticum aestivum* L.). New Phytol 166:917–932
Van Riet L, Nagaraj V, Van den Ende W, Clerens S, Wiemken A, Van Laere A (2006) Purification, cloning and functional characterization of a fructan 6-exohydrolase from wheat (*Triticum aestivum* L.). J Exp Bot 57:213–223
Van Riet L, Altenbach D, Vergauwen R, Clerens S, Kawakami A, Yoshida M, Van den Ende W, Wiemken A, Van Laere A (2008) Purification, cloning and functional differences of a third fructan 1-exohydrolase (1-FEHw3) from wheat (*Triticum aestivum*). Physiol Plant 133:242–253
Yoshida M, Abe J, Moriyama M, Shimokawa S, Nakamura Y (1997) Seasonal changes in the physical state of crown water associated with freezing tolerance in winter wheat. Physiol Plant 99:363–370
Yoshida M, Abe J, Moriyama M, Kuwabara T (1998) Carbohydrate levels among winter wheat cultivars varying in freezing tolerance and snow mold resistance during autumn and winter. Physiol Plant 103:8–16
Yoshida M, Kawakami A, Sato Y (2009) Growth and characteristics of fructan-accumulating transgenic rice: potential for utilization in forage. Dyn Biochem Process Biotechnol Mol Biol 3(SI 1):78–84
Yoshida M, Kawakami A, Van den Ende W (2007) Graminan metabolism in cereals: wheat as a model system (Chapter 10). In: Shiomi N, Benkeblia N, Onodera S (eds) Recent advances in fructooligosaccharides research. Research Signpost Publisher, Kerala, ISBN: 81–308–014609 pp 201–212

Yoshida M, Tamura K (2011) Research on fructan in wheat and temperate grasses in Japan. Jpn Agric Res Q 45:9–14

Yukawa T, Kobayashi M, Watanabe Y, Yamamoto S (1995) Studies on fructan accumulation in wheat (*Triticum aestivum* L.): IV. Fructan accumulation under cold treatment and its varietal difference in relation to the activities of sucrose-sucrose fructosyltransferase and fructan exo-hydrolase. Jpn J Crop Sci 64:801–806

Do Growth Kinetics of Snow-mold Fungi Explain Exponential CO_2 Fluxes Through the Snow?

Steven K. Schmidt, S. R. Frankel, R. L. Wagner and Ryan C. Lynch

Introduction

Due to the insulating properties of snow, the subnivean environment is a surprisingly hospitable habitat for the growth and development of fungal communities, especially in the late winter and early spring (Schadt et al. 2003; Schmidt and Lipson 2004; Schmidt et al. 2009). Plant pathogenic snow-mold fungi have received much attention for many years due to their economic importance (Hartig 1888; Hoshino et al. 2009; Hsiang et al. 1999; Simms 1967; Tojo and Newsham 2012), but the identity and activity of nonpathogenic fungi of the under-snow environment have received much less study (Schadt et al. 2003; Freeman et al. 2009a, b; Schmidt et al. 2009). In polar and alpine areas of the world, mats of saprophytic snow molds are commonly observed as snow melts in both forested and tundra biomes (Nemergut et al. 2005; Schmidt et al. 2007, 2008). Recent biogeochemical studies in high-elevation and high-latitude environments have shown conclusively that microbial activity under the snow pack contributes significantly to gas fluxes and to carbon and nitrogen cycling (Brooks et al. 1998; Grogan and Jonasson 2006; Larsen et al. 2007; Lipson et al. 1999; Schmidt and Lipson 2004; Monson et al. 2006). In addition, Schmidt et al. (2008, 2009, 2012) have demonstrated that nonpathogenic snow-mold fungi are among the dominant organisms of the sub-nivean environment and carry out important ecosystem functions in tundra and forested ecosystems. Snow-mold fungi are especially important in coniferous forests of the Colorado Rocky Mountains, where dry conditions limit microbial decomposition during much of the summer. In contrast, the relatively mild and moist conditions under alpine snow packs are conducive to decomposition and fungal growth resulting in high rates of CO_2 flux through the late-winter snowpack. At these sites, fluxes of respired CO_2 through the snow pack can be high, accounting for a significant fraction of the total annual

S. K. Schmidt (✉) · S. R. Frankel · R. L. Wagner · R. C. Lynch
Department of Ecology and Evolutionary Biology, University of Colorado,
Boulder, Colorado 80309, USA
e-mail: steve.schmidt@colorado.edu

R. Imai et al. (eds.), *Plant and Microbe Adaptations to Cold in a Changing World*,
DOI 10.1007/978-1-4614-8253-6_21,

ecosystem CO_2 flux (Monson et al. 2006). In addition, late-winter and early-spring CO_2 fluxes have been observed to increase exponentially through the snow pack (Brooks et al. 1997; Mast et al. 1998; Monson et al. 2006; Sommerfeld et al. 1996). These exponential kinetics result in unexpectedly high Q_{10} values for sub-nivean CO_2 fluxes when calculated across the seasonal range of soil temperatures (Monson et al. 2006; Schmidt et al. 2009). However, it is still a mystery whether these high Q_{10} values are due to physical or biological phenomena or a combination of both working in concert (Lipson et al. 2009; Schmidt et al. 2009).

The question addressed in this chapter is whether isolated snow-mold fungi from the same sites studied by Monson et al. (2006) have growth kinetics and low-temperature Q_{10} values similar to those for CO_2 fluxes observed in the field. This chapter builds on the findings of Schmidt et al. (2009), which showed that several snow-mold fungi from this site could grow exponentially at sub-zero temperatures and that Q_{10} values for sub-zero changes in exponential growth rates were orders of magnitude higher for sub-zero temperatures compared to Q_{10} values at temperatures above 0 °C. Here, we test whether Q_{10} values calculated from *linear growth rates* show the same extreme changes at low temperatures as those reported by Schmidt et al. (2009) for *exponential growth rates*.

Materials and Methods

Field data and isolates were obtained from the intensively studied forest sites described in detail elsewhere (Bowling et al. 2009; Monson et al. 2006; Porter et al. 2008; Schmidt et al. 2007; Weintraub et al. 2007). The study area is at 3,050 m above sea level (40° 1' 58"N; 105° 32' 47"W) and is fully instrumented with eddy-flux towers, data loggers (e.g., for soil water and temperature), and equipment for measuring year-round gas fluxes and gas fluxes through the snow pack (Bowling et al. 2009; Monson et al. 2006). This mixed forest is dominated by *Pinus contorta*, *Picea engelmannii*, and *Abies lasiocarpa* and the soils are sandy inceptisols on top of granite moraine material; an organic horizon ranging from 0 to 6 cm covers most of the soils at the site (Weintraub et al. 2007).

The snow-mold fungi were isolated from sub-nivean fungal mats (see Schmidt et al. 2007, 2009 for photos) using cold temperatures (< 0 °C) and media that mimics the contents of the forest floor (Schmidt et al. 2008, 2009). Inocula for experiments were grown at low temperatures on the same media used in each experiment; uniform plugs for inoculation were obtained by using a sterile 6-mm-diameter Biopsy Punch (AcuPunch, Acuderm Inc., Florida, USA). All plugs for inoculating experiments were taken at the outer edge of expanding colonies to ensure that the fungi used for all treatments were at same metabolic state at the beginning of the experiments (Schmidt et al. 2008).

Growth rates at different temperatures (− 3, − 2, − 0.3, 3.8, 7, and 13 °C) were estimated using the approach of Kerry (1990) and Schmidt et al. (2008). Briefly, at each sampling time, we marked the extent of radial growth along four axes (to obtain four estimates of radius at each sampling interval) and the mean of the four measurements was used at each time point. All temperature treatments were done in triplicate in low-temperature incubators (Sheldon Manufacturing, Cornelius, OR, USA). Incubators were calibrated according to the manufacturer's instructions and temperatures were measured every 15 min for the duration of the experiments using data loggers, model UA-002-64 (Onset, Pocasset, MA, USA).

The media used for growth experiments involving isolates in the former Zygomycota (now Mucoromycotina according to Hibbett et al. 2007) contained (per liter of water): 5 g of inulin, 0.5 g yeast extract, 0.5 g of KCl, 1 g of KH_2PO_4, 1 g of $(NH_4)_2SO_4$, 2 g of $NaNO_3$, 12.3 g of $MgSO_4{\cdot}7H_2O$, 20 g of agar, 50 mg of $CaCl_2$, 10 mg of $FeSO_4$, 10 mg of $CuSO_4$, 5 mg of $MnSO_4$, 1 mg of $ZnSO_4$, and 1 ml of soil extract solution to provide other growth factors and trace minerals (Schmidt et al. 2009). The media used for growth experiments with Ascomycetes and Basidiomycetes contained (per liter of water): 5 g glucose, 5 g of glycerol, 1 g finely ground ponderosa pine leaf litter, 0.5 g yeast extract, 0.5 g of KCl, 1 g of KH_2PO_4, 1 g of $(NH_4)_2SO_4$, 2 g of $NaNO_3$, 0.5 g of $MgSO_4{\cdot}7H_2O$, 1 g of yeast extract, 1 g of peptone, and 20 g of agar.

Kinetic Considerations

Linear growth rates were calculated using linear regression of the increase in radial growth versus time as described elsewhere (Kerry 1990; Schmidt et al. 2008). Exponential growth rates (for the data of Haines 1931) were estimated using the exponential growth equation

$$L_t = L_0 e^{\mu t} \tag{1}$$

where μ is exponential growth rate with units of h^{-1} and L_t and L_0 represent the total length of fungal hyphae (millimeters) at time *t* and time 0, respectively (Schmidt et al. 2009). To obtain parameter estimates from each equation, growth curves were analyzed using the nonlinear regression package of Kaliedagraph (Synergy Software Co., Reading, PA, USA).

The effects of temperature on rates of biological processes were evaluated using the Q_{10} relationship for Arrhenius-like behavior of enzymes and organisms (Hochachka and Somero 1984)

$$Q_{10} = (r_2 / r_1)^{10/T_2 - T_1} \tag{2}$$

where r_1 is the growth rate at temperature 1 (T1) and r_2 is the growth rate at temperature 2 (T2).

Table 1 Phylogenetic affinities of the snow-mold fungi used in this study. NCBI accession numbers for our isolates and the closest match to each isolate are given in parentheses. All these fungi were isolated from the research sites discussed in the text and by Monson et al. (2006) and Schmidt et al. (2008, 2009)

Isolate no.	Fungal group	Closest match	Ref. for closest match
316	Mucorales *Helicostylum elegans*	(EU428767) (AF157139)	O'Donnell et al. 2001
317	Mortierellales *Mortierella alpina*	(EU428769) (AJ271630)	Mackenzie et al. 2000
319	Mucorales *Mucor mucedo*	(EU428766) (X89434)	O'Donnell et al. 2001
*2AB**	Atheliales *Athelia bombacina*	(M55638)	Illingworth et al. 1991
A3	Melanommataceae *Herpotrichia juniperi*	(U42483)	Berbee (1996)

Results and Discussion

In a previous study, we found that the Q_{10} for exponential growth rates of snow-mold fungi in the Mucoromycotina increased substantially at sub-zero temperatures compared to Q_{10} values at temperatures above 0 °C (Schmidt et al. 2009). For example, isolates 316, 317, and 319 had Q_{10} values (comparing exponential growth rates) averaging 2.6 at temperatures above 0 °C compared to an average of 210 at temperatures below 0 °C (Table 2 in Schmidt et al. 2009). In this study, we extend that work to test whether these patterns hold up for linear growth rates using a broader range of fungi including Basidiomycetes and Ascomycetes from our field sites and using data for other fungi from the literature.

The phylogenetic affinities of the snow-mold fungi discussed in this chapter are shown in Table 1. All these fungi have been isolated numerous times from under-snow fungal mats on the forest floor at our research sites. The fungi in the Mucoromycotina discussed here have been identified phylogenetically in our previous studies as falling into the Mortierellales (Isolate 317) and Mucorales (Isolates 316 and 319) Orders of the Mucoromycotina (Schmidt et al. 2008). Recently, we have also identified two deeply melanized isolates: a relative (Isolate A3) of the Basidiomycete, *Athelia bombacina*, and an Ascomycete (Isolate 2AB*) that is phylogeneticaly very similar to *Herpotrichia juniperi* (Table 1).

We conducted experiments to determine if snow molds could grow across the range of surface-soil temperatures (−3 to 0 °C) commonly observed under the snow in the late winter and early spring at our research sites (Monson et al. 2006) and at higher temperatures encountered at this site in the early summer months. For example, Fig. 1 shows the growth curves for isolates 316 and 317 across the environmentally relevant temperature range. These data clearly show that these isolates can grow across the temperature range that corresponds to our temperature measurements at the field sites. The two isolates of the Mucoromycotina

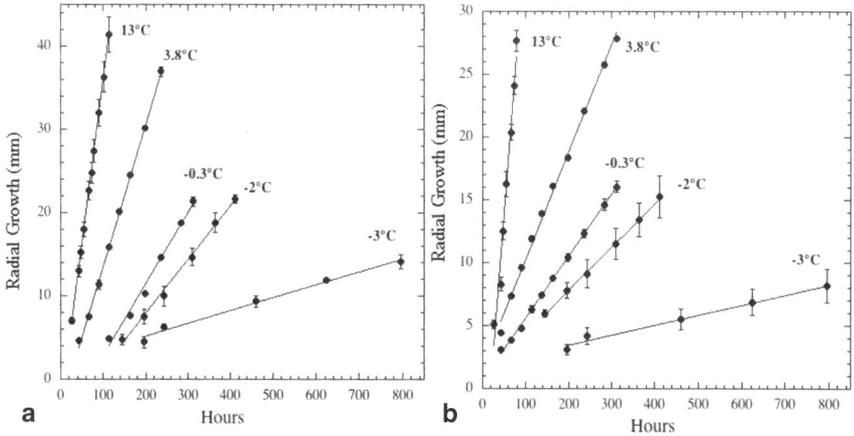

Fig. 1 Linear (*radial*) growth of isolates 316 **a** and 317 **b** at temperatures of −3 to 13 °C. Each point is the mean of three replicates and error bars are one standard deviation of the mean. Lines are linear regression fits to the data with R^2 values of greater than 0.96 for all regressions. The data show the extreme change in rate that occurs as temperatures drop from −2 to −3 °C resulting in the extreme Q_{10} values for this temperature interval (see Table 2)

(isolaters 316 and 317) grew at higher rates at all temperatures tested than did the Ascomycete and isolates. For example, at −2 °C isolates 316 and 317 exhibited growth rates of 0.035 and 0.065 mm/h, respectively, whereas isolates 2AB* and A3 both grew at an average rate of 0.01 mm/h (data not shown). All our isolates grew at higher rates than did a group of five Antarctic fungi that exhibited growth rates of 0.002 mm/h at −2 °C (Hughes et al. 2003).

More important for this study is that the growth rates across this temperature range change in a fashion that is analogous to changes in CO_2 flux rates through the snowpack that we observed in our field studies. That is, Q_{10} values are dramatically higher at lower temperature ranges compared to higher temperatures for these fungi just as we have observed for our field respiration data (cf Monson et al. 2006; Schmidt et al. 2009). Table 2 shows Q_{10} values for all five of our isolates. The most dramatic Q_{10} values we observed were for isolates 316 and 317 when temperatures increased from −3 to −2 °C. Across this 1 °C change in temperature, the rate of linear growth increased fourfold for both isolates, which translates into a Q_{10} of over 1×10^6 (Table 2). All other calculated Q_{10} values were less dramatic but, nonetheless, all isolates showed a pattern of higher Q_{10} values at lower temperatures compared to temperatures above 0 °C (Table 2).

We also calculated the Q_{10} values for a fungus that spoils meat at sub-zero temperatures. These data were previously published by Haines (1931) and the growth curves from which growth rates (using Eq. 1) were obtained are shown in Fig. 2. As with our soil isolates, Q_{10} values were higher at temperatures below 0 °C compared to temperatures above 0 °C; the Q_{10} was 3.9 for 0 to 5 °C compared to a Q_{10} of 111 for −5 to 0 °C. The growth data of Haines (1931) are exponential (Fig. 2) because

Table 2 Q_{10} values calculated using linear growth rates and Eq. 2 for different temperature intervals. These calculations demonstrate the extreme sensitivity of apparent Q_{10} values to sub-zero changes in temperature and are similar to Q_{10} values observed for through-snow CO_2 fluxes (Schmidt et al. 2009)

Temperature interval for the Mucoromycotina				
Isolate no.	−3 to −2 °C	−2 to −0.3 °C	−0.3 to +3.8 °C	3.8 to 13 °C
Isolate 316	1.7×10^6	5.4	4.1	2.1
Isolate 317	2.3×10^6	7.7	3.8	4.3
Isolate 319	N. D.	38.8	7.1	2.1
Temperature interval for the Ascomycete and Basidiomycete				
Isolate no.	−2 to 0 °C	0 to 2 °C	2 to 7 °C	
*Isolate 2AB**	6.1	3.0	2.7	
Isolate A3	3.4	1.8	1.7	

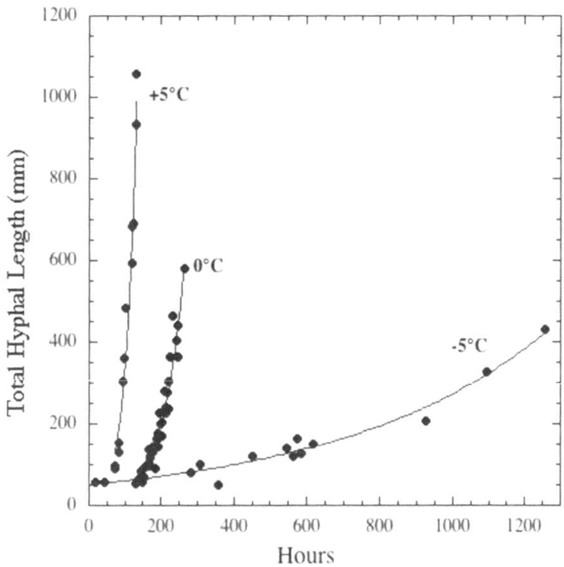

Fig. 2 Exponential growth of *Sporotrichum carnis* at temperatures of −5, 0, and 5 °C. Lines are fits of the exponential growth equation (Eq. 1) to the data from Haines (1931) with R^2 values of greater than 0.94 for all curve fits. The data for growth at −5 °C were truncated to remove the lag phase. Q_{10} values were 3.9 for the temperature interval of 0 to 5 °C compared to a Q_{10} of 111 for the temperature interval of −5 to 0 °C

they measured total length of hyphae produced rather than the radial growth of colonies as was done in this study. A better comparison to Haines' data (Fig. 2) would be to plot the areal growth of fungal colonies as was done by Schmidt et al. (2009). Areal growth of the fastest growing isolates (316, 317, and 319) studied here also shows exponential growth during the "early" phase of colony growth (i.e., the first 400 and 800 h at −2 and −3 °C, respectively), and Q_{10} values calculated from those exponential growth rates are very similar to those calculated using linear growth rates in this study (*cf.* Schmidt et al. 2009).

Overall, the work discussed above shows conclusively that the behavior of Q_{10} values for through-snow CO_2 flux rates observed by many investigators (Monson et al. 2006 and references therein) *could* potentially be explained by the growth behavior of the dominant denizens of the subnivean environment, that is, saprophytic snow-mold fungi. This is not to say that we have proven beyond a doubt that this is the case and it is likely that a combination of biological and physical factors are involved in producing the exponential CO_2 flux kinetics observed in many low-temperature filed studies. In the next section, we speculate on why Q_{10} values are so different at subzero temperatures.

Why Are Q_{10} Values So High At Sub-zero Temperatures?

In this section, we will present some of the ideas that can explain why Q_{10} (and other kinetic parameters) change so dramatically at low temperatures. We do not mean for this section to be an exhaustive discussion of this topic, but rather as an introduction to some ideas that are discussed in much more detail (and much more articulately) in the papers cited below.

First, it must be stated that even in the classical derivation of the Q_{10} model (as an Arrhenius function) it was pointed out that Q_{10} values should increase as temperatures decrease (Davidson and Janssens 2006). This shift is presumably due to changes in the fraction of molecules that have energy less than the required activation energy for a reaction as temperature decreases. (Please see Davidson and Janssens (2006) and references therein for a more thorough discussion of this idea.)

It has also been pointed out by many workers that substrate availability can limit microbial growth and respiration rates at low temperatures under the snow (Lipson et al. 2000) and that substrate availability decreases rapidly and nonlinearly as temperatures drop (Ley et al. 2004; Mikan et al. 2002). There are several reasons as to why substrate availability likely decreases with decreasing temperature. The most often stated explanation for exponential changes in substrate availability as temperatures decrease below the freezing point is the fact that water availability (and therefore nutrient availability) decreases exponentially as soil temperatures drop from 0 to −3 °C (see data and discussion in Ley et al. 2004; Romanovsky and Osterkamp 2000). This physical phenomenon could help explain dramatic rate changes in soils (and fungal cultures) as temperatures drop from 0 to −3 °C, but at least in the study of Monson et al. (2006) the exponential change in soil water availability happened at a much faster rate than did the exponential change in soil respiration rates (see Schmidt et al. 2009 for more discussion). Finally, there is evidence that growth rate-limiting enzymes show decreased substrate affinity as temperatures decrease (Davidson and Janssens 2006; Nedwell 1999), which would result in decreased growth rates and increased Q_{10} values as temperatures drop. Thus, there are many possible explanations as to why we see large changes in Q_{10} values as temperatures decrease, however much more work is needed to test the mechanistic basis of these ideas in pure cultures of fungi and in soils.

Acknowledgments Thanks to M. M. Gebauer, K. L. Wilson, and A. F. Meyer for assistance in the laboratory and D. R. Bowling, R. K. Monson, D. A. Lipson, and N. Trahan for helpful discussions. This work was supported by a grant (MCB-0455606) from the Microbial Observatories Program of the National Science Foundation of the USA.

References

Berbee ML (1996) Loculoascomycete origins and evolution of filamentous ascomycete morphology based on 18S rRNA gene sequence data. Mol Biol Evol 13:462–470

Bowling DR, Massman WJ, Schaeffer SM, Burns SP, Monson RK, Williams MW (2009) Biological and physical influences on the carbon isotope content of CO_2 in a subalpine forest snowpack, Niwot Ridge, Colorado. Biogeochemistry 95:37–59

Brooks PD, Schmidt SK, Williams MW (1997) Winter production of CO_2 and N_2O from alpine tundra: environmental controls and relationship to inter-system C and N fluxes. Oecologia 110:403–413

Brooks PD, Williams MW, Schmidt SK (1998) Inorganic N and microbial biomass dynamics before and during spring snowmelt. Biogeochemistry 43:1–15

Davidson EA, Janssens IA (2006) Temperature sensitivity of soil carbon decomposition and feedbacks to climate change. Nature 440:165–173

Freeman KR, Martin AP, Karki D, Lynch RC, Mitter MS, Meyer AF, Longcore JE, Simmons DR, Schmidt SK (2009a) Evidence that chytrids dominate fungal communities in high-elevation soils. Proc Natl Acad Sci U S A 106:18315–18320

Freeman KR, Pescador MY, Reed SC, Costello EK, Robeson MS, Schmidt SK (2009b) Soil CO_2 flux and photoautotrophic community composition in high-elevation, "barren" soils. Environ Microbiol 11:674–686

Grogan P, Jonasson S (2006) Ecosystem CO_2 production during winter in a Swedish subarctic region: the relative importance of climate and vegetation type. Glob Chang Biol 12:1479–1495

Haines RB (1931) The influence of temperature on the rate of growth of *Sporotrichum carnis*, from − 10 °C to + 30 °C. J Exp Biol 8:379–388

Hartig R (1888) *Herpotrichia nigra* n. sp. Allg Forst Jagdztg 64:15–17

Hibbett DS et al. (2007) A higher-level phylogenetic classification of the Fungi. Mycol Res 111:509–547

Hochachka PW, Somero GN. 1984. Biochemical adaptation. Princeton University Press, Princeton

Hoshino T, Xiao N, Tkachenko OB (2009) Cold adaptation in phytopathogenic fungi causing snow molds. Mycoscience 50:26–38

Hsiang T, Matsumoto N, Millett SM (1999) Biology and management of *Typhula* snow molds of turfgrass. Plant Dis 83:788–798

Hughes KA, Lawley B, Newsham KK (2003) Solar UV-B radiation inhibits the growth of Antarctic terrestrial fungi. Appl Environ Microbiol 69:1488–1491

Illingworth CA, Andrews JH, Bibeau C, Sogin ML (1991) Phylogenetic placement of *Athelia bombacina*, *Aureobasidium pullulans* and *Colletotrichum gloeosporioides* inferred from sequence comparisons of small-subunit ribosomal RNAs. Exp Mycol 15:65–75

Kerry E (1990) Effects of temperature on growth rates of fungi from sub-Antarctic Macquarie Island and Casey, Antarctica. Polar Biol 10:293–299

Larsen KS, Grogan P, Jonasson S, Michelsen A (2007) Respiration and microbial dynamics in two subarctic ecosystems during winter and spring thaw: effects of increased snow depth. Arct Alp Res 39:268–276

Ley RE, Williams MW, Schmidt SK (2004) Microbial population dynamics in an extreme environment: controlling factors in talus soils at 3750 m in the Colorado Rocky Mountains. Biogeochemistry 68:313–335

Lipson DA, Schmidt SK, Monson RK (1999) Links between microbial population dynamics and nitrogen availability in an alpine ecosystem. Ecology 80:1623–1631

Lipson DA, Schmidt SK, Monson RK (2000) Carbon availability and temperature control the post-snowmelt decline in alpine soil microbial biomass. Soil Biology Biochemistry 32:441–448

Lipson DA, Monson RK, Schmidt SK, Weintraub MN (2009) The trade-off between growth rate and yield in microbial communities and the consequences for soil respiration in a high elevation coniferous forest. Biogeochemistry 95:23–35

Mackenzie DA, Wongwathanarat P, Carter AT, Archer DB (2000) Isolation and use of a homologous histone H4 promoter and a ribosomal DNA region in a transformation vector for the oil-producing fungus *Mortierella alpina*. Applied Environmental Microbiology 66:4655–4661

Mast MA, Wickland WP, Striegl RT, Clow DW (1998) Winter fluxes of CO_2 and CH_4 from subalpine soils in Rocky Mountain National Park, Colorado. Global Biogeochem Cycles 12:607–620

Mikan CJ, Schimel JP, Doyle AP (2002) Temperature controls of microbial respiration in arctic tundra soils above and below freezing. Soil Biology Biochemistry 34:1785–1795

Monson RK, Lipson DL, Burns SP, Turnipseed AA, Delany AC, Williams MW, Schmidt SK (2006) Winter forest soil respiration controlled by climate and microbial community composition. Nature 439:711–714

Nedwell DB (1999) Effect of low temperature on microbial growth: lowered affinity for substrates limits growth at low temperature. FEMS Microbiol Ecol 30:101–111

Nemergut DR, Costello E, Meyer AF, Pescador MY, Weintraub MW, Schmidt SK (2005) Structure and function of alpine and arctic soil microbial communities. Res in Microbiol 156:775–784

O'Donnell K, Lutzoni FM, Ward TJ, Benny GL (2001) Evolutionary relationships among mucoralean fungi (Zygomycota): evidence for family polyphyly on a large scale. Mycologia 93:286–296

Porter TM, Schadt CW, Rizvi L, Martin AP, Schmidt SK, Scott-Denton L, Vilgalys R, Moncalvo J-M. (2008) Widespread occurrence and phylogenetic placement of a soil clone group adds a prominent new branch to the fungal tree of life. Mol Phylogenet Evol 46:635–644

Romanovsky VE, Osterkamp TE (2000) Effects of unfrozen water on heat and mass transport processes in the active layer of permafrost. Permafrost Periglac Process 11:219–239

Schadt CW, Martin AP, Lipson DA, Schmidt SK (2003) Seasonal dynamics of previously unknown fungal lineages in tundra soils. Science 301:1359–1361

Schmidt SK, Lipson DA (2004) Microbial growth under the snow: implications for nutrient and alleochemical availability in temperate soils. Plant Soil 259:1–7

Schmidt SK, Costello EK, Nemergut DR, Cleveland CC, Reed SC, Weintraub MN, Meyer AF, Martin AP (2007) Biogeochemical consequences of rapid microbial turnover and seasonal succession in soil. Ecology 88:1379–1385

Schmidt SK, Wilson KL, Gebauer MM, Meyer AF, King AJ (2008) Phylogeny and ecophysiology of opportunistic "snow molds" from a sub-alpine forest ecosystem. Microb Ecol 56:681–687

Schmidt SK, Wilson KL, Monson RK, Lipson DA (2009) Exponential growth of "snow molds" at sub-zero temperatures: an explanation for high beneath-snow respiration rates and Q_{10} values. Biogeochemistry 95:13–21

Schmidt SK, Naff CS, Lynch RC (2012) Fungal communities at the edge: Ecological lessons from high alpine fungi. Fungal Ecology 5:443–452

Simms HR (1967) On the ecology of *Herpotrichia nigra*. Mycologia 59:902–909

Sommerfeld RA, Massman WJ, Musselman RC, Mosier AR (1996) Diffusional flux of CO_2 through snow: Spatial and temporal variability among alpine-subalpine sites. Global Biogeochem Cycles 10:473–482

Tojo M, Newsham KK (2012) Snow moulds in polar environments. Fungal Ecology 5:395–402

Weintraub MN, Scott-Denton LE, Schmidt SK, Monson RK (2007) The effects of tree rhizodeposition on soil exoenzyme activity, dissolved organic carbon, and nutrient availability in a subalpine forest ecosystem. Oecologia 154:327–338

Change in Snow Mold Flora in Eastern Hokkaido and its Impact on Agriculture

Naoyuki Matsumoto and Tamotsu Hoshino

Introduction

A group of fungi cause disease on dormant plants under snow. They are referred to as snow molds. Snow mold fungi are taxonomically diverse and vary ecologically (Matsumoto 2009). Snow cover protects plants from freezing but provides snow molds with optimal conditions to prevail (Matsumoto 1994). The environment under snow is characterized by constant low temperatures at about 0 °C, darkness, and high moisture, which bring about the following features of the habitat of snow molds: (1) low temperature restricts species diversity and only psychrophiles can grow (Matsumoto and Hoshino 2009), (2) reserve carbohydrates are depleted through respiration and plant resistance decreases with time (Nakajima and Abe 1994), and (3) resources, i.e., plant tissues, do not increase under snow (Matsumoto 1994).

The life cycle of snow mold is clearly divided into active and dormant stages based on the presence or absence of snow cover (Hsiang et al. 1999). The life cycle of two major snow molds, *Typhula ishikariensis* and *Sclerotinia borealis* is schematically presented along with factors critical for disease incidence (Fig. 1). The duration of persistent snow cover is most important and changing winter climate concerns us most seriously. The timing of persistent snow cover affects plant resistance through plant cold hardiness. Humid and cloudy summer climate reduces snow mold resistance of alfalfa through foliar disease in some areas.

Tomiyama (1955) divided Hokkaido into two, according to the major snow mold fungi (Fig. 2). The *Sclerotinia* snow mold fungus, *S. borealis* was prevalent in eastern Hokkaido characterized by severe soil frost and thin snow cover, whereas little or slight soil frost and deep snow cover favored the occurrence of *Typhula* spp.

N. Matsumoto (✉)
Graduate School of Agriculture, Hokkaido University, Sapporo, Japan
e-mail: nowmat@for.agr.hokudai.ac.jp

T. Hoshino
National Institute of Advanced Industrial Science and Technology (AIST), Higashi-Hiroshima, Japan

R. Imai et al. (eds.), *Plant and Microbe Adaptations to Cold in a Changing World*,
DOI 10.1007/978-1-4614-8253-6_22,

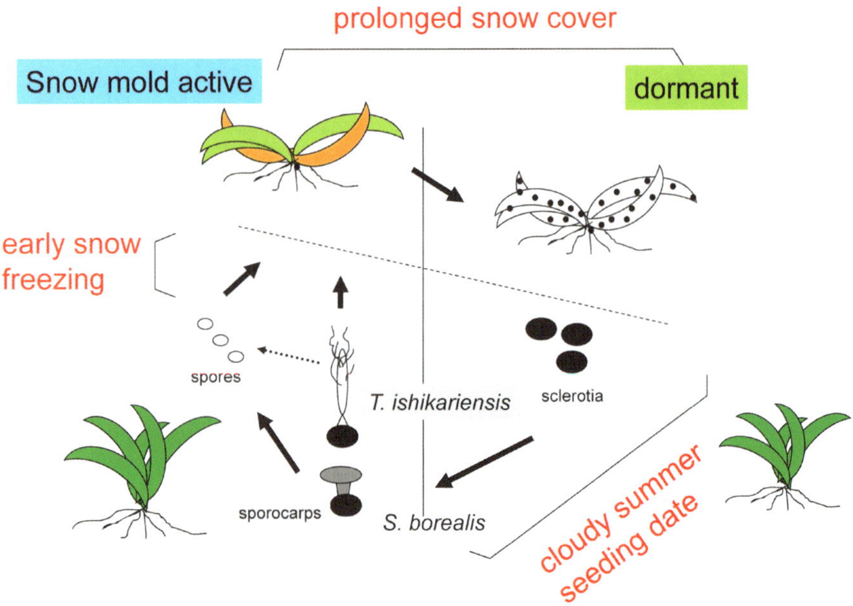

Fig. 1 Life cycle of the two major snow mold fungi, *Sclerotinia borealis* and *Typhula ishikariensis*. Cloudy summers predispose forage crops to snow mold through the incidence of leaf disease. Seeding date is critical for both winter cereals and forage crops

in western Hokkaido. This pattern is not absolute and simply represents the outcome of host–parasite interactions, especially in the case of *Sclerotinia* snow mold; the disease could occur even in Sapporo, western Hokkaido when less hardy plants such as perennial ryegrass (*Lolium perenne*) were grown (Matsumoto and Araki 1982). Matsumoto et al. (1982) found that *T. ishikariensis* biotype B occurred in less snowy areas, including eastern Hokkaido. They further revealed that biotype A was prevailing in eastern Hokkaido presumably due to the change in winter climate (Matsumoto et al. 2000).

Sclerotinia borealis

Persistent snow cover occurred much later than usual in eastern Hokkaido in the 1974–1975 winter with frozen soil more than 30 cm deep, which predisposed orchardgrass (*Dactylis glomerata*) to *Sclerotinia* snow mold (Araki 1975). To make matters worse, deep snowfall in late March prolonged thawing for about a month. Nearly, a half of grasslands in eastern Hokkaido suffered from *Sclerotinia* snow mold and 10% of the affected fields required reseeding and planting with other crops or were abandoned.

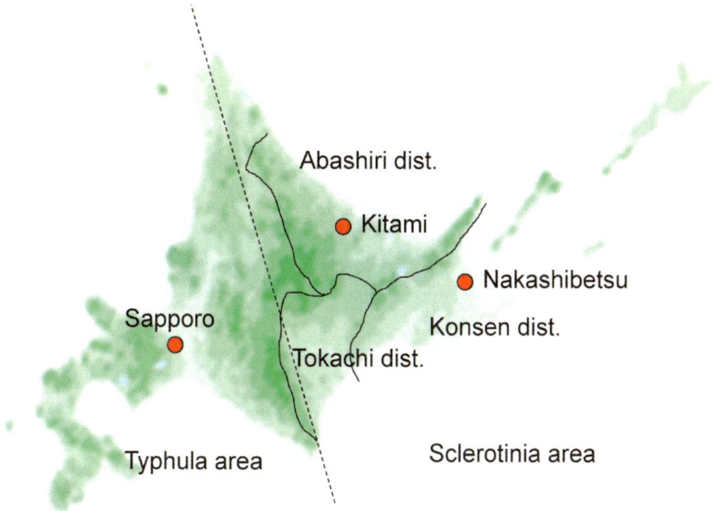

Fig. 2 Hokkaido is divided into the *Typhula* and *Sclerotinia* areas. Districts in eastern Hokkaido are shown in the map. Summer in districts of Abashiri and Tokachi is warm enough for upland crops such as wheat, beans, etc., whereas forage grasses and legumes are exclusively cultivated in the Konsen district due to its damp and cool summer climate. Breeding and experimental cites are indicated by *red* dots

Sclerotinia borealis is an epidemic snow mold fungus and occurs irregularly (Fig. 3). Recent climate change has reduced soil frost indicative of the *Sclerotinia* area in eastern Hokkaido (Hirota et al. 2006). Also, orchardgrass was replaced with more hardy, timothy (*Phleum pratense*). These facts, along with improved cultivation methods, minimized the occurrence of *Sclerotinia* snow mold.

Seeding date is critical to *Sclerotinia* snow mold on grasses as is the case with *Typhula* snow mold on winter wheat (Bruehl et al. 1975). Delay in seeding significantly impairs overwintering of timothy (Nissinen 1996; Sato et al. 2009). The critical issue has been ignored, inciting outbreak of *Sclerotinia* snow mold on first-year timothy in the Tokachi district, eastern Hokkaido in 2008. Timothy is sown after corn to renovate grasslands in Tokachi, but corn plants are left unharvested in the field as late as possible to attain full maturity, which results in the delay of seeding timothy. Timothy should have been sown at least by early September (Sato et al. 2009). Thus, cultivation methods established for snow mold control are liable to be ignored or forgotten due to changes in agroeconomic situation.

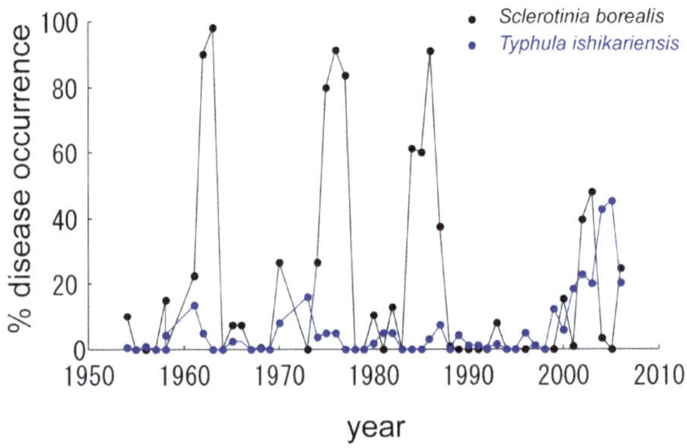

Fig. 3 Occurrence of *Typhula ishikariensis* and *Sclerotinia borealis* on winter wheat in Kitami. No chemical was sprayed. Note that *T. ishikariensis* has been increasing during the last decade. *T. ishikariensis* biotype B, originally prevalent there, is considered to have been replaced by biotype A before that. The assumption was supported by the yield decrease of alfalfa between 1982 and 1993 (see Fig. 4)

Fig. 4 Yield of Kita-wakaba alfalfa in 1982 and 1993 in Kitami and Sapporo. Note that yield decreased between 1982 and 1993 in Kitami. (Ueda et al.1985; Yamaguchi et al. 1995)

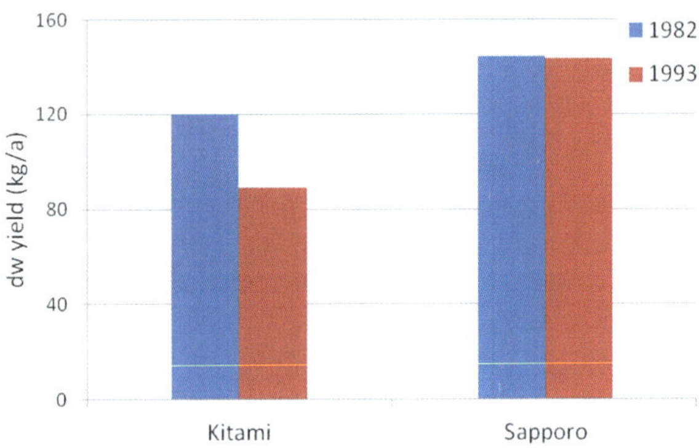

Typhula ishikariensis

Tomiyama (1955) did not recognize intraspecies differentiation in the *Typhula ishikariensis* complex in Japan (Matsumoto et al. 1982). *T. ishikariensis* that he referred was considered to be biotype A with a broader host range. Biotype A can attack dicots such as canola and alfalfa, as well as monocots and prevails in snowy western Hokkaido. Biotype B exists in less snowy areas, including eastern Hokkaido and is excluded from monocots when biotype A coexists (Matsumoto and Sato 1983). Two examples are illustrated below that relate to the change in snow mold flora from *T. ishikariensis* biotype B to biotype A in eastern Hokkaido.

Typhula ishikariensis occurred consistently at low levels till 1998 in eastern Hokkaido (Fig. 3) and biotype B was regarded as the principal taxon of the *T. ishikariensis* complex (Matsumoto et al. 1982). Biotype B predominated over

Fig. 5 Severe damage of *Typhula ishikariensis* biotype A on alfalfa lines selected for Lept leaf spot resistance in Nakashibetsu, Konsen

biotype A mainly due to shallow snow cover, and possibly foliage application of fungicides favored biotype B, which could also attack wheat roots. However, snow cover occurred 1 month earlier in eastern Hokkaido in the winter of 1998–1999. Farmers were unable to spray fungicides on winter wheat. Consequently, biotype A caused serious damage and 30 % of wheat crop had to be abandoned (Matsumoto et al. 2000).

Alfalfa (*Medicago sativa*) used to grow well in the districts of Abashiri (Fig. 4) and Tokachi except mountainous areas in Tokachi where *T. ishikariensis* biotype A caused significant damage (Komatsu 1983). Scientists tried to extend alfalfa cultivation to the Konsen district where freezing tolerance was at first considered most critical. Sixty cultivars, including Canadian cultivars were tested in the field. Many Canadian cultivars with strong fall dormancy failed to survive the first winter due to soil heaving before winter (Takeda and Nakajima 1997a). Second-year experiments revealed that Lepto leaf spot caused by *Leptosphaerulina briosiana* reduced growth and winter survival (Takeda and Nakajima 1997b). Damp, cool summer climate in the Konsen district favored the leaf disease and alfalfa plants were unable to harden enough. *T. ishikariensis* biotype A was not observed.

Consequently, breeding objectives shifted to Lept leaf spot resistance. Freezing-tolerant cultivars were susceptible to the disease since they were bred under dry summer conditions (Takeda and Nakajima 1997c). Takeda et al. (1998) found resistant alfalfa plants among commercially available cultivars and lines. They conducted further field screening. Unusually, snowy winter in 1998–1999 favored the outbreak of *T. ishikariensis* biotype A even in Nakashibetsu, Konsen, and all the experimental lines were badly damaged by snow mold (Fig. 5).

Sapporo represents one of the alfalfa breeding cites and is located in snowy western Hokkaido (Fig. 2). Field experiments there naturally select snow mold resistance. They were also aware of the significance of Lepto leaf spot and released "Hisa-wakaba" in collaboration with the breeding cites in eastern Hokkaido (Yamaguchi et al. 1995). "Hisa-wakaba" alfalfa improved the productivity in Kitami and made alfalfa cultivation possible in the Konsen district (Fig. 6).

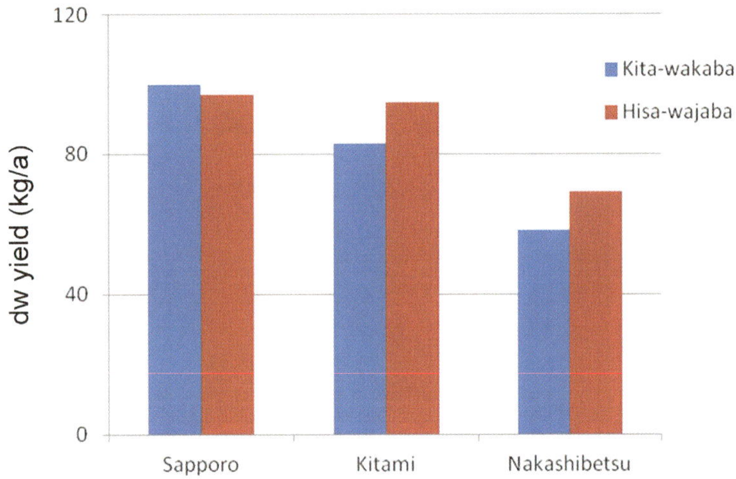

Fig. 6 Yield of Kita-wakaba and Hisa-wakaba in three alfalfa cites, Sapporo in western Hokkaido, Kitami in the Abashiri district, and Nakashibetsu in the Konsen district (see Fig. 2). Figures indicate average annual yield during 1990–1993.(Yamaguchi et al. 1995)

Conclusion

Persistent snow cover occurs much earlier than before in eastern Hokkaido, which has alleviated severe soil frost (Hirota et al. 2006). Contrasting winter climate in Hokkaido is no longer obvious in terms of snow mold flora. Farmers and scientists have established agricultural methods suitable for eastern Hokkaido and some of them are not effective. We illustrated some examples.

Winter climate, as well as summer climate, is likely to fluctuate more seriously than ever. Snow mold fungi are dependent on snow cover and physiologic conditions of plants may vary every year. These parameters affect host–parasite interactions, resulting in the need for novel strategies against snow molds.

Breeding is doubtlessly most effective and the diversity in agro-ecosystem should also be remembered. These two issues are not mutually exclusive but difficult to harmonize. Multidisciplinary collaboration is essential to cope with unpredictable climate change.

References

Araki T (1975) Outbreak of snow mold on forage grasses in Hokkaido. Shokubutsu-boueki 29:484–488

Bruehl GW, Kiyomoto R, Peterson C, Nagamitsu M (1975) Testing winter wheats for snow mold resistance in Washington. Plant Disease Reporter 51:815–819

Hirota T, Iwata Y, Hayashi M, Suzuki S, Hamasaki T, Sameshima R, Takayabu I (2006) Decreasing soil-frost depth and its relation to climate change in Tokachi, Hokkaido, Japan. J Meteor Soc Japan 84:821–833

Hsiang T, Matsumoto N, Millett SM (1999) Biology and management of Typhula snow molds of turfgrass. Plant Dis 83:788–798

Komatsu T (1983) Problems and perspectives on alfalfa cultivation in terms of winter climate. Tokachi Nogaku Danwakaishi 24:92–101

Matsumoto N (1994) Ecological adaptations of low temperature plant pathogenic fungi to diverse winter climates. Can J Plant Pathol 16:237–240

Matsumoto N (2009) Snow molds: a group of fungi that prevail under snow. Microbes Environ 24:14–20

Matsumoto N, Araki T (1982) Field observation of snow mold pathogens of grasses under snow cover in Sapporo. Research Bulletin of the Hokkaido National Agricultural Experiment Station 135:1–10

Matsumoto N, Sato T, Araki T (1982) Biotype differentiation in the *Typhula ishikariensis* complex and their allopatry in Hokkaido. Ann Phytopathol Soc Jpn 48:275–280

Matsumoto N, Sato T (1983) Niche separation in the pathogenic species of *Typhula*. Ann Phytopathol Soc Jpn 49:293–298

Matsumoto N, Kawakami A, Izutsu S (2000) Distribution of *Typhula ishikariensis* biotype A isolates belonging to a predominant mycelia compatibility group. J Gen Plant Pathol 66:103–108

Matsumoto N, Hoshino T (2009) Fungi in snow environments: phychrophilic molds—a group of pathogens affecting plants under snow. In: Misra JK, Deshmukh SK (Eds.) Fungi from different environments. Science Publishers, Enfield, pp.169–188

Nakajima T, Abe J (1994) Development of resistance to *Microdochium nivale* in winter wheat during autumn and decline of the resistance under snow. Can J Bot 72:1211–1215

Nissinen O (1996) Analyses of climatic factors affecting snow mould injury in first-year timothy (*Phleum pretense* L.) with special reference to Sclerotinia borealis. Acta Universitatis Ouluensis A289, pp 115.

Sato T, Abe T, Etori R, Tani H, Yamakawa M, Morimoto M (2009) Winter killing of timothy at Tokachi district in 2008. Journal of Hokkaido Society of Grassland Science 43:43

Takeda Y, Nakajima K (1997a) Characteristics of alfalfa (*Medicago satava* L.) cultivars adapted to Konsen district. 1. Differences among cultivars in winter hardiness of the first winter and associated traits. Grassland Science 43:144–149

Takeda Y, Nakajima K (1997b) Characteristics of alfalfa (*Medicago satava* L.) cultivars adapted to Konsen district. 2. Differences among cultivars in winter hardiness after the 2nd year and associated traits. Grassland Science 43: 150–156

Takeda Y, Nakajima K (1997c) Characteristics of alfalfa (*Medicago satava* L.) cultivars adapted to Konsen district. 3. Differences among cultivars in Lepto-leaf spot under the natural infection. Grassland Science 43:157–163

Takeda Y, Uchiyama K, Nakajima K, Yamaguchi H (1998) Individual variation in Lepto-leaf spot resistance and effects of phenotypic recurrent selection on its resistance in alfalfa (*Medicago sativa* L.) cultivars. Grassland Science 44:73–79

Tomiyama K (1955) Studies of the snow blight disease of winter cereals. Hokkaido National Agricultural Experiment Station Bulletin 47:234pp.

Ueda S, Gau M, Matsu-ura M, Suginobu K, Maki Y, Sato H, Hayakawa R, Miyashita Y, Nishimura N, Kaneko K, Murakami K (1985) Breeding of 'Kitawakaba' alfalfa and its characteristics. Research Bulletin of the Hokkaido National Agricultural Experiment Station 143:1–21

Yamaguchi H, Uchiyama K, Sawai A, Gau M, Ueda S, Maki Y, Matsu-ura M, Suginobu K, Sato R, Takeda Y, Nakajima K, Chiba I, Ochi H, Sawada Y, Tamakake H (1995) Breeding of 'Hisawakaba' alfalfa and its characteristics. Research Bulletin of the Hokkaido National Agricultural Experiment Station 161:17–31

Phytopathogenic Fungi and Fungal-Like Microbes in Svalbard

Motoaki Tojo, Shota Masumoto and Tamotsu Hoshino

Introduction

Plant pathogens can affect individual growth and community structure in many wild plants (Gilbert 2002). Phytopathogenic fungi and fungal-like microbes occur frequently on moss and vascular plant species in the polar regions (Hoshino et al. 2001; Tojo and Newsham 2012). Svalbard, a high Arctic archipelago, has been investigated for phytopathogenic fungi and fungal-like microbes since the late nineteenth century (Karsten 1872; Lind 1928). Among the 578 species of recorded fungi and fungal-like microbes in Svalbard (Elvebakk et al. 1996; Singh et al. 2012; Tojo and Newsham 2012; Tojo et al. 2012), at least 176 species (100 Ascomycetes, 45 Deuteromycetes [Anamorphic fungi], 25 Basidiomycetes, 3 Chytridiomycetes, and 3 Oomycetes) have been found from plants (Table 1). They are thought to be pathogens or potential pathogens of plants. However, there are very few reports regarding their distribution and pathogenic capacity in natural polar ecosystems. In this review, we describe the taxonomical and ecological features of several plant pathogenic fungi and fungal-like microbes recently reported in Svalbard.

M. Tojo (✉)
Graduate School of Life and Environmental Sciences, Osaka Prefecture University, Sakai, Japan
e-mail: tojo@plant.osakafu-u.ac.jp

S. Masumoto
The Graduate University for Advanced Studies (SOKENDAI), Tachikawa, Japan

T. Hoshino
National Institute of Advanced Industrial Science and Technology (AIST),
Higashi-Hiroshima, Japan
e-mail: tamotsu.hoshino@aist.go.jp

R. Imai et al. (eds.), *Plant and Microbe Adaptations to Cold in a Changing World*,
DOI 10.1007/978-1-4614-8253-6_23,

Table 1 Fungi or fungal-like microbes reported from plants in Svalbard

Fungi or fungal-like microbe	Host or suspected host	Distribution	References
Ascomycetes			
Anthostroma polaris	*Dryas octopetala*	Ny-Ålesund and Longyearbyen	Holm and Holm (1993a)
Atopospora betulina	*Betula nana*	Many localities in Svalbard	Lind (1928)
Blumeria graminis	*Poa* spp. and *Phippsia algida*	Isfjorden, Wijdefjorden, and Longyearbyen	Lind (1928), Hagen (1952)
Bricookea sepalorum	*Luzula arctica* and *Juncus* species	Gråhuken	Lind (1928, 1934)
Bryochiton miroscopicus	*Gymnomitrion* spp.	Amsterdamøya, Barentsøya, Edgeøya, and Kvalhovden	Döbbeler (1978)
Bryochiton monascus	*Racomiotrium lanuginosum*	Amsterdamøya	Döbbeler (1978)
Bryochiton perpusillus	*Polytrichastrum alpinum, Polytrichastrum hyperboreum*, and *Ptilidium ciliare*	Amsterdamøya and Longyearbyen	Döbbeler (1978)
Cainiella borealis	*Cassiope tetragona*	Longyearbyen and Kongsfjorden	Holm (1975), Holm and Holm (1994)
Cainiella johansonii	*Dryas* species	Grønfjorden and Longyearbyen	Holm (1979), Holm and Holm (1993a)
Chamaeascus arcticus	*Carex misandra* and *C. rupestris*	Kongsfjorden and Gipsvika	Holm and Holm (1993c, 1994)
Ciboria aschersoniana	*Carex subspathacea*	Kapp Wijk	Schumacher and Kohn (1985)
Ciborinia ciborium	*Eriophorum* spp. especially *Eriophorum scheuchzeri*	Hornsund, Van Mijenfjorden, Bockfjorden, and Brøggerhalvøya	Elvebakk et al. (1996)
Cilioplea coronata	Unknown (Four different host species)	Scattered localities on Spitsbergen	Holm and Holm (1993b)
Clathrospora deflectens	*Poa pratensis* and *Trisetum spicatum*	Liefdefjorden	Elvebakk et al. (1996)
Clathrospora heterospora	*Carex nardina* and other graminoids	Many localities in Svalbard	Karsten (1872), Lind (1928), Holm and Holm (1994)

Table 1 (continued)

Fungi or fungal-like microbe	Host or suspected host	Distribution	References
Clathrospora planispora	*Puccinellia angustata* and *Eriophorum scheuchzeri*	Gipsvika	Holm and Holm (1994)
Clathrospora verruculosa	*Poa glauca*	Billefjiorden	Eriksson (1967)
Crocicreas culmicola	*Carex saxatilis* and grass culms	Adventfjorden and Longyearbyen	Lind (1928), Huhtinen (1987)
Crocicreas cyathoideum	*Poa pratensis* ssp. *alpingena*	Longyearbyen	Huhtinen (1987)
Crocicreas gramineum	*Poa* species, especially *Poa arctica*	Grønfjorden and north-western Spitsbergen	Karsten (1872), Lind (1928), Carpenter (1981)
Didymella glacialis	*Poa* species	Billefjiorden	Lind (1928)
Didymosphaeria futilis	*Dryas* sp.	Blomstrandhalvøya and Grønsteinfjorden	Holm and Holm (1993a)
Diplocarpon polygoni	*Bistorta vivipara*	Isfjorden	Lind (1928)
Diplonaevia circinata	*Juncus biglumis*	Bellsund	Lind (1928)
Diplonaevia hyperborea	*Juncus castaneus* and *J. biglumis*	Søkapp Land and Longyeardalen	Nannfeldt (1984b)
Diplonaevia savilei	*Luzula arctica*	Grønfjorden and Bellsund	Nannfeldt (1984b)
Discostroma hyperborea	*Cassiope tetragona*	Grønfjorden	Karsten (1872)
Duplicaria empetri	*Empetrum* sp.	Grønfjorden	Karsten (1872), Lind (1928)
Epibryon polysporum	*Ptilidum ciliare*	Amsterdamøya	Döbbeler (1978)
Epibryon absconditum	*Dryas* leaves	Ny-Ålesund and Adventfjorden	Holm and Holm (1993a)
Gibbera barriae	*Cassiope hypnoides*	Grønfjorden	Holm and Holm (1980)
Gibbera latispora	*Dryas* leaves and *Cassiope tetragona*	Ny-Ålesund and Longyearbyen	Holm and Holm (1993a, 1994)
Glomerella amenti	*Salix polaris*	Longyeardalen, Kapp Thordsen, and Endalen	Lind (1928), Holm and Holm (1994)

Table 1 (continued)

Fungi or fungal-like microbe	Host or suspected host	Distribution	References
Gnomonia dryadis	*Dryas octopetala*	Blomstrandhalvøya and Endalen	Holm and Holm (1993a, 1994)
Gnomoniella vagans	*Dryas octopetala*	Blomstrandhalvøya and Endalen	Holm and Holm (1993a)
Graphyllium pentamerum	Several graminoid species and *Dryas* sp.	Several localities in Svalbard	Karsten (1872), Lind (1928), Eriksson (1967), Holm and Holm (1994)
Guignardia graminicola	Various grasses	Kongsfjorden, Gipsvika, and Longyearbyen	Holm and Holm (1994)
Hypoderma degenerans	*Vaccinium uliginosum*	Colesbukta	Lind (1928)
Hysteronaevia advena	*Eriophorum angustifolium* ssp. *triste, Luzula arctica*, and *L. arctica* ssp. *confusa*	Adventdalen and Ekmanfjorden	Karsten (1872)
Hysteronaevia clavulifera	*Juncus biglumis*	Sassendalen	Nannfeldt (1984a)
Hysteronaevia kobayasii	*Eriophorum* spiecies and several other host plants	Ny-Ålesund and Sassendalen	Nannfeldt (1984a)
Hysteronaevia luzulicola	*Luzula arctica*	Many localities of Svalbard	Nannfeldt (1984a)
Hysteronaevia lyngei	*Fesutuca rubra*	Sørkapp Land and Kapp Thordsen	Nannfeldt (1932, 1984a)
Hysteropezizella fuscella	*Festuca rubra* and *Luzula arctica*	Sørkapp Land, Edgeøya, and Kavadehuken	Lind (1928, 1934), Nannfeldt (1932)
Isothea rhytismoides	Leaves of *Dryas* spp.	Wijdefjorden, Bellsund, Isfjorden, and Adventdalen	Lind (1928), Lid (1967), Holm (1979)
Kalmusia coniothyrium	*Salix polaris*	Sassendalen	Lind (1928)
Lachnum palearum	*Phippsia aldida*	Hopen	Elvebakk et al. (1996)
Laetinaevia erythrostigma	*Cerastium* spp.	Many localities in Svalbard	Lind (1928), Nannfeldt (1932), Hagen (1941)

Table 1 (continued)

Fungi or fungal-like microbe	Host or suspected host	Distribution	References
Laetinaevia stellariae	*Stellaria longipes*	Bellsund, Colesbukta, and Adventfjorden	Lind (1928, 1934)
Leptosphaeria brachyasca	*Saxifraga oppositifolia*	Bjørnøya	Lind (1928)
Leptosphaeria monotis	*Saxifraga oppositifolia*	Ny-Ålesund	Holm and Holm (1994)
Leptotrochila cerastiorum	*Cerastium* sp.	Adventfjorden	Elvebakk et al. (1996)
Lophiostoma winteri	*Dryas* spp.	Many localities in Svalbard	Holm and Holm (1993a)
Lophiotrema vagabundum	*Luzula arctica* ssp. *confusa* and *Deschampsia alpina*	Ny-Ålesund and Blomstrandhalvøya	Holm and Holm (1994)
Lophodermium caricinum	*Eriophorum scheuchzeri* and *Eriophorum angustifolium*	Adventdalen	Lind (1928)
Lophodermium culmigenum	Large number of grasses	All over many localities in Svalbard	Karsten (1872), Lind (1928)
Massarina balnei-ursi	*Dryas* spp.	Ny-Ålesund, Gipsvika, and west of Sassendalen	Holm and Holm (1993a)
Massariopsis wulffii	Grasses	Ekmanfjorden, Brøgger- halvøya, and Longyearbyen	Elvebakk et al. (1996)
Melanomma dryadis	*Dryas* spp.	Kongsfjorden and Longyearbyen	Holm and Holm (1994)
Melaspilea lecideopsida	*Dryas* spp.	Ny-Ålesund, Blomstrandhalvøya, and Longyearbyen	Holm and Holm (1993a)
Microthyrium holmiae	*Dryas* spp.	Ny-Ålesund	Holm and Holm (1993a)
Microthyrium microscopicum	*Carex misandra*	Gluudneset	Holm and Holm (1994)
Mollisia graminis	Grasses	Bellsund, Grønfjorden, and Bohemannesset	Lind (1928)

Table 1 (continued)

Fungi or fungal-like microbe	Host or suspected host	Distribution	References
Montagnula spinosella	Graminoids and *Deschampsia cespitosa*	Ny-Ålesund and Endalen	Holm and Holm (1993b, 1994)
Mycosphaerella arthopyrenoides	*Papaver dahlianum*	Longyearbyen	Holm and Holm (1994)
Mycosphaerella cassiopes	*Cassiope tetragona*	Many localities in Svalbard	Holm (1975)
Mycosphaerella confinis	*Brassjcaceae plants*	Krossfjorden, Kongsfjorden, Adventfjorden, and Wijdefjorden	Holm and Holm (1994)
Mycosphaerella densa	different herbaceous host species	Many localities in Svalbard	Lind (1928)
Mycosphaerella equiseti	*Equisetum scirpoides*	Endalen	Holm and Holm (1994)
Mycosphaerella equiseticola	*Equisetum scirpoides*	Endalen	Holm and Holm (1994)
Mycosphaerella halophila	*Honkenya peploides*	Unknown place in Svalbard	Holm and Holm (1994)
Mycosphaerella lycopodii	*Huperzia selago*	Ny-Ålesund and Longyearbyen	Holm and Holm (1994)
Mycosphaerella minor	*Dryas* spp. and many dicotyledons	Isfjorden, Bellsund, and Bjørnøya	Lind (1928), Holm and Holm (1993a, 1994)
Mycosphaerella octopetalae	*Dryas* sp.	Many localities in Svalbard	Holm and Holm (1993a)
Mycosphaerella pachyasca	*Eutrema edwardsii*	Wijdefjorden	Wulff (1902)
Mycosphaerella pedicularidis	*Pedicularis hirsuta* and *P. lanata* ssp. *dasyatha*	Grønfjorden and many localities	Karsten (1872), Lind (1928)
Mycosphaerella perexigua	*Juncus, Luzula*, and Cyperaceae plants	Many localities in Svalbard	Holm and Holm (1994)
Mycosphaerella polaris	*Salix polaris* and *S. polaris* × *herbacea*	Lomfjorden, Bellsund, Wijdefjorden, and Sørkapp Land	Elvebakk et al. (1996)
Mycosphaerella pusilla	*Carex* spp.	Isfjorden	Lind (1928)
Mycosphaerella ranunculi	*Ranunculus* spp.	Many localities in Svalbard	Karsten (1872), Lind (1928)

Table 1 (continued)

Fungi or fungal-like microbe	Host or suspected host	Distribution	References
Mycosphaerella salicicola	*Salix polaris*	Wijdefjorden and Sorgfjorden	Wulff (1902)
Mycosphaerella recutita	Various monocotyledons	Many localities in Svalbard	Elvebakk et al. (1996)
Naemacyclus lambertii	*Dryas octopetala*	Many localities in Svalbard	Holm and Holm (1993a)
Naviopsis primulae	*Taraxacum arcticum*	Grønfjorden	Lind (1928)
Nimbomollisia eriophori	*Carex misandra* and *C. subspathacea*	Kongsfjorden, Bellsund, and Adventpynten	Lind (1928)
Octospora melina	*Bryaceae* mosses	Longyearbyen	Huhtinen (1987)
Odontotrema cassiopes	*Cassiope tetragona*	Mimerdalen and Longyeardalen	Lind (1928)
Otthia dryadis	*Dryas* sp.	Ny-Ålesund and Adventdalen	Holm and Holm (1993a, 1994)
Phaeosphaeria caricinella	*Leptosphaeria caricinella. L. junciseda* and *L. vagans*	Many localities in Svalbard	Karsten (1872), Lind (1928), Holm and Holm (1994)
Phaeosphaeria culmorum	*Luzula* sp.	Ny-Ålesund	Holm and Holm (1994)
Phaeosphaeria equiseti	*Equisetum arvense* and *E. scirpoides*	Ny-Ålesund and Endalen	Holm and Holm (1994)
Phaeosphaeria herpotrichoides	Graminoids	Kongsfjorden, Gipsvika, and Longyearbyen	Holm and Holm (1994)
Phaeosphaeria insignis	Variety of grasses	Many localities in Svalbard	Karsten (1872), Lind (1928), Holm (1957), Eriksson (1967)
Phaeosphaeria juncina	*Luzula arctica* ssp. *confusa*	Ny-Ålesund	Holm and Holm (1994)
Phaeosphaeria lindii	*Equisetum arvense* and *E. scirpoides*	Ny-Ålesund and Longyearbyen	Holm and Holm (1994)
Phaeosphaeria marcyensis	*Huperzia selago*	Ny-Ålesund and Longyearbyen	Holm and Holm (1994)
Phaeosphaeria microscopica	Grasses	Many localities in Svalbard	Karsten (1872), Lind (1934), Holm (1957), Eriksson (1967)

Table 1 (continued)

Fungi or fungal-like microbe	Host or suspected host	Distribution	References
Phaeosphaeria nigrans	*Carex parallela* and *Festuca* species	Billefjorden and Dicksonfjorden	Elvebakk et al. (1996)
Phaeosphaeria silenes-acaulis	*Silene acaulis, S. humifsa*, and *S. longipes*	Grønfjorden and Ny-Ålesund	Karsten (1872), Lind (1928), Holm and Holm (1994)
Phaeosphaeria stellariae	*Stellaria longipes*	Longyearbyen	Holm and Holm (1994)
Phaeosphaeria vagans	*Phippsia algida, Dupontia fisheri*, and *Eriophorum scheuchzeri*	Brennevinsfjorden and Gipsvika	Lind (1928), Holm and Holm (1994)
Phaeosphaeria weberi	*Ranunculus pygmaeus* and *R. sulphureus*	Sørkapp Land, Kapp Thordsen, Bohemannesset, Grønfjorden, and Endalen	Lind (1928), Holm and Holm (1994)
Potriphila navicularis	*Polytrichastrum alpinum*	Spitsbergen and Barentsøya	Döbbeler (1996)
Pseudopeziza drabae	*Draba* spp.	Bohemannesset and Kongsfjorden	Lind (1928), Elvebakk et al. (1996)
Pseudopeziza svalbardensis	*Saxifraga hirculus*	Adventfjorden	Lind (1928)
Pseudorhytisma bistortae	*Bistorta vivipara*	Kapp Thordsen, Dickson- fjorden, Tempefjorden, and Adventdalen	Lind (1928), Hagen (1941)
Psilachnum acutum	Grass culms	Ny-Ålesund and Longyearbyen	Huhtinen (1993)
Psilachnum inquilinum	*Equisetum arvense*	Bjørndalen	Huhtinen (1987)
Pyrenopeziza atrata	*Potentilla* sp.	Many localities in Svalbard	Elvebakk et al. (1996)
Pyrenophora raetica	Several grasses species	Scattered Spitsbergen localities	Holm and Holm (1993b)
Pyrenophora schroeteri	*Arctophila fulva* and *Hierochloe alpina*	Sørkapp Land and Colesbukta	Lind (1928), Holm and Holm (1994)
Rhytisma salicinum	*Salix polaris* and *S. polaris* × *S. herbacea*	Many localities in Svalbard	Lind (1928), Hagen (1941), Lid (1967)
Ronnigeria arctica	*Potentilla pulchella*	Adventdalen	Lind (1928)
Scleropleella hyperborea	*Cassiope* and *Dryas* species	Many localities in Svalbard	Holm and Holm (1993a)

Table 1 (continued)

Fungi or fungal-like microbe	Host or suspected host	Distribution	References
Sclerotinia borealis (syn. *Myriosclerotinia borealis*)	*Poa hartzii*	Barentsburg	Hoshino et al. (2003)
Sphaerotheca erigeronis-canadensis	*Taraxacum arcticum*	Adventfjorden and Helvetiadalen	Holm and Holm (1994)
Sydowiella dryadis	*Dryas* spp.	Many localities in Svalbard	Holm and Holm (1993a)
Taphrina carnea	*Betula nana*	Colesbukta	Elvebakk et al. (1996)
Taphrophila argyllensis	*Deschampsia alpina*	Endalen	Holm and Holm (1994)
Trichopezizella nidulus	*Carex lachenalii*	Carolinedalen	Huhtinen (1987)
Trichothyrina salicis	*Salix reticulata*	Ny-Ålesund	Holm and Holm (1994)
Venturia oxyriae	*Oxyria digyna*	Ny-Ålesund and Longyearbyen	Holm and Holm (1994)
Venturia potentillae	*Potentilla pulchella*	Bellsund, Adventdalen, Billefjorden, and Gipsvika	Lind (1928), Holm and Holm (1994)
Wettsteinina andromedae	*Cassiope tetragona*	Many localities in Svalbard	Holm and Holm (1994)
Wettsteinina distincta	*Phippsia* sp. and *Puccinellia* species	Many localities in Svalbard	Holm and Holm (1994)
Wettsteinina eucarpa	*Bistorta vivipara*	Many localities in Svalbard	Elvebakk et al. (1996)
Wettsteinina junci	*Juncus biglumis*	Gluudneset and Blomstrandhalvøya	Holm and Holm (1994)
Wettsteinina macrotheca	*Carex saxatilis* and *C. subspathacea*	Bellsund and Kongsfjorden	Elvebakk et al. (1996)
Wettsteinina salicicola	*Salix reticulata*	Ny-Ålesund	Holm and Holm (1994)
Wettsteinina savilei	*Carex misandra* and *Juncus biglumis*	Ny-Ålesund	Holm and Holm (1994)
Deuteromycotes (Anamorphic fungi)			
Arthinium puccinioides	*Juncus triglumis* and *Carex misandra*	Adventfjorden and Longyeardalen	Lind (1928)

Table 1 (continued)

Fungi or fungal-like microbe	Host or suspected host	Distribution	References
Ascochyta arctica	*Poa* sp. and *Festuca rubra*	Olsokflyan, Grønfjorden, and Krossfjorden	Lind (1928)
Ascochyta dianthi	*Stellaria humifusa*	Klovningen and Grønfjorden	Lind (1928)
Ascochyta arctica	*Alopecurus borealis*, *Festuca* sp., and *Poa arctica*	Several localities in the northwest Spitsbergen	Lind (1928)
Dendryphion fumosum	*Eutrema edwardsii*	Dicksonfjorden	Lind (1928)
Diplodia bessimyanii	*Poa pratensis* ssp. *alpigena*	Sørkapp Land	Lind (1928)
Diplodia simmonsii	*Poa* sp. and *Luzula arctica* ssp. *confusa*	Billefjiorden and Bjørnøya	Lind (1928)
Diplodia euphrasiae	*Pedicularis hirsuta*	Grønfjorden	Lind (1928)
Diplodia bessimyanii	14 different dicotyledoneous	Many localities in Svalbard	Lind (1928)
Diplodia pedicularidis	*Pedicularis hirsuta*	Adventfjorden	Lind (1928)
Eriospora leucostoma	*Carex misandra*	Adventfjorden	Lind (1928)
Gloeossporium roaldii	*Polemonium boreale*	Adventfjorden and Grønfjorden	Lind (1928)
Hendersonia arabidis	*Draba* sp.	Several localities in Svalbard	Lind (1928)
Hendersonia arundinacea	Monocotyledoneous plants	Many localities in Svalbard	Lind (1928)
Hendersonia gigantea	*Juncus arcticus*, *Carex saxatilis*, and *C. bigelowii*	Central parts of Isfjorden	Lind (1928)
Hendersonia rostrupii	*Hierochloe alpine* and *Poa abbreviata*	Adventfjorden, Sassen- fjorden, and Billefjiorden	Lind (1928)
Hendersonia arabidis	*Carex misandra*	Kongsfjorden and Krossfjorden	Lind (1928)
Leptothyrium arcticun	*Potentilla hyparctica* and *P. nivea*	Many localities in Svalbard	Lind (1928)
Leptothyrium palustre	*Potentilla hirsuta*	Sørkapp Land	Lind (1928)

Table 1 (continued)

Fungi or fungal-like microbe	Host or suspected host	Distribution	References
Marssonina obscura	*Salix polaris* and *S. polaris* × *S. herbacea*	Sørkapp Land	Lind (1928)
Mastigosporium album	*Poa* sp.	Sørkapp Land and Forlandssundet	Lind (1928)
Microdochium bolleyi	Unknown	Unknown	Väre et al. (1992)
Microsphaeriopsis olivacea	*Draba oblongata* (probably *D. micropetala*)	Sassenfjorden	Lind (1928)
Phoma alpina	Numerous plants	Many localities in Svalbard	Karsten (1872), Lind (1928)
Phoma caricis	*Carex glareosa* and *C. saxatilis*	Adventfjorden and Sassenfjorden	Lind (1928)
Phoma complanata	*Pedicularis hirsuta*	Kapp Thordsen	Lind (1928)
Phoma graminis	*Festuca* sp. and *Poa abbreviata*	Sassenfjorden	Lind (1928)
Phoma herbarum	16 species of plant	Many localities in Svalbard	Lind (1928)
Phoma nebulosa	*Stellaria longipes*	Adventfjorden, Colesbukta	Lind (1928)
Phoma oudemansii	*Polemonium boreale*	Bellsund and Isfjorden	Lind (1928)
Phoma ranunculi	*Ranunculus* species	Sørkapp Land, Tempel- fjorden, Krossfjorden, and Gråhuken	Lind (1928)
Phoma sceptri	*Pedicularis hirsute* and *P. lanata* ssp. *dasyantha*	Sorgfjorden, Dicksonfjorden	Wulff (1902), Lind (1928)
Phyllosticta saxifragarum	*Saxifraga hirculus*	Bellsund	Lind (1928)
Plenodomus svalbardensis	*Draba alpina*	Magdalenefjorden	Lind (1928)
Ramularia alborosella	*Cerastium* species	Many places of Spitsbergen and Bjørnøya	Lind (1928)
Rhabdospora campanulae	*Campanula uniflora*	Krossfjorden	Lind (1928)
Rhabdospora pleosporoides	*Oxyria digyna* and *Draba* sp.	Many localities in Svalbard	Lind (1928)

Table 1 (continued)

Fungi or fungal-like microbe	Host or suspected host	Distribution	References
Rhizoctonia solani	Unknown	Unknown	Väre et al. (1992)
Selenophoma drabae	*Carex*, *Draba*, and four species within Caryophyllaceae	Many localities in Svalbard	Lind (1928)
Septoria lychnidis	*Silene uralensis*	Tempelfjorden	Lind (1928)
Septoria polaris	*Ranunculus lapponicus*	Adventfjorden	Karsten (1884)
Septoria punctoidea	*Carex misandra*, *Luzula arctica*, and *L. arctica* ssp. *confusa*	Adventfjorden, numerous localities of Spitsbergen	Karsten (1884), Lind (1928)
Septoria saxifragae	*Saxifraga cespitosa*	Bjørnøya	Lind (1928)
Sphaeronaema foliicolum	*Salix polaris*	Kapp Thordsen	Lind (1928)
Stagonospora eriophorella	*Juncus arcticus*	Dicksonfjorden	Lind (1928)
Basidiomycetes			
Anthraconidea altera	*Carex misandra*	Teist Fjellet	Elvebakk et al. (1996)
Anthraconidea elynae	*Carex nardina*	Ekmanfjorden, Blomstrandhalvøya, and Kongs- fjorden	Kukkonen (1963)
Anthraconidea lindebergiae	*Kobresia simpliciuscula*	Mimerdalen	Elvebakk et al. (1996)
Anthraconidea altera	*Carex misandra*	Bohemannesset	Elvebakk et al. (1996)
Entyloma dactylidis	Species of *Dupontia* and *Poa*	Adventfjorden, Bjørnøya, Bellsund, Grønfjorden, Colesbukta, Lomf-jorden, and Bohemannesset	Karsten (1872), Lind (1928)
Melampsora epitea	*Saxifraga cespitosa*, *S. oppositifolia*, and *Salix polaris*	Bjørnøya, Bellsund, Isfjorden, and Krossfjorden	Lind (1928), Hagen (1941, 1950, 1952)
Puccinia arenariae	*Cerastium alpinum* (probably *C. arcticum*)	Adventfjorden and Wijdefjorden	Wulff (1902)

Table 1 (continued)

Fungi or fungal-like microbe	Host or suspected host	Distribution	References
Puccinia bistortae	*Bistorta vivipara*	Bellsund, Isfjorden, and Sabine Land	Lind (1928), Hagen (1941), Elvebakk et al. (1996)
Puccinia curciferarum	*Cardamine bellidifolia*	Isfjorden and Vesleraudfjorden	Lind (1928), Hagen (1941)
Puccinia drabae	*Draba micropetala, D.* "*alpina* × *oblongata,*" *D. arctica*, and *D. corymbosa*	Dicksonfjorden, Advent- fjorden, Skansbukta, Lovénø- yane, and Kongsfjorden	Lind (1928), Hagen (1941), Elvebakk et al. (1996)
Puccinia eutremae	*Cochlearia groenlandica* and *Eutrema edwardsii*	Sassendalen, Bjørnbeinflya, Bohemannesset, and Kapp Linné	Lind (1928, 1934), Nannfeldt (1933), Lid (1967), Hagen (1941)
Puccinia gibberulosa	*Rununculus* sp.	Bjonahamna at Tempelfijorden	Jørstad I (1950)
Puccinia heucherae	Six *Saxifraga* spp.	Most parts in Svalbard	Karsten (1872), Lind (1928), Hagen (1941, 1950, 1952), Lid (1967)
Puccinia hieracii	*Taraxacum cymbifolium*	Bjørnøya	Gjærum (1991)
Puccinia oxyriae	Unknown	Adventfjorden and Adventdalen	Lind (1928), Hagen(1941)
Puccinia pazschkei	*Saxifraga aizoides*	Lyckholmdalen and Myggdalen	Elvebakk et al. (1996)
Schizonella melanogramma	*Carex rupestris*	Sorgfjorden and Adventfjorden	Wulff (1902), Lind (1928)
Tolyposporium junci	*Juncus biglumis*	Adventfjorden	Lind (1928)
Typhula isikariensis	*Poa hartzii*	Barentsburg	Hoshino et al. (2003)
Ustilago bistortarum	*Bistorta vivipara*	Numerous localities in Svalbard	Karsten (1872), Lind (1928,1934), Hagen (1941, 1950, 1952)
Ustilago hyperborea	*Luzula arctica* spp. *confusa*	Adventdalen	Lind (1928)
Ustilago nivalis	*Sagina nivalis*	Adventfjorden and Blåhuken	Lind (1928), Hagen(1941)
Ustilago picacea	*Koenigia islandica*	Adventdalen	Lind (1928)
Ustilago striiformis	*Poa arctica*	Brentskardet	Hagen (1941)

Table 1 (continued)

Fungi or fungal-like microbe	Host or suspected host	Distribution	References
Ustilago vinosa	*Oxyria digyna*	Numerous localities in Svalbard	Karsten (1872), Lind (1928,1934), Hagen (1941, 1952)
Ustilago violacea	*Silene acaulis* and *Stellaria longipes*	Isfjorden, Engelskbukta, Wijdefjorden, Lomfjorden, and Tempelfjorden	Wulff (1902), Lind (1928,1934), Hagen(1941, 1950, 1952)
Chytridiomycetes			
Olpidium brassicae	*Saxifraga* sp.	Spitsbergen	Väre et al. (1992)
Synchytrium groenladicum	*Saxifraga cernua, S. rivularis*	Colesbukta and Sørkapp land	Lind (1928, 1934)
Synchytrium potentiallae	*Potentilla* sp.	Adventifjorden	Hagen (1941)
Oomycetes			
Peronospora alsinearum	*Cerastium alpinum* and Caryophylaceous plants	Central Spitsbergen	Hagen (1941)
Peronospora parasitica	*Cochlearia groenlandica* and Brassicaceous plants	Bellsund and Isfjorden	Hagen (1941)
Pythium polare	*Sanionia uncinata*	Ny-Ålesund and Longyearbyen	Tojo et al. (2012)

Fig. 1 Morphology of the smut fungus *Microbotryum bistortarum* on *Polygonum viviparum* in Svalbard. (**a**) Bulbils and flowers of the healthy plant; (**b**) signs of smut on bulbils; (**c**) optical micrograph of teliospores; (**d**) scanning electron micrograph showing the fine and dense verruculose pattern of a teliospore surface; and (**e**) teliospores that have germinated to produce a basidium with basidiospores

Taxonomic and Ecological Features of Representative Fungi and Fungal-Like Microbes

Smut Fungus on Polygonum viviparum

Smut fungus (*Microbotryum bistortarum*) is a common pathogen on *P. viviparum*, a polygonaceous perennial plant common in the arctic and alpine tundra (Lindeberg 1959; Vánky 1994; Vánky and Oberwinkler 1994). This pathogen systemically infects *P. viviparum*, spreads throughout the plant, and replaces all the bulbils and flowers with teliospore-bearing structures (Vánky 1994; Vánky and Oberwinkler 1994). The fungus can cause severe damage to *P. viviparum* plants in Svalbard (Tojo and Nishitani 2005; Fig. 1). The diseased plants had visually smaller leaves and inflorescences compared to the healthy plants. The survival rate among smut-infected

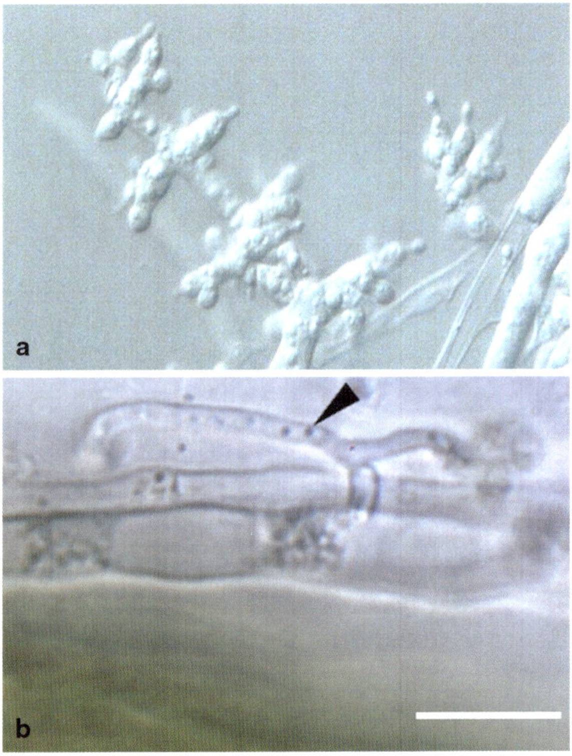

Fig. 2 Morphology of *Trichoderma polysporum* isolated from Sanionia moss in Svalbard. (**a**) Conidiophores with conidia. (**b**) Hyphae (*arrowhead*) formed in and around dead epidermal cells of Sanionia moss. Bars = 10 μm

plants was significantly lower than that of healthy plants ($P<0.001$). The smut-infected plants had a significantly lower number of bulbils and flowers than the healthy plants, and the size of the largest leaf of each infected plant was smaller than that of a healthy plant. This suggests that the smut fungus has a negative effect on the growth and survival of individual *P. viviparum* in the high Arctic region. The high mortality of the infected *P. viviparum* implies that other factors, such as cold stress, may enhance the damage of the disease.

Trichoderma polysporum on Sanionia moss

Trichoderma polysporum is one of the psychrotrophic *Trichoderma* species present in temperate (Goldfarb et al. 1989; Lu et al. 2004) and Arctic environments (Zabawski 1982). Svalbard is the only Arctic environment where this species has been recorded (Zabawski 1982). *T. polysporum*-like species isolated from a moss (*Sanionia uncinata*) were recently obtained in high Arctic wetlands in Svalbard and examined for species identification, the effects of temperature on growth, pathogenicity to the moss, and polygalacturonase (PGase) activity (Yamazaki et al. 2011). The species were identified as *T. polysporum* based on, morphology (Fig. 2a), the sequence of the rDNA-ITS regions, and the growth response to temperature. *T. polysporum* infected epidermal tissues of the moss (Fig. 2b), but did not cause any symptoms in in vitro inoculation tests at 0°C. One of the isolates of *T. polysporum* from Spitsbergen showed high activity of PGase, an enzyme produced by many saprophytic and pathogenic fungi, in a pectin liquid medium. The enzyme activity was evident even at 0 °C. The epidermal infection of the moss and the PGase production at 0 °C of the *T. polysporum* suggest that the species likely inhabits a niche as a saprophyte or an endotroph in the Arctic environments.

Pythium polare from *Sanionia* moss

Pythium species are cosmopolitan and many species of this genus can infect a variety of host plants. There has been only one report on their distribution and pathogenic capacity in natural ecosystems in polar regions (Bridge et al. 2008). *Pythium polare* is a heterothallic, newly described oomycete (Tojo et al. 2012). This fungal-like microbe has been isolated from fresh water and moss in various locations of both the Antarctic and Arctic including Svalbard (Fig. 3), King George Island, Greenland, the Norwegian alpine region, and Baffin Island. In vitro inoculation tests showed that *P. polare* is able to infect stems and leaves of Sanionia moss (*S. uncinata*) and cause brown discoloration.

P. polare is characterized by globose sporangia with discharge tubes of various lengths that release zoospores and aplerotic oospores, usually with 1–5 antheridia. The sexual structures are only produced in a dual culture of antheridial and oogonial isolates. Phylogenetic analysis, based on internal transcribed spacer (ITS) sequencing, places all isolated strains of *P. polare* in a unique new clade, hence it is considered a novel species. *Pythium canariense* and *Pythium violae* are the species most closely related to *P. polare* based on both morphology and the phylogenetic analysis. *P. polare* is also closely related to *P. iwayamai*, *P. okanoganense*, and *P. paddicum*, based on morphology and growth temperature, but is distinguished from these species by heterothallism and a longer discharge tube (5.5–88.0 μm) compared to *P. iwayamai* (3–35 μm, Tojo unpublished), *P. okanoganense* (3–25 μm, Van der Plaats-Niterink 1981), and *P. paddicum* (3–25 μm, Tojo unpublished).

The weak pathogenicity but strong colonization ability of *P. polare* on *S. uncinata* was elucidated by an in vitro inoculation test. *P. polare* was isolated from both nonsymptomatic moss and discolored moss, although the isolation frequency was lower in the former. The results suggest that *P. polare* is a weak pathogen in its natural environment.

Bridge et al. (2008) isolated a *Pythium* species from Signy Island almost identical to *P. polare* based on ITS sequences, and demonstrated that this species is the causative agent of snow rot on Antarctic hairgrass (*Deschampsia antarctica*). Its asexual morphology (Bridge et al. 2008) corresponds with *P. polare* and it is thought to be the same species. A *Pythium* isolate belonging to the same monophyletic group of *P. polare* has been recorded from alpine haircap moss (*Polytrichastrum alpinum*) on King George Island. (Yu and Hur unpublished). These isolates suggest that *P. polare* may have a wider host range, which includes other mosses and monocotyledons.

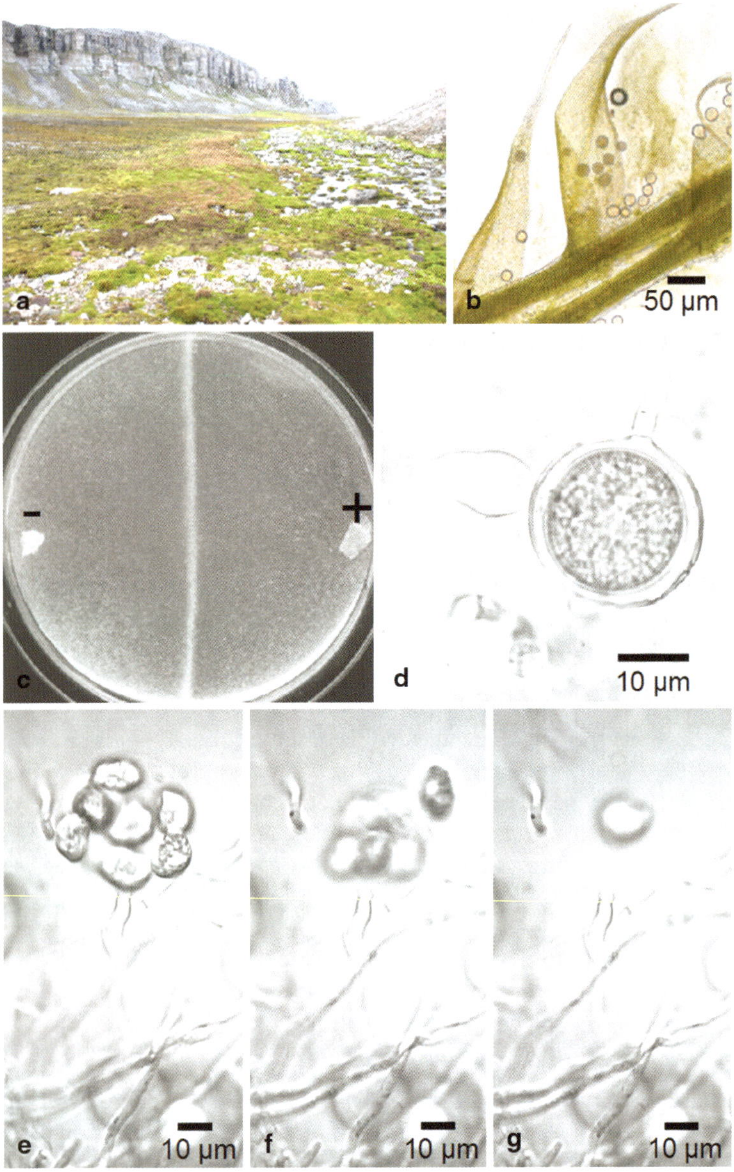

Fig. 3 Habitat and morphology of *Pythium polare* in Svalbard. (**a**) Brown discoloration of mosses occurs along low and wet area where water collected during snowmelt; (**b**) Sporangial-like structure of *P. polare*, which caused discoloration on the stem-leaf tissue of Sanionia moss; (**c**) Contact line of the heterothallic reaction of oogonial "+" and antheridial "–" types of *P. polare* isolates on potato carrot agar. Petri dish = 9 cm diameter; (**d**) Sexual organs of P. polare formed in corn fish meal agar. Oogonia with aplerotic oospores and antheridia formed in a dual culture of isolates "–" and "+"; (**e–g**) Development of zoospores of *P. polare* in a pond water culture

Fig. 4 Signs of tar spot (*left*) and rust (*right*) diseases on leaves of polar willow in Svalbard.

Although the reason for the limited distribution of *P. polare* within polar regions is unclear, this species may require a specific habitat that is only available in these environments. For example, its main host, Sanionia moss, has a bipolar distribution (Smith 1996; Virtanen et al. 1997) and is predominantly found in locations with a steady supply of melt water (Davey 1997; Victoria et al. 2009). This means that the distribution of *P. polare* is highly dependent on favorable habitats for Sanionia moss and other potential host plants, such as Antarctic hairgrass (Bridge et al. 2008), in the polar regions. Polar regions are currently warming at rapid rates (Fleming et al. 1997; Hansen et al. 2006; Turner et al. 2009). As *P. polare* can grow at warmer temperatures as well as in temperate regions (Tojo et al. 2012), it has a potential impact on plant survival in polar regions.

Tar Spot and Rust Fungi on Polar Willow

Tar spot disease caused by *Rhytisma* sp. and rust disease caused by *Melampsora* sp. are commonly found on polar willow (*Salix polaris*) in Svalbard (Fig. 4). Although these fungi have been recorded as *R. salicinum* and *M. epitea* in the herbal catalogs (Hagen 1941; Lid 1967; Lind 1928), details of their taxonomical characteristics have remained unknown. Morphological characterization and molecular taxonomic identification are being performed on Svalbard specimens (Masumoto, unpublished).

Concluding Remarks

Svalbard is the polar region most investigated and well described for fungi or fungal-like microbes. At least 176 species of fungi or fungal-like microbes have been recorded as pathogens or potential plant pathogen from the archipelago. Taxonomic and ecological diversity of the pathogens shown in this review suggests that more detailed research on these fungi or fungal-like microbes is required. The potential impact of effects of climate warming on them and their host plant survival is also warranted.

References

Bridge PD, Newsham KK, Denton GJ (2008) Snow mould caused by a *Pythium* sp.: a potential vascular plant pathogen in the maritime Antarctic. Plant Pathol 57:1066–1072

Carpenter SE (1981) Monograph of Crocicreas (Ascomycetes, Helotiales, Leotiaceae). New York Botanical Garden, New York, pp 33:1–290

Davey MC (1997) Effects of continuous and repeated dehydration on carbon fixation by bryophytes from the maritime Antarctic. Oecologia 110:25–31

Döbbeler P (1978) Moosbewohnende Ascomyceten I. Die Pyrenocarpen, den Gametophyten besiedelen Arten, vol 14. Botanische Staatssammlung, München, pp 1–360

Döbbeler P (1996) *Potriphila navicularis* gen. et sp. nov. (Ostropales, Ascomycetes), ein bipolar verbreiteter Parasit von *Polytrichum alpinum*. Nova Hedw 62:61–77

Elvebakk A, Gjærum HB, Sivertsen S (1996) Part 4. Fungi II. Myxomycota, Oomycota, Chytridiomycota, Zygomycota, Ascomycota, Deutromycota, Basidiomycota: Uredinales and Ustilaginales. In: Elvebakk A, Prestrud P (eds) A catalogue of Svalbard plants, fungi, algae and cyanobacteria. Norsk Polarinstitutt, Oslo, pp 207–259

Eriksson OE (1967) On graminicolous pyrenomycetes from Fennoscandia. 1. Dictyosporous species. Ark Bot 2:339–379

Fleming KM, Dowdeswell JA, Oerlemans J (1997) Modelling the mass balance of northwest Spitsbergen glaciers and responses to climate change. Ann Glaciol 24:203–210

Goldfarb B, Nelson EE, Hansen EM (1989) *Trichoderma* spp.: growth rates and antagonism to Phellinus weirii *in vitro*. Mycologia 81:375–381

Gilbert GS (2002) Evolutionary ecology of plant disease in natural ecosystems. Annu Rev Phytopathol 40:13–43

Gjærum H (1991) Rust- og sotsopper på Svalbard. Blyttia 49:167–169

Hagen A (1941) Micromycetes from Vestspitsbergen collected by dr. Emil Hadač in 1939 vol 49. Dybwad i Komm, Norway, pp 1–11

Hagen A (1950) Notes on arctic fungi. Norsk Polarinst Srk 93:1–25

Hagen A (1952) Plants collected in Vestspitsbergen in the summer of 1933. Norsk Polarinstitutt Medd 70:1–14

Hansen J, Sato M, Ruedy R, Lo K, Lea DW, Medina-Elizade M (2006) Global temperature change. Proc Nat Acad Sci USA 103:14288–14293

Holm K, Holm L (1993a) Dryadicolous ascomycetes from Svalbard. In: Petrini O, Laursen GA (eds) Arctic and alpine mycology 3–4. Bibl Mycol 150:53–62

Holm L (1957) Études taxonomiques sur les pléosporacées. Symb Bot upsal 14:1–188

Holm L (1975) Taxonomic notes on ascomycetes. VII. Microfungi on *Cassiope tetragona*. Svensk Bot Tidskr 69:143–160

Holm L (1979) Microfungi on *Dryas*. Bot Not 132:77–92

Holm L, Holm K (1980) Microfungi on *Cassiope* (*Harrimanella*) *hypnoides*. Norweg J Bot 27:179–184
Holm L, Holm K (1993b) The genus Pleospora s. l. from Svalbard. Sydowia 45:167–187
Holm L, Holm K (1993c) Two new northern pyrenomycetes. Blyttia 51:121–123
Holm L, Holm K (1994) Svalbard Pyrenomycetes. An annotated checklist. Karstenia 34:65–78
Hoshino T, Tojo M, Kanda H, Tronsmo AM (2001) Ecological role of fungal infections of moss carpet in Svalbard. Mem Nat Inst Polar Res 54:507–513
Hoshino T, Saito I, Tronsmo AM (2003) Two snow mold fungi from Svalbard. Lidia 6:30–32
Huhtinen S (1987) New Svalbard Fungi. In: Laursen GA, Ammirati JF (eds) Arctic and Alpine Mycology II (= Proceedings ISAM 2). Plenum, New York, pp 123–151
Huhtinen S (1993) Some hyaloscyphaceous fungi from tundra and taiga. Sydowia 45:188–198
Jørstad I (1950) *Puccinia blyttiana*, a new member of the East Arctic rust flora. Blyttia 8:81–90
Karsten P (1872) Fungi in insulis Spitsbergen et Beeren Eiland collecti. Examinat, enumerate. Öfvers K Vet-Akd Förth 2:91–108
Karsten P (1884) Fragmenta mycological VI. Hedwigia 23:37–39
Kukkonen I (1963) Taxonomic studies on the genus Anthracoidea (Ustilaginales). Ann Bot Soc Zool Bot Fenn Vanamo 34:1–122
Lid J (1967) Synedria of twenty vascular plants from Svalbard. Bot Jahrb 86:481–493
Lind J (1928) The Micromycetes of Svalbard. Skr Svalbard Ishavet 13:1–61
Lind J (1934) Studies on the geographical distribution of arctic circumpolar micromycetes. K Danske Vid Selsk Biol Medd 11:1–152
Lindeberg B (1959) Ustilaginales of Sweden. Symb Bot Upsal 14:1–175
Lu B, Druzhinina I, Fallah P, Chaverri P, Gradinger C, Kubicek CP, Samuels GJ (2004) *Hypocrea/ Trichoderma* species with *pachybasium*-like conidiophores: teleomorphs for T. *minutisporum* and *T. polysporum*, and their newly discovered relatives. Mycologia 96:310–342
Nannfeldt JA (1932) Studien über die Morphologie und Systematik der nicht lichenisierten, inoperculaten Discomyceten.—Nova Acta Reg Soc Sei Upsaliensis. Ser 4(8):1–376
Nannfeldt JA (1933) *Puccinia eutremae* Lindr. Found in Spitsbergen. Svensk Bot Tidskr 27:116–117
Nannfeldt JA (1984a) *Hysteronaevia*, a new genus of mollisioid Discomycetes. Nord J Bot 4:225–247
Nannfeldt JA (1984b) Notes on *Diplonaevia* (Discomycetes inoperculati), with special regard to the species on Juncaceae. Nord J Bot 4:791–815
Schumacher T, Kohn LM (1985) A monographic revision of the genus Myriosclerotinia. Can J Bot 63:1610–1640
Singh SM, Singh SK, Yadav LS, Ravindra R (2012) Filamentous soil fungi from Ny-Ålesund, Spitsbergen, and screening for extracellular enzymes. Arctic 65:45–55
Smith RIL (1996) Terrestrial and freshwater biotic components of the western Antarctic Peninsula. In: Ross RM, Hofmann EE, Quentin LB (eds) Foundations of Ecological Research West of the Antarctic Peninsula Antarctic Research Series vol 70. American Geophysical Union, Washington DC, pp 15–59
Tojo M, Newsham KK (2012) Snow moulds in polar environments. Fungal Ecol 5:395–402
Tojo M, Nishitani S (2005) The effects of the smut fungus *Microbotryum bistortarum* on survival and growth of *Polygonum viviparum* in Svalbard, Norway. Can J Bot 83:1513–1517
Tojo M, Van West P, Hoshino T, Kida K, Fujii H, Hakoda H, Kawaguchi Y, Mühlhauser HA, Van den Berg AH, Küpper FC, Herrero ML, Klemsdal SS, Tronsmo AM, Kanda H (2012) *Pythium polare*, a new heterothallic Oomycete causing brown discoloration of *Sanionia uncinata* in the Arctic and Antarctic. Fungal Biol 116:756–768
Turner J, Bindschadler R, Convey P, di Prisco G, Fahrbach E, Gutt J, Hodgson D, Mayewski P, Summerhayes C (2009) Antarctic climate change and the environment. Scientific Committee on Antarctic Research, Scott Polar Research Institute, Cambridge, UK.
Van der Plaats-Niterink AJ (1981) Stud Mycol 21:1–242
Vánky K (1994) European smut fungi. Gustav Fischer Verlag, Stuttgart

Vánky K, Oberwinkler F (1994) Ustilaginales on Polygonaceae, a taxonomic revision. Nova Hedwigia 107:1–96

Väre H, Vestberg M, Eurola S (1992) Mycorrhiza and root-associated fungi in Spitsbergen. Mykorrhiza 1:93–104

Victoria FC, Pereira AB, Costa DP (2009) Composition and distribution of moss formations in the ice-free areas adjoining the Arctowski region, Admiralty Bay, King George Island, Antarctica. Iheringia, Série Botânica 64:81–91

Virtanen RJ, Lundberg PA, Moen J, Oksanen L (1997) Topographic and altitudinal patterns in plant communities on European Arctic islands. Polar Biol 17:95–113

Wulff T (1902) Botanische Beobachtungen aus Spitzbergen. Kessinger, Montana

Yamazaki Y, Tojo M, Hoshino T, Kida K, Sakamoto T, Ihara H, Yumoto I, Tronsmo AM, Kanda H (2011) Characterization of *Trichoderma polysporum* from Spitsbergen, Svalbard archipelago, Norway, on species identity, infectivity to moss, and polygalacturonase activity. Fungal Ecol 4:15–21

Zabawski J (1982) Soil microfungi of peats in the Hornsund region (West Spitsbergen). Acta Universitatis Wratislaviensis 525:269–279

Ecological Strategies of Snow Molds to Tolerate Freezing Stress

Tamotsu Hoshino, Nan Xiao, Yuka Yajima, Kenichi Kida, Katsuyuki Tokura, Ryo Murakami, Motoaki Tojo and Naoyuki Matsumoto

What is Snow Mold?

Snow molds attack dormant plants, such as forage crops, winter cereals, and conifer seedlings, at low temperatures under snow cover (e.g., Hoshino et al. 2009). The first observation was probably made on rotted winter barley under snow cover in Toyama, Japan in 1788 (Hori 1934), and Eriksson (1879) published a scientific report on winter injury of winter crops in Sweden. "Snow molds" or "snow mold fungi" is a generic name including diverse taxonomic groups of fungi. Bruehl et al. (1966) found 55 fungi including unidentified species from winter wheat in early spring, and Årsvoll (1975) described 33 fungi from forage grasses just after snowmelt; however, most of these fungi are considered not pathogenic, or simply remained inactive under snow. Important pathogens of agricultural crops are oomycetes: *Pythium iwayamai*, *P. okanoganense*, and *P. paddicum*; ascomycetes: *Microdochium nivale* (syn.: *Fusarium nivale*), *Sclerotinia borealis* (syn.: *Myriosclerotinia borealis*), *S. nivalis*, and *S. trifoliorum*; and basidiomycets: *Athelia* sp., *Coprinus psychromorbidus*, *Typhula incarnata*, *T. ishikariensis* (syn.: *T. idahoensis*), and *T. trifolii*. Ascomycetes attacking conifer seedlings include *Phacidium infestans*, *P. abietis*, and *Racodium therryanum*.

T. Hoshino (✉) · Y. Yajima
National Institute of Advanced Industrial Science and Technology (AIST), Higashi-Hiroshima, Japan
e-mail: tamotsu.hoshino@aist.go.jp

N. Xiao
Kochi Institute for Core Sample Research, Japan Agency Marine-Earth Science and Technology (JAMSTEC), Nankoku, Japan

K. Kida · K. Tokura · M. Tojo
Graduate School of Life and Environmental Sciences, Osaka Prefecture University, Sakai, Japan

R. Murakami
Graduate School of Life Sciences, Hokkaido University, Sapporo, Japan

N. Matsumoto
Graduate School of Agriculture, Hokkaido University, Sapporo, Japan

R. Imai et al. (eds.), *Plant and Microbe Adaptations to Cold in a Changing World,*
DOI 10.1007/978-1-4614-8253-6_24,

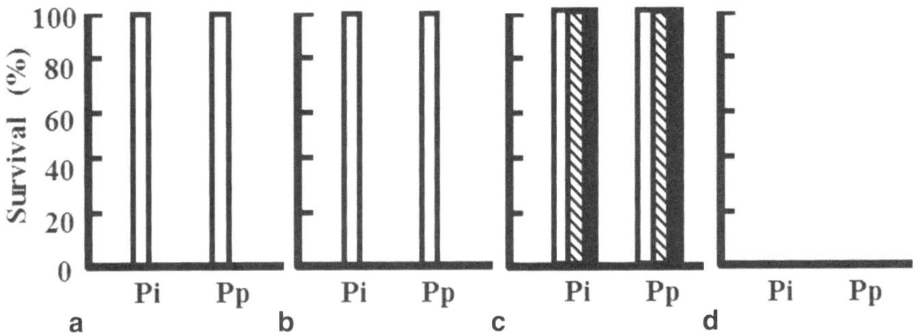

Fig. 1 Freeze tolerance of oomycetous snow molds (Murakami 2012). **A** *mycelia*, **b** hyphal swellings, **c** fungi in living host plants, **d** fungi in dead host plants. Open column: control (nonfreezing), shaded column: one freeze–thaw treatment, closed column: two cycles of freeze–thaw. *Pi P. iwayamai* and *Pp P. padicum*. Mycelial discs free of sexual and nonsexual organs were frozen at −20 °C in a program freezer at a rate of 10 °C/h in test tubes. They were then thawed at 2 °C for 8 h, transferred to fresh PDA plates, and incubated at 15 °C. Hyphal swellings were prepared according to Kusunoki and Ichitani (1982). Fungi in the host: mycelial discs were inoculated to living or autoclaved bentgrass and incubated at 10 °C for 1 month in the darkness. Ten infected bentgrass leaves were placed in sterilized test tubes and frozen at −20 °C under the same condition as above. After freezing, infected bentgrass leaves were plated on corn meal agar plates and incubated at 20 °C for 3 days

Ecophysiological Strategies of Snow Mold to Adapt Frozen Environments

Snow molds are taxonomically diverse and can grow at subzero temperatures and tolerate freeze–thaw event. Many studies have been carried out on snow molds from plant protection view point. New findings regarding their ecophysiological and biochemical characteristics have been reported during the past decade by a few researchers, including the authors group. In this chapter, we show various strategies of snow molds against freezing tolerance.

Oomycetes

Oomyceteous snow molds, *Pythium* spp. were reported to be less tolerant to chilling and freezing temperatures than other snow mold taxa (Takamatsu 1989; Hoshino et al. 2002). However, *Pythium* species are often found pathogenic on mosses in polar regions (e.g., Tojo and Newsham 2012). Tokura (2009) and Murakami (2012) investigated the freezing resistance of *Pythium* spp. from temperate and polar regions. Free mycelia and hyphal swellings, structures for survival of *P. iwayamai* and *P. paddicum* lost viability after 3 cycles of freezing–thawing; however, mycelia survived the treatment when they were in the host tissues (Figs. 1 and 2; Tokura 2009; Murakami 2012).

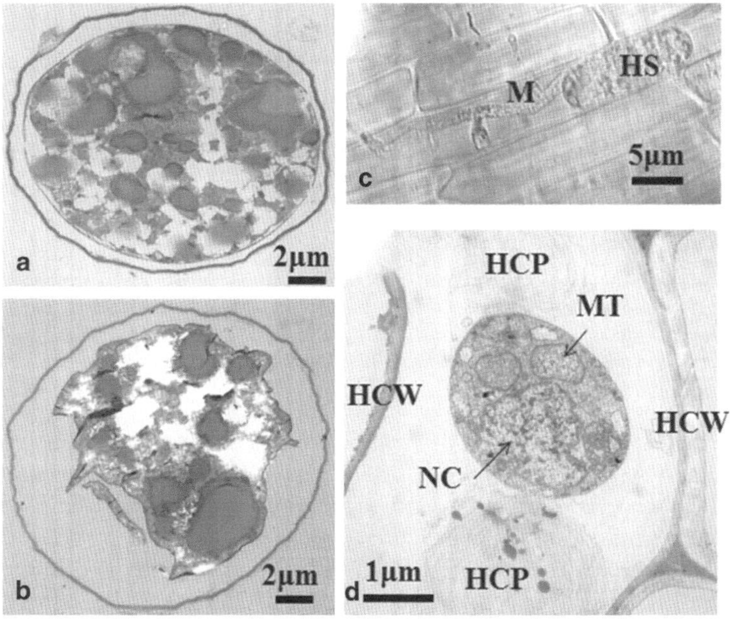

Fig. 2 Electron microscopy of hyphal swellings and mycelia of *Pythium*. Hyphal swellings before (**a**) and after freezing (**b**). light microscopy of hyphal swelling of *Pythium* sp. germinated in bentgrass leaf before freezing (**c**). *HS* hyphal swelling, *M* mycelia, (**d**) electron microscopy of mycelium of *P. iwayamai* in infected bentgrass leaf after freezing, *HCP* host chloroplast, *HCW* host cell wall, *MT* mitochondrion, *NC* nucleus

Fungi in permafrost are characterized by both the presence of natural cryoprotectants such as plant substrates or derivatives in these ecotopes and the ability to utilize their inherent mechanisms of protection (Ozerskaya et al. 2009). Stakhov et al. (2008) demonstrated that ancient seeds of higher plants constituted a specific habitat for microorganisms in frozen ground, which favored their survival for millennia.

Cold-accumulated plants also store cryoprotectants in their cytosol and *Pythium* spp. in infected tissues, probably to utilize host cryoprotectants to tolerate freezing. Our results are the first to prove the hypothesis of Ozerskaya et al. (2009) experimentally.

Basidiomycetes

Duman and Olsen (1993) first reported antifreeze activities from fruit bodies of four kinds of mushrooms, i.e., *Trametes versicolor*, *Flammulina velutipes*, *Pleurotus ostreatus*, and *Stereum* sp. Basidiomycetes produce extracellular antifreeze proteins (AFPs), which attach to the surface of ice crystal to inhibit its growth (Hoshino et al. 2009). Extracellular antifreeze activity is the mechanism unique to basidiomycetous snow molds but not in other snow molds belonging to oomycetes or ascomycetes (Snider et al. 2000; Hoshino et al. 2003a, b, 2009).

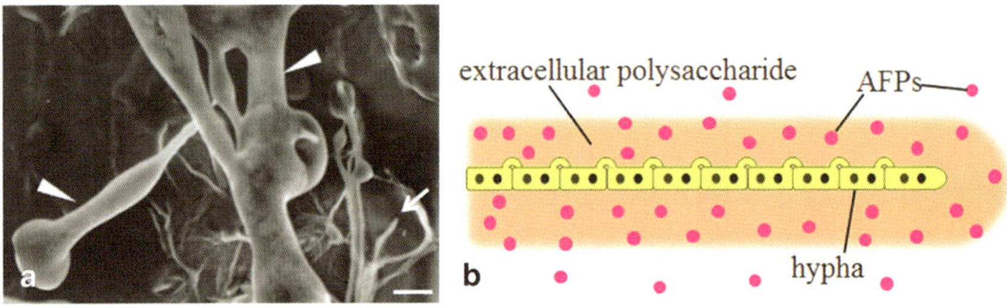

Fig. 3 Extracellular polysaccharide and AFPs. **a** SEM image of extracellular polysaccharide covering mycelia of *Typhula ishikariensis* on a leaf of *Arabidopsis thaliana*. *Arrow* normal mycelium, *triangles* extracellular cellulose covering mycelia Bar is 10 μm. **b** Schematic image: extracellular polysaccharide concentrate secreted AFPs around hypha

We purified and cloned AFPs from basidiomycetous snow molds *Coprinus psychromorbidus* and *T. ishikariensis* (Hoshino et al. 2003a, b). Cloned genes of AFPs from *T. ishikariensis* did not have any similarity with known proteins when we found these genes. After our publications, isogenes were found in bacteria, algae, and animals (these isogenes without fungi will be discussed later). Therefore, these fungal AFPs were considered to be a new class of AFPs. AFP of *T. ishikariensis* (TisAFP) exhibited bursting ice growth normal to the c-axis of the ice crystal and high thermal hysteresis activity (approximately 2 °C), as in the case of insect hyperactive AFPs (Xiao et al. 2010b). TisAFP was crystallized and its crystal structure was analyzed (Kondo et al. 2012). The structure of TisAFP6 (one isoform of AFPs from *T. ishikariensis*) and that of a recently determined homolog from an Arctic yeast *Leucosporidium* sp. (Lee et al. 2012) are the only β-helices that deviate from sequential stacking of β-helical loops. A carbonic anhydrase from *Methanosarcina thermophila* (Protein Data Bank ID code = 1THJ) bears a resemblance in having six 18-residue triangular loops with an α-helix alongside. However, the β-helical core of 1THJ is left-handed, whereas that of TisAFP6 is right-handed. The helical structure of TisAFP6 is surprisingly irregular in comparison with the other β-helical AFPs.

Diffusion of AFPs secreted in the extracellular environment does not support mycelial growth under subzero temperatures. AFPs and thermal hysteresis activity at concentrations are found in the extracellular polysaccharide, which surround Arctic diatoms (Krembs et al. 2002). *T. incarnata* and *T. ishikariensis* also produce extracellular polysaccharides (Fig. 3; Hoshino et al. 2009) and they probably bind the AFP molecules they secrete. *M. nivale* does not produce AFPs but produces extracellular polysaccharides such as cellulose (Schweiger-Hufnagel et al. 2001) and fructan (Cairns et al. 1995). These extracellular polysaccharides of *M. nivale* are considered to bind plant antifungal peptides and reduce activities of these peptides (Koike et al. 2003; Hoshino et al. 2009).

Similar kinds of AFP genes were found from other edible mushrooms (*Lentinula edodes* and *F. velutipes*; Raymond and Janech 2009) and Arctic basidiomycetous yeast *Leucosporidium* sp. (Lee et al. 2010). AFPs were also cloned from an ice diatom (Janech et al. 2005) and a bacterium in Antarctica (Raymond et al. 2007), and these proteins showed high similarities to fungal AFPs. These findings suggest that fungal AFP homologs are widely distributed in different kingdoms, implying the possibility of horizontal gene transfer between eukaryotic microbes and prokaryotes (Raymond et al. 2007; Sorhannus 2011; Raymond and Kim 2012).

Ascomycetes

S. borealis prevails where soil freezing is severe (Tomiyama 1955; Röed 1960; Nissinen 1996). The fungus, however, does not produce extracellular AFPs (Snider et al. 2000; Hoshino et al. 2003a, b, 2009) and has another physiologic strategy to adapt to the frozen environment. Tomiyama (1955) cultured *S. borealis* and *T. incarnata* on both frozen and unfrozen potato dextrose agar (PDA) plates that were kept outside in Sapporo, northern Japan; mycelial growth of *T. incarnata* was inhibited on frozen plates, but *S. borealis* grew faster on frozen plates than on unfrozen ones. His experiments were not made under controlled conditions and have not been reproduced by others. Under controlled conditions, we confirmed his results that *S. borealis* grew on frozen PDA (Hoshino et al. 2009, 2010). *S. borealis* showed normal mycelial growth under the frozen condition and mycelial growth rate on frozen plates at −1 °C was faster than that on unfrozen PDA at its optimal growth temperature of 4–10 °C. *S. borealis* can grow at low water potential on plates containing twice the concentration of medium ingredients (Tomiyama 1955), sucrose and KCl (Bruehl and Cunfer 1971) as well as D-mannitol (Namikawa et al. 2004). An increase in intracellular osmosis enhanced mycelial growth and shifted the optimal mycelial growth temperature from 10–15 °C to 4 °C. However, mycelial growth of other snow molds such as *S. nivalis* and *S. trifoliorum* was inhibited at low water potential (Hoshino et al. 2009, 2010). Therefore, osmophilism of *S. borealis* was probably a unique feature to adapt severe freezing.

Antifreeze activities were detected in various taxa including oomycete, blastocladiomycete, ascomycetes, and basidiomycetous yeasts from Antarctica (Xiao et al. 2010a). AFP from Antarctic ascomycete *Antarctomyces psychrotrophicus* (AnpAFP) was glycosylated and AnpAFP and Tis AFP were similar in that they both inhibit recrystallization (Xiao et al. 2010b). Though AnpAFP was not as active as TisAFP, *A. psychrotrophicus* had higher freeze tolerance than *T. ishikariensis* and survived 15 freeze–thaw cycles (Xiao 2012). AnpAFP may inhibit recrystallization more effectively than TisAFP. Xiao (2012) cloned AnpAFP gene to find that it had low similarity (ca. 25 % in amino acid sequence) to TisAFP gene. These two AFP genes from basidiomycete and ascomycetes might have evolved independently from the same ancestor. Ascomycetes are largest fungal taxon in Kingdom Fungi and probably have more variable mechanisms to adapt cold environment in the cryosphere than basidiomycetes.

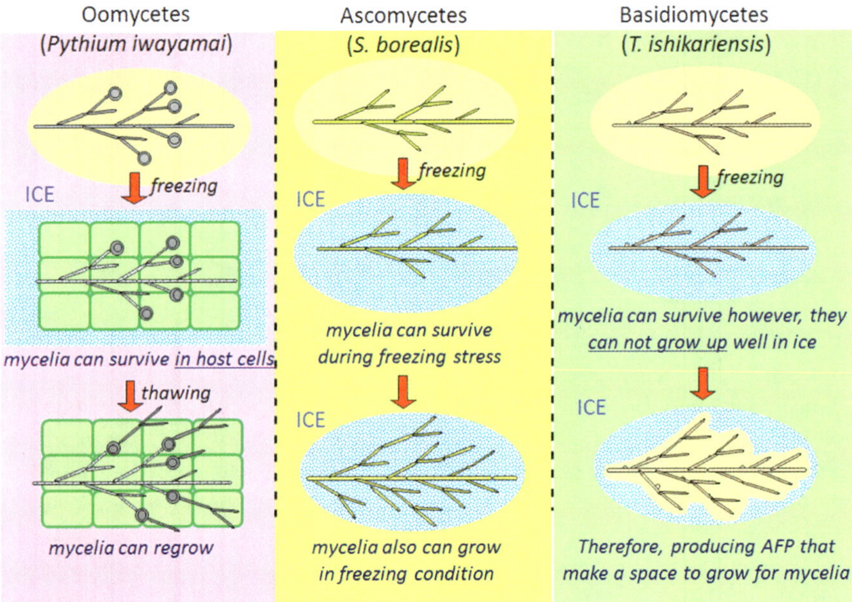

Fig. 4 Ecophysiological strategies of snow molds against freezing stress. *Sb* mycelia can survive freezing stress. Mycelia can also grow under freezing condition. *Tish* mycelia can survive; however, they cannot grow well in ice

Concluding Remarks

These findings suggested that snow molds have cold adaptation mechanisms that differ at the phylum level (summarized in Fig. 4). Oomycetous and basidiomycetous snow molds avoid extracellular frozen environments. However, ascomycete, *S. borealis* adapted well to frozen condition. Our results supported previous reports that snow mold pathogens were different in snow environment and it is important for the proper usage of pesticides that control each snow mold species.

References

Årsvoll K (1975) Fungi causing winter damage on cultivated grasses in Norway. Meld Norg Landbrukshøgsk 54(9):1–49

Bruehl GW, Cunfer BM (1971) Physiologic and environmental factors that affect the severity of snow mold of wheat. Phytopathology 61:792–799

Bruehl GW, Sprague R, Fischer WB, Nagamitsu M, Nelson WL, Vogel OA (1966) Snow mold of winter wheat in Washington. Wash Agric Stn Bull 677:1–21

Cairns AJ, Howarth CJ, Pollock CJ (1995) Submerged batch culture of the psychrophile *Monographella nivalis* in a defined medium; growth, carbohydrate utilization and responses to temperature. New Phytol 129:299–308

Duman JA, Olsen TM (1993) Thermal hysteresis protein activity in bacteria, fungi, and phylogenetically diverse plants. Cryobiology 30:322–328

Eriksson J (1879) En ny parasitsvamp å hvete, *Typhula graminum* Karst. Kungl Landtbr-Acad Handl o Tidskr 3:1–6
Hoshino T, Tojo M, Kanda H, Herrero ML, Tronsmo AM, Kiriaki M, Yokota Y, Yumoto I (2002) Chilling resistance of isolates of *Pythium ultimum* var. *ultimum* from the Arctic and Temperate Zones. Cryo Lett 23:151–156
Hoshino T, Kiriaki M, Nakajima T (2003a) Novel thermal hysteresis proteins from low temperature basidiomycete, *Coprinus psychromorbidus*. Cryo Lett 24:135–142
Hoshino T, Kiriaki M, Ohgiya S, Fujiwara M, Kondo H, Nihimiya Y, Yumoto I, Tsuda S (2003b) Antifreeze proteins from snow molds. Can J Bot 81:1175–1181
Hoshino T, Xiao N, Tkachenko OB (2009) Cold adaptation in the phytopathogenic fungi causing snow molds. Mycoscience 50:26–38
Hoshino T, Terami F, Tkachenko OB, Tojo M, Matsumoto N (2010) Mycelial growth of the snow mold fungus, *Sclerotinia borealis* improved at low water potentials: an adaptation to frozen environment. Mycoscience 51:98–102
Hori S (1934) Old record of snow molds of winter wheat (in Japanese). J Plant Protection 21:165–166
Janech MG, Krell A, Mock T, Kang J-S, Raymond JA (2005) Icebinding proteins from sea ice diatoms (Bacillariophyceae). J Phycol 42:410–416
Koike M, Isobe A, Imai R (2003) A cotton jacket for winter survival: cellulose production of the snow mold fungus *Microdochium nivale*, is induced by the wheat defensing, TAD1, which accumulates during winter hardening. In: Functional Genomics and Breeding Strategies for Cold Tolerance in Plants. National Agricultural Research Center for Hokkaido Region, Sapporo, pp46
Kondo H, Hanada Y, Sugimoto H, Hoshino T, Garnham CP, Davies PL, Tsuda S (2012) Ice-binding site of snow mold fungus antifreeze protein deviates from structural regularity and high conservation. Proc Natl Acad Sci U S A 109:9360–9365
Krembs C, Eicken H, Junge K, Deming JW (2002) High concentrations of exoplymeric substances in Artic winter sea ice: implications for the polar ocean carbon cycle and cryoprotection of diatoms. Deep Sea Res I 49:2163–2181
Kusunoki M, Ichitani T (1982) Preparation of mycelium-free oospores of *Pythium bulteri* by a freezing method. Ann Phytopath Soc Jpn 48:695–698
Lee JK, Park KS, Park S, Park H, Song YH, Kang S-H, Kim HJ (2010) An extracellular ice-binding glycoprotein from an Arctic psychrophilic yeast. Cryobiology 50:222–228
Lee JH, Park AK, Do H, Park KS, Moh AH, Chi YM, Kim HJ (2012) Structural basis for the anti-freeze activity of an ice-binding protein from an Arctic yeast. J Biol Chem 287:11460–11468
Murakami R (2012) Studies on freezing resistance of *Pythium* species from Polar Regions. Master thesis, Graduate School of Life Science, Hokkaido University, Sapporo
Namikawa Y, Watanabe T, Saito I, Takasawa T (2004) Growth of the psychrophilic snow mold *Sclerotinia borealis* on the agar under xerophilic conditions. Res Bull Obihiro Univ 25:23–26
Nissinen O (1996) Analyses of climatic factors affecting snow mould injury in first-year timothy (*Phleum pratense* L.) with special reference to *Sclerotinia borealis*. Acta Univ Oul A 289:1–115
Ozerskaya S, Kochkina G, Ivanushkina N, Gilichinsky DA (2009) Fungi in Permafrost. In: Margesin R (ed) Permafrost Soils, Soil Biology 16. Springer, Berlin, pp 85–95
Raymond JA, Janech MG (2009) Ice-binding proteins from enoki and shiitake mushrooms. Cryobiology 58:151–156
Raymond JA, Kim HJ (2012) Possible Role of Horizontal Gene Transfer in the Colonization of Sea Ice by Algae. PLoS ONE 7(5):e35968. doi:10.1371/journal.pone.0035968
Raymond JA, Fritsen C, Shen K (2007) An ice-binding protein from an Antarctic sea ice bacterium. FEMS Microbiol Ecol 61:214–221
Röed H (1960) *Sclerotinia borealis* Bub. & Vleg., a cause of winter injuries to winter cereals and grasses in Norway. Acta Agric Scand 10:74–82
Schweiger-Hufnagel U, Ono T, Izumi K, Hufnagel P, Morita N, Kaga H, Morita M, Hoshino T, Yumoto I, Matsumoto N, Yoshida M, Sawada MT, Okuyama H (2000) Identification of the extracellular polysaccharide produced by snow mold fungus *Microdochium nivale*. Biotechnol Lett 22:183–187

Snider CS, Hsiang T, Zhao GY, Griffith M (2000) Role of ice nucleation and antifreeze activities in pathogenesis and growth of snow molds. Phytopathology 90:354–361

Sorhannus U (2011) Evolution of antifreeze protein genes in the diatom genus *Fragilaripsis*: Evidence for horizontal gene transfer, gene duplication and episodic diversity selection. Evol Bioinformatics 7:279–289

Stakhov VL, Gubin SV, Maksimovich SV, Rebrikov DV, Savilova AM, Kochkina GA, Ozerskaya SM, Ivanushkina NE, Vorobyva EA (2008) Microbial communities of ancient seeds derived from permanently frozen Pleistocene deposits. Mikrobiologia 77:348–355

Takamatsu S (1989) Snow molds in winter wheat: studies on occurrence of *Pythium* snow rot (in Japanese with English summary). Spec Bull Fukui Agric Exp Stn 9:1–135

Tojo M, Newsham KK (2012) Snow moulds in polar environments. Fungal Ecol 5:379–480

Tokura K (2009) Species composition and ecology of *Pythium* spp. in a moss colony in the high Arctic region (in Japanese). Master thesis, Graduate School of Life and Environmental Sciences, Osaka Prefecture University, Sakai

Tomiyama K (1955) Studies of the snow blight disease of winter cereals. Hokkaido Natl Agric Exp Stn Rep 47:1–234

Xiao N, Inaba S, Tojo M, Degawa Y, Fujiu S, Hanada Y, Kudoh S, Hoshino T (2010a) Antifreeze activities of various fungi and Stramenopila isolates from Antarctica. North American Fungi 5:215–220

Xiao N, Suzuki K, Nishimiya Y, Kondo H, Miura A, Tsuda S, Hoshino T (2010b) Comparison of functional properties of two fungal antifreeze proteins from *Antarctomyces psychrotrophicus* and *Typhula ishikariensis*. FEBS Lett 277:394–403

Xiao N (2012) Study on antifreeze proteins from cold adapted fungi and diatoms. Ph.D. thesis, Graduate School of Life Science, Hokkaido University, Sapporo

Snow Mold Fungi in Russia

Oleg B. Tkachenko

Russia is a very large country and occupies 10 % of world's arable lands. Four-fifths of our arable lands are located in Central Volga, North Caucasus, Ural, and West Siberia. Areas with permafrost in Siberia and the Far East occupy 65 % of Russian territory, representing about 70 % of risky agriculture zone of the country. Therefore, most Russian territory is unfit for agriculture.

My Japanese colleagues and I studied sclerotial snow mold fungi in Russian territories at the beginning of this century. We investigated both typical agricultural areas and areas where agriculture was difficult or impossible because our interest also included fungi inhabiting natural ecosystem. The system of Russian botanical gardens helped us with our survey. We investigated sclerotial snow mold fungi in botanical gardens. Where botanical gardens were absent, we visited these regions by ourselves.

Investigations of snow molds have, as a rule, agricultural purposes. As in other countries, our snow mold research was first focused on winter cereals.

Sclerotinia spp.

The first record of snow mold fungi appeared in Russia at the beginning of last century. Trusova was the first to isolate and describe *Sclerotinia* sp. from winter rye as a separate taxon in 1901 (Khokhrjakov 1935). This fungus prevails on tissues of winter cereals severely damaged by freezing. Elenev made more detailed description of the fungus and named it *Sclerotinia graminearum* in 1918, and then Elenev and Solkina made complete description in Latin in 1939. Japanese scientists used the name *S. graminearum* for a long time (Tomiyama 1955). Bubác and Vleugel separately described *S. borealis* in 1917 in North Sweden (Saccardo 1925–1928).

O. B. Tkachenko (✉)
Plant Protection Department with Quarantine Service, Federal State Budgetary Institution for Science, Main Botanical Gardens n. a. N. V. Tsitsin of Russian Academy of Sciences (MBG RAS), Botanicheskaya ul. 4, Moscow, Russia 127276
e-mail: otkach@postman.ru

R. Imai et al. (eds.), *Plant and Microbe Adaptations to Cold in a Changing World*,
DOI 10.1007/978-1-4614-8253-6_25, © Springer Science+Business Media New York 2013

Fig. 1 Expedition routes

S. borealis was judged to have priority over *S. graminearum* in IX International Botanical Congress (Montreal, Canada) in 1959. Some Russian scientists still use the name *S. graminearum*.

Our expeditions showed that this fungus is widespread throughout Russia (Fig. 1). *S. borealis* is a necrotrophic fungus that colonizes damaged plant tissues in frost regions, such as Middle Volga, Siberia, North Russian Far East, and other Russian North regions. In these regions, gramineous plants grow despite severe frost. We could not find any low-temperature fungi in Yakutsk, where winter temperature falls as low as −45 to −55 °C. *Sclerotinia* snow mold is rare in Moscow region (approximately 1 %) on winter cereals (Anonymous 1968, 1969).

Saito, who made detailed investigations on *S. borealis* in Japan, was the first to find nongramineous host plants, including Asteraceae, Campanulaceae, Fabaceae, Iridaceae, and Liliaceae (Saito 1998). Gulaev found *S. graminearum* (*S. borealis*) even on 1-year Scots pine seedlings in Russia (Gulaev 1948). Pine plants were not killed but became freakish. We have no information on the *Sclerotinia* pine disease in countries other than Russia.

Sclerotinia nivalis was described by Saito in Hokkaido on several plants in 1997 (Saito 1997). I met him in the meeting next year in Sapporo and told him that this species existed in Russia. I sent him several specimens from Moscow, Saint Petersburg, and Novosibirsk, the following year. These isolates appeared to be *S. nivalis* (Tkachenko et al. 2003a). We later found this fungus in other Russian regions (Tkachenko 2006; Fig. 2).

Kohn made a monographic revision of the genus *Sclerotinia* in 1979 (Kohn 1979), in which she included only three pathogenic species of this genus: *S. sclerotiorum*, *S. trifoliorum*, and *S. minor*. Many low-temperature *Sclerotiniaceae*

Fig. 2 Place of detection of *Sclerotinia borealis* and *S. nivalis* during the expeditions

fungi have been recognized as synonyms of mesophile *S. minor*, such as taxonomically unknown *S. bulborum* on hyacinth from Holland (Wakker 1889), *S. sativa* on tulip and sweet clover from Canada (Drayton and Groves 1943), and *S. intermedia* on salsify from the USA (Ramsey 1924). *S. bulborum* was described in Main Botanical Gardens, Russian Academy of Science on tulips by Dr. Procenko in 1968 (Procenko 1968), but our study on this fungus showed that it was typical of *S. nivalis*. Saito and I visited Polar-Alpine Botanical Garden-Institute in Apatity (Kola Peninsula), where Shavrova had identified *Sclerotinia* fungus on 21 wintering plants, first as *Sclerotinia* sp. (Shavrova and Kislykh 1987) and then as *S. minor*, according to Kohn's revision (Shavrova 1989). We confirmed that these species were *S. nivalis*.

There are a large number of introduced plants in botanical gardens. These plants are usually not fully adapted to local conditions and are often affected by endemic diseases that are absent in their original habitat. Thus, the study of diseases of introduced plants may predict their success and difficulties on introduction. Cultivated plants are also often less resistant to pathogens than wild ones. All the above applies to snow mold pathogens, especially to *S. nivalis*. Saito's first record of *S. nivalis* was made on edible burdock, *Arctium lappa* var. *sativa*. Wild tulips and hyacinths, for example, cannot be damaged by this fungus in their original habitats. Most of the new host plants of *S. nivalis* were introduced plants. I can give only one example of wild plant damaged by this fungus—it is *Tripleurospermum perforatum* (Merat) M. Lainz (syn. *Matricaria inodora* Merat.). Our investigation and literature data have shown that *S. nivalis* infected 97 plant species of 54 genera and 19 families (Tkachenko 2006).

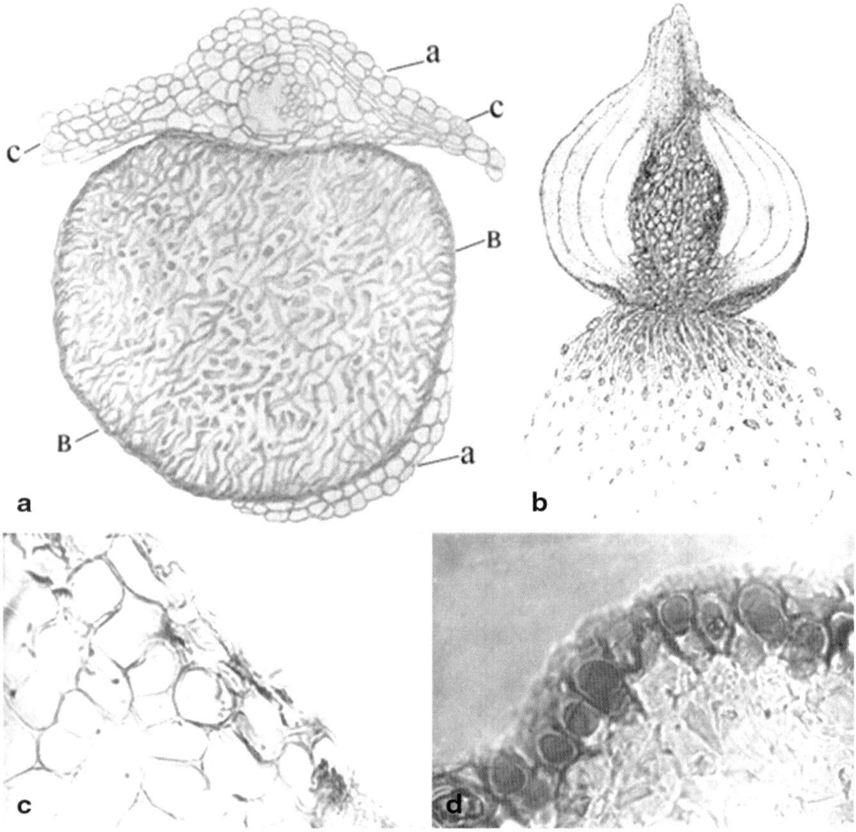

Fig. 3 **a** Cross section of sclerotia of *Sclerotium tuliparum* Kleb. through root (×80). *a* root tissue, *b* sclerotial tissue, *c* hyphae in root tissue cells (Elenkin 1911). **b** Longitudinal section of tulip bulb with sclerotia of *Sclerotium tuliparum* Kleb. (Elenkin 1911). **c** Cross section of pseudosclerotium of *Rhizoctonia tuliparum*. **d** Cross section of sclerotium of *Typhula ishikariensis* (Tkachenko 1983)

Typhula ishikariensis

Typhula ishikariensis S. Imai is the most devastating pathogen in the genus *Typhula*. I found a first description of *T. ishikariensis* tulip bulb rot in a Russian journal *Diseases of Plants* of 1911. Elenkin, mycologist of the Central Phytopathological Station of Imperial Saint Petersburg Botanical Garden (Elenkin 1911), described the sclerotial disease of tulip from the Botanical Garden in detail. He considered that tulip was damaged by *Sclerotium tuliparum* Kleb. (*Rhizoctonia tuliparum* (Kleb.) Whetz. et Arth.; Fig. 3a, b). The latter, in contrast to *T. ishikariensis*, never damaged roots of tulips (Gladders and Coley-Smith 1978). Sclerotia of *Rhizoctonia tuliparum* are really pseudosclerotia (resting bodies that do not have basic structure of sclerotia—cortex outside and medulla inside; Fig. 3c), consisting of large cells without distinct cell rind (only cell walls are pigmented) in contrast to the dense medulla cells and distinct rind cells (cortex) of *T. ishikariensis* sclerotia (Tkachenko 1983; Fig. 3d). These facts indicate that the fungus in Elenkin's drawings is *T. ishikariensis* S. Imai.

This fungal species has a large history of descriptions. H. Ekstrand described two species in Sweden: *Typhula borealis* and *T. hyperborea*; first based on morphological characters of anamorpha (Ekstrand 1934) and then using teleomorpha (Ekstrand 1955). Closely related *T. idahoensis* was described by R. Remsberg in the USA (Remsberg 1940). It turns out that S. Imai in Hokkaido (Japan) had described the fungus as *T. ishikariensis* in Japanese in 1929 (McDonald 1961), then he published in English (Imai 1930). Røed (1956) and Jamalainen (1957) compared the morphological characteristics of Japanese isolates of *T. ishikariensis*, North American *T. idahoensis*, and Finnish-Scandinavian *T. borealis* to conclude that they were synonymous. Investigations in Canada revealed three varieties of *T. ishikariensis*: *T. ishikariensis* var. *ishikariensis*, *T. ishikariensis* var. *idahoensis*, and *T. ishikariensis* var. *canadensis* (Årsvoll and Smith 1978). Matsumoto divided this species in Japan into three biotypes, i.e., biotypes A, B, and C (Matsumoto et al. 1982, 1983), then biotype C was considered as an ecotype of biotype B and referred to as small sclerotium (ss) form of biotype B (Matsumoto and Tajimi 1991). Matsumoto's conclusion is that the *T. ishikariensis* complex consists of two groups, species I and II (Matsumoto 1997; Matsumoto et al. 2001). In our work, we support his classification of the fungus.

Several *Typhula* species described in Russia, whose epithets were given from their host plants, appeared to be synonyms of *T. ishikariensis*. They are *T. humulina* from hop *Humulus lupulus* L. by Kuznetzova (1953) and *T. graminearum* by V.V. Gulaev (1948; this fungus was described on Scots pine young seedling, in addition to gramineous weeds). Potatosova (1960) was first, who considered that *T. humulina* and *T. graminearum* were synonymous with *T. idahoensis*. We confirmed that these fungi belonged to *Typhula ishikariensis* species I (Hoshino et al. 2004).

T. ishikariensis species I was known in Russia only in European part of the country before our expeditions in the last century (Tkachenko et al. 1997). We found the fungus during our expeditions in different parts of Russia (Fig. 4), even where we did not expect to find it due to deep frozen soil in winter, such as in Novosibirsk and Tomsk (West Siberia) (Tkachenko et al. 2003b). Hoshino et al. (2001) compared the survival of *T. ishikariensis* sclerotia after freeze–thaw cycles using Moscow and Novosibirsk isolates. Moscow isolates were killed after three cycles; significant deterioration of mycelial growth was observed. Novosibirsk isolates survived seven treatment cycles with a slight deterioration of mycelial growth, and Hoshino gave up further experiments.

We could not find *T. ishikariensis* in East Siberia near Irkutsk. This fungus was found near Baikal Lake at the foothill of Khamar-Daban. Khamar-Daban is, as some other Baikal regions (Salair ridge, Kuznetsk Alatau, and Western Sayan), a refugium of the glacial period (Epova 1956; Zarubin 2000; Krapivkina 2009). These areas were not affected by continuous glaciation. *T. ishikariensis* from Khamar-Daban belonged to species II designated by Matsumoto (1997). I think that this

Fig. 4 Place of detection of *Typhula ishikariensis* species I and II during the expeditions

species is the most primitive taxon of *T. ishikariensis* with only gramineous host plants.

We were surprised to find this fungus in the Chukotka region at Anadyr, which is situated at 64°4′N. It is a region of the continental tundra with almost no grasses (Poaceae), but *Poa ochotensis* Trin. damaged by *T. ishikariensis* was found on the shores of the Bering Sea, Anadyr Gulf.

Remarkably, *T. ishikariensis* species I damages underground plant tissues of hops (Kuznetzova 1953) and tulip (Procenko 1967) in Russia. Few reports of *Typhula* tulip disease are available from Canada and Sweden, but there are no reports of *Typhula* hop damage in areas other than Russia. The fungus produces secondary sclerotia (Tkachenko 1995), which may represent adaptation to environmental change.

Our investigation and literature data have shown that *T. ishikariensis* infected 97 plant species of 54 genera and 20 families (Tkachenko 1983, 2006, 2012; Seraya 2001).

Other Snow Mold Fungi

Lastly, I want to mention other snow mold fungi, which were subsidiary to our expedition. These are *Microdochium nivale*, *Phacidium infestans*, *Pythium* spp., *Sclerotinia trifoliorum*, *Sclerotium nivale*, *Typhula incarnata*, and *T. phacorrhiza*.

While *T. ishikariensis* is a biotroph, *T. incarnata* is a facultative saprophyte (Jacobs and Bruehl 1986). We did not see *T. incarnata* in areas with severe winters (Siberia), but the fungus was common even near Yamal Peninsula (Shiryaev 2006).

We found it in South Russian Far East (Vladivostok, Khabarovsk) and even in Caucasus, both at foots and at uplands in Adygea Republic. We have also recorded this fungus above the Arctic Circle at the Kola Peninsula in Polar-Alpine Botanical Garden-Institute at 67°7′N.

Typhula phacorrhiza is apparently widely spread in Russia. It was found in south tundra zone, the Polar Urals, and Novaya Zemlya (Shiryaev 2006), but detailed research or practical achievements using this species as biocontrol agent were not made. There was only one PhD thesis of Tazina (2005), where she successfully used this fungus against *T. ishikariensis* on winter wheat.

A cause of pink snow mold is a psychrotolerant fungus *Microdochium nivale*. The fungus has been much studied in Russia regarding its distribution and disease severity (Dmitriev and Saulich 2004). The disease is severe in wet areas. Russian scientists compared *Fusarium* rot and pink snow mold diseases (Ovsyankina 2000; Kostenko 2012). The term "snow mold" was applied only to the disease caused by *Fusarium* (*Microdochium*) *nivale* for a long time in Russia. We have never seen this fungus during our expeditions in Siberia.

Clover is, of course, widely grown in Russia and this plant is damaged by low-temperature pathogen *S. trifoliorum*. The fungus is found in most places where clover is grown.

Although not included in our investigations, snowy pine Schütte caused by *Phacidium infestans* is important. This disease is widely spread in Russia: European part of Russia, the Urals, and Siberia. Snowy pine Schütte damages Scots pine, rarely Siberian cedar, spruce, and juniper (Kuz'mina and Kuz'min 2009). The fungus can grow under the snow; hence, the name of the disease is snowy pine Schütte. The disease poses a great threat to seedlings both in nurseries and in natural ecosystem as well as regrowth of young trees, as the disease occurs almost exclusively in the winter under the snow. Plant loss in nurseries, in some cases, is up to 80%.

Pythium sp. has been recognized as a low-temperature pathogen only once on the Kola Peninsula in the 1980s in the former USSR. Petrov (1983) revealed that prevalence of *S. graminearum* (*S. borealis*) was largely dependent on the preconditioning by infection of *Pythium*, and that the fungus had no appreciable effect on perennial grasses.

Finally, about the fungus described by Elenev—*Sclerotium nivale* Elenev. I translate Russian name of this disease "Снеговая крупка" in English as "Snow tiny pellets" or "Snow tiny grain." I have not found the record on this fungus from abroad, although its record was cited in *Phytopathological Dictionary* (Diakova 1969) and the Internet in "Encyclopedias, dictionaries, reference books (Search)" (in Russian)[1] with translation into English (Sclerotium disease of rye), German (Sclerotium-Krankheit, Roggen), and French (sclérotiose du seigle). Although the disease is often found in more southern Russian areas of cultivation of winter cereals (Fig. 5a), it is also found in other gramineous plants (Fig. 5b, c). Symptoms of this low-temperature disease are characterized by white web-like mold with small, round, white sclerotia with a diameter of 0.5–1 mm on leaves and stems. The cause of disease is the imperfect

[1] Encyclopedias, dictionaries, reference books (Search) available at www.cnshb.ru/AKDiL/0040/k307340.shtm (in Russian).

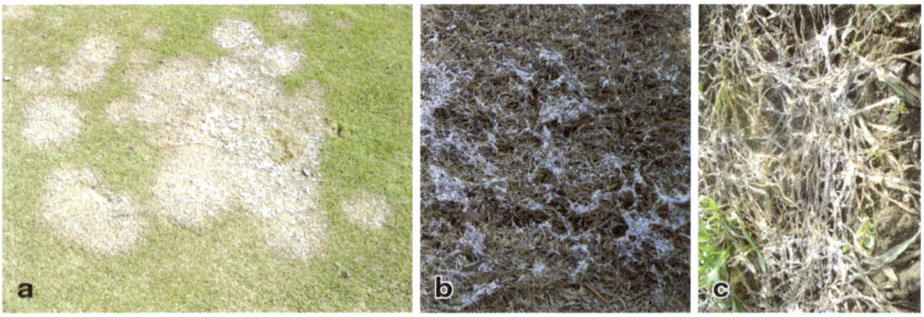

Fig. 5 *Sclerotium nivale* **a** on winter triticale, **b** on grasses, and **c** on turf. (Photographs by L. M. Sarycheva)

fungus *Sclerotium nivale*. The fungus forms only mycelium and sclerotia in its cycle. Sclerotia remain until an autumn and develop a new mycelium that infects plants in the presence of abundant moisture, causing rot and plant death. No resistant varieties are available. Obviously, this fungus has never been investigated deeply. Perhaps, this fungus exists in Russia (Khokhryakov et al. 1966; Semenkova and Sokolova 2003) but is regarded as "illegal" because it was not included even in the *Index Fungorum*.[2]

Conclusion

Global warming is likely to occur in Russia. Winters in Moscow district used to arrive about 1 month later in 1970s–1980s. Siberian climate has become much warmer. Weather forecast predicts that weather in European Part of Russia may become the same as in West Europe and that Siberian climate may become as in the European part. It means that large Siberian area will be developed for growing of winter cereals and that we need to work on snow mold there in the future. I welcome any collaboration with participants here.

But I doubt the reality of global warming using this site http://planeta.moy.su/blog/globalnoe_poteplenie_lozh_veka/2012-02-18-14885. According to expert evaluation of the National Oceanic and Atmospheric Administration (NOAA), USA, the average global temperature of the planet in 2011 is not among the 10 warmest.

Russian geographer Prof. Andrei Kapitsa (he had died in August 2011) said that global warming does not exist. On the contrary, it is slowly cooling for more than 30 years.

I think that there is warming in Russian territory, but Gulf Stream has become colder. I am not sure if warming is a universal issue of the whole globe, as I saw snowy

[2] Editors' note: Matsumoto et al. (1996) described a similar fungus in Norway, "Group II isolates from the Oppland locality often had abundant aerial mycelium. Their sclerotia were small and covered with thick mycelium. Those sclerotia developed normal sporocarps and produced basidiospores. Their mycelia produced on infected plants were usually, but not always, very fluffy."

Siberian pictures in a TV program of West European countries. I would say that different types of climate change should not be simply termed as global warming, but we should identify the differences to find countermeasures for sustainable agriculture.

Acknowledgments Expeditions across Russia were impossible without Japanese collaborators, Dr. Tamotsu Hoshino, Dr. Izumi Saito, Dr. Motoaki Tojo, and, certainly, Dr. Naoyuki Matsumoto. I appreciate their support. I also thank Dr. N. Matsumoto for help in translating article, Dr. E. V. Tkacheva for help with illustrations, and Dr. L. M. Sarycheva for the photographs of the fungus *Sclerotium nivale*.

References

Anonymous (1968) Distribution of pests and diseases of crops in the USSR in 1966 and the forecast of their appearance in 1967 [Распространение вредителей и болезней с.-х. культур в СССР в 1967 г. и прогноз их появления в 1968 г.]. Rossel'hozizdat, Moscow (in Russian)

Anonymous (1969) Distribution of pests and diseases of crops in the USSR in 1967 and the forecast of their appearance in 1968 [Распространение вредителей и болезней с.-х. культур в СССР в 1966 г. и прогноз их появления в 1967 г.]. Rossel'hozizdat, Moscow (in Russian)

Årsvoll K, Smith JD (1978) *Typhula ishikariensis* and its varieties, var. *idahoensis* comb. nov. and var. *canadensis* comb. nov. Can J Bot 56:348–364

Diakova GA (1969) Phytopathological dictionary: Russian-English-German-French. Nauka, Moscow

Dmitriev AP, Saulich MI (2004) Distribution of rye pink snow mold. Interactive agricultural Ecological atlas of Russia and neighboring countries. http://www.agroatlas.ru/en/content/diseases/Secalis/Secalis_Microdochium_nivale/map/

Drayton FL, Groves JW (1943) A new *Sclerotinia* causing a destructive disease of bulbs and legumes. Mycologia 35:517–528

Ekstrand H (1934) Trådkiubba på wintersåd. *Typhula* on winter cereals. Växtskyddsnotiser Stat. Växtskyddsanst. Stockholm. 1:3–4 (in Sweden, English summary)

Ekstrand H (1955) Overwintering of winter cereals and forage grasses [Höstädens och välgrasens övervintering] Medd. Stat. Växstkyddanst. 67:125 (in Sweden, English summary)

Elenkin AA (1911) About fungal diseases of tulip bulbs [О грибных болезнях луковиц тюльпанов]. Diseases of Plants [Болезни растений]. 5:105–124 (in Russian, German summary)

Epova NA (1956) Relicts of deciduous forests in Khamar-Daban flora [Реликты широколиственных лесов во флоре Хамар-Дабана]. Proceedings of the Biological-Geological Research Institute [Изв. БГНИИ] 16:25–61 (in Russian)

Gladders P, Coley-Smith JR (1978) *Rhizoctonia tuliparum* a winter-active pathogen. Trans Brit Micol Soc 71:129–139

Gulaev VV (1948) Damping off Pine seedlings in forest nurseries [Выпревание сеянцев сосны в лесных питомниках]. Proceedings/ Tatar forest experiment station [Труды/ Татарская лесная опытная станция] Kazan [Казань] 9:44–49 (in Russian)

Hoshino T, Tkachenko OB, Tronsmo AM, Kawakami A, Morita N, Ohgiya S, Ishizaki K (2001) Temperature sensitivity and freezing resistance among isolates of *Typhula ishikariensis* from Russia Icel. Agr Sci 14:61–65

Hoshino T, Tkachenko OB, Kiriaki M, Yumoto I, Matsumoto N (2004) Winter damage caused by *Typhula ishikariensis* biological species I on conifer seedlings and hop roots collected in the Volga–Ural regions of Russia. Can J Plant Path 26:391–396

Imai S (1930) On the Clavariacea of Japan. Transact Sapporo Nat His Soc 11:70–77

Jacobs DL, Bruehl GW (1986) Saprophytic ability of *Typhula incarnata*, *T. idahoensis*, and *T. ishikariensis*. Phytopathology 76:695–698

Jamalainen EA (1957) Overwintering Gramineae-plants and parasitic fungi. II. On the *Typhula* sp. fungi in Finland. J Sci Agr Soc Fin 29:76–81
Khokhrjakov M (1935) A little known disease of winter corn crops (*Sclerotinia*). Plant Protection 4:94–97 (in Russian, English summary)
Khokhryakov MK, Dobrozrakova TL, Stepanov KM, Letova MF (1966) Determinant of plant diseases [Определитель болезней растений]. Kolos, Leningrad (in Russian)
Kohn LM (1979) A monographic revision of the genus *Sclerotinia*. Mycotaxon 9:365–444
Kostenko ES (2012) Improved methods of pest monitoring and protection of turf grass stands from the major pests (snow mold and click beetles) [Совершенствование приёмов фитосанитарного мониторинга и защиты грибных травостоев от основных вредных организмов (снежной плесени и жуков щелкунов)]. Dissertation, Moscow Agricultural Academy (in Russian)
Krapivkina ED (2009) Nemoral relicts in flora of Chernevaya taiga in Mountain Shoria [Неморальные реликты во флоре черневой тайги Горной Шории]. Publishing House of Siberian Branch of Russian Academy of Sciences, Novosibirsk (in Russian)
Kuz'mina NA, Kuz'min SR (2009) Selection of Scots pine for resistance to fungal pathogens in the geographic cultures. Conifers Boreal Zone 1:76–81 (in Russian, English summary)
Kuznetzova AP (1953) New fungal species *Typhula humulina* A. Kuzn. on underground stems of hop [Новый вид гриба *Typhula humulina* A. Kuzn. на подземных стеблях хмеля]. Botanical Materials of Department of Cryptogamic Plants of Botanical Institute of Academy of Sciences of the USSR [Бот. мат. отд. споровых раст. Бот. ин-та АН СССР] 9:142–145 (in Russian)
Matsumoto N (1997) Evolution and adaptation in snow mold fungi. Soil Microorg 50:13–19 (in Japanese)
Matsumoto N, Tajimi A (1991) *Typhula ishikariensis* biotypes B and C, a single biological species. Trans Mycol Soc Jpn 32:273–281
Matsumoto N, Sato T, Araki T (1982) Biotype differentiation in the *Typhula ishikariensis* complex and their allopatry in Hokkaido. Ann Phytopathol Soc Jpn 48:275–280
Matsumoto N, Sato T, Araki T, Tajimi A (1983) Genetic relationships within the *Typhula ishikariensis* complex. Trans Mycol Soc Jpn 24:313–318
Matsumoto N, Tronsmo AM, Shimanuki T (1996) Genetic and biological characteristics of *Typhula ishikariensis* isolates from Norway. Eur J Plant Pathol 102:431–439
Matsumoto N, Tkachenko OB, Hoshino T (2001) The pathogenic species of *Typhula*. Low temperature plant microbe interactions under snow, pp 49–59
McDonald WC (1961) A review of the taxonomy and nomenclature of some low-temperature forage pathogens. Can Plant Dis Surv 41:256–260
Ovsyankina AV (2000) The structure of the populations of root rot pathogens and snow mold of winter rye and selection of source material for breeding resistant cultivars [Структура популяций возбудителей корневой гнили и снежной плесени озимой ржи и отбор исходного материала для селекции устойчивых сортов]. Dissertation, All-Russian Research Institute of Phytopathology, Bolshye Vezyomy Moscow District (in Russian)
Petrov VP (1983) Pathogenic microflora of root system of perennial gramineous grasses in Khibiny [Патогенная микофлора корневой системы многолетних злаковых трав в Хибинах]. Proceedings in Applied Botany, Genetic, and Breeding [Труды по прикладной ботанике, генетике и селекции] 82:38–45 (in Russian)
Potatosova EG (1960) Fungi of *Typhula* genus in the USSR. Bot J 45:567–572 (in Russian, English summary)
Procenko EP (1967) *Typhula borealis* Ekstrand on tulips in the USSR. Mycol Phytopathol 1:107–109 (in Russian, English summary)
Procenko EP (1968) New data of tulip sclerotial diseases. Mycol Phytopathol 2:252–253 (in Russian, English summary)
Ramsey GB (1924) *Sclerotinia intermedia* n. sp. a cause of decay of salsify and carrots. Phytopathology 14:323–328
Remsberg R (1940) Studies in the genus *Typhula*. Mycologia 40:52–96
Røed H (1956) Parasitic winter injury on pasture crops and sown cereals in Norway [Parasittaere vinterskader påengvekster og høstsaed i Norge]. Nord Jordbrugsforsk 38:428–432

Saccardo PA (1925–1928) Sylloge Fungorum XXIV:1179
Saito I (1997) *Sclerotinia nivalis*, sp. nov., the pathogen of snow mold of herbaceous dicots in Northern Japan. Mycoscience 38:227–236
Saito I (1998) Non-gramineous hosts of *Myriosclerotinia borealis*. Mycoscience 39:145–153
Semenkova IG, Sokolova ES (2003) Phytopathology: textbook for university students [Фитопатология: Учебник для студентов вузов]. Publishing House "Academia" [издательский центр «Академия»], Moscow (in Russian)
Seraya LG (2001) A cause of grey (speckled) snow mold fungus *Typhula ishikariensis* S. Imai: biology, ecology, and pathogenesis and study method of protection [Возбудитель серой (пятнистой) снежной плесени гриб *Typhula ishikariensis* S. Imai: биология, экология, патогенез и обоснование приемов защиты]. Dissertation, Moscow Agricultural Academy (in Russian)
Shavrova LA (1989) Parasitic fungi of introduced plants in Kola Peninsula conditions [Паразитные грибы интродуцированных растений в условиях Кольского полуострова]. Nauka, Leningrad (in Russian)
Shavrova LA, Kislykh EE (1987) Control of sclerotial rot of perennial plants [Борьба со склероциальной гнилью многолетников]. Plant Protection [Защита растений] 3:39 (in Russian)
Shiryaev AG (2006) Clavarioid fungi of Urals. III. Arctic zone. Mycol Phytopathol 40:294–304
Tazina SV (2005) Rationale for the protection of winter crops from the infection loss of plants [Обоснование защиты озимых зерновых культур от инфекционного выпадения растений]. Dissertation, All-Russian Agricultural University/Moscow Agricultural Academy (in Russian)
Tkachenko OB (1983) Biological basis of sclerotial rot control [Биологические основы борьбы со склероциальными гнилями]. Dissertation, Moscow Agricultural Academy (in Russian)
Tkachenko OB (1995) Adaptation of the fungus *Typhula ishikariensis* Imai to the soil inhabitance. Mycol Phytopathol 29:14–19 (in Russian, English summary)
Tkachenko OB (2006) Low temperature sclerotial fungi is a limiting factor in wintering of introduced herbaceous plants [Низкотемпературные склероциальные грибы – лимитирующий фактор зимующих интродуцируемых травянистых растений]. Dissertation, Main Botanical Gardens n.a. N.V. Tsitsin of Russian Academy of Sciences (in Russian)
Tkachenko OB (2012) Distribution and plant-host range of the most dangerous agents of snow molds, sclerotial fungi *Sclerotinia borealis*, *S. nivalis* and *Typhula ishikariensis*. Bull Main Botanical Garden 198(4):65–72 (in Russian, English summary)
Tkachenko OB, Matsumoto N, Shimanuki T (1997) Mating patterns of east European isolates of *Typhula ishikariensis* with isolates from distant regions. Mycol Phytopathol 31:68–72 (in English, Russian summary)
Tkachenko OB, Saito I, Novozhilova OA (2003a) A new snow mold *Sclerotinia* fungus in Russia. J Russ Phytopathol Soc 4:53–67 (in Russian, English elect. version)
Tkachenko OB, Hoshino T, Chikin YuA (2003b) A cause of speckled (grey) snow mold in Western Siberia [Возбудитель крапчатой (серой) снежной плесени в Западной Сибири]. Protection and Quarantine of Plants [Защита и карантин растений] 10:43 (in Russian)
Tomiyama K (1955) Studies on the snow blight disease of winter cereals. Hokkaido Nat Agric Exp Stn Rep 47:1–234 (in Japan, English summary)
Wakker JH (1889) La morve noire des jacinthes et plantes analoques, produite per le *Peziza bulborum*. Arch Neerland XXIII(II):25–45 (in French)
Zarubin AM (2000) Botanical refugia Baikal Siberia and the need to protect them [Ботанические рефугиумы Байкальской Сибири и необходимость их охраны]. Thesis of All-Russian Scientific-Practical Symposium on Youth Oct 24–27 Irkutsk. 8–9

Snow Moulds in a Changing Environment—A Scandinavian Perspective

Anne Marte Tronsmo

Introduction

The sub-Arctic area is assumed to be greatly affected by climate change. For plant production, this represents both an opportunity and a challenge. Rising temperatures, longer growth season, elevated CO_2 and increased precipitation represent a potential for increased biomass production of several crops and ought to be exploited for a more sustainable food production (Flæte et al. 2010). Climatic change models for Norway have projected temperature increase permitting 2 months' longer growth season (Fig. 1). However, diseases previously restricted by a hostile climate may prevail due to warmer and shorter winters and changes in precipitation patterns. Endemic diseases may appear more aggressive due to the same changes (Gaudet et al. 2012). In the northern parts of Scandinavia, plant production is carried out under combination of climate factors found at no other place on the earth. In northern Norway, crops are produced up to latitude 70° N and with 24-h day length during summer and very low irradiance during winter (Tromsø at 69° N has 2 months with midnight sun and 2 months without sun).

With a climatic change, plants will be facing temperature x day length combinations never experienced before. Current cultivars of crop plants will no longer be suitable: The new environment may pose a significant stress factor and these cultivars do not harbour the maximal yield potential under the new environment. Cultivars bred for southern latitudes are able to exploit the higher temperature, but are not adapted to the long days in northern Scandinavia. The main agricultural production/crops in northern Scandinavia are fodder crops for animal production, such as forage grasses and legumes;cereals and potatoes are also important crops. The most severe diseases on the main crops, perennial grasses and winter cereals in the sub-Arctic are caused by snow mould fungi. Even though winter temperatures may

A. M. Tronsmo (✉)
Department of Plant and Environmental Sciences, Norwegian University of Life Sciences,
1432 Ås, Norway
e-mail: anne-marte.tronsmo@umb.no

R. Imai et al. (eds.), *Plant and Microbe Adaptations to Cold in a Changing World*,
DOI 10.1007/978-1-4614-8253-6_26, © Springer Science+Business Media New York 2013

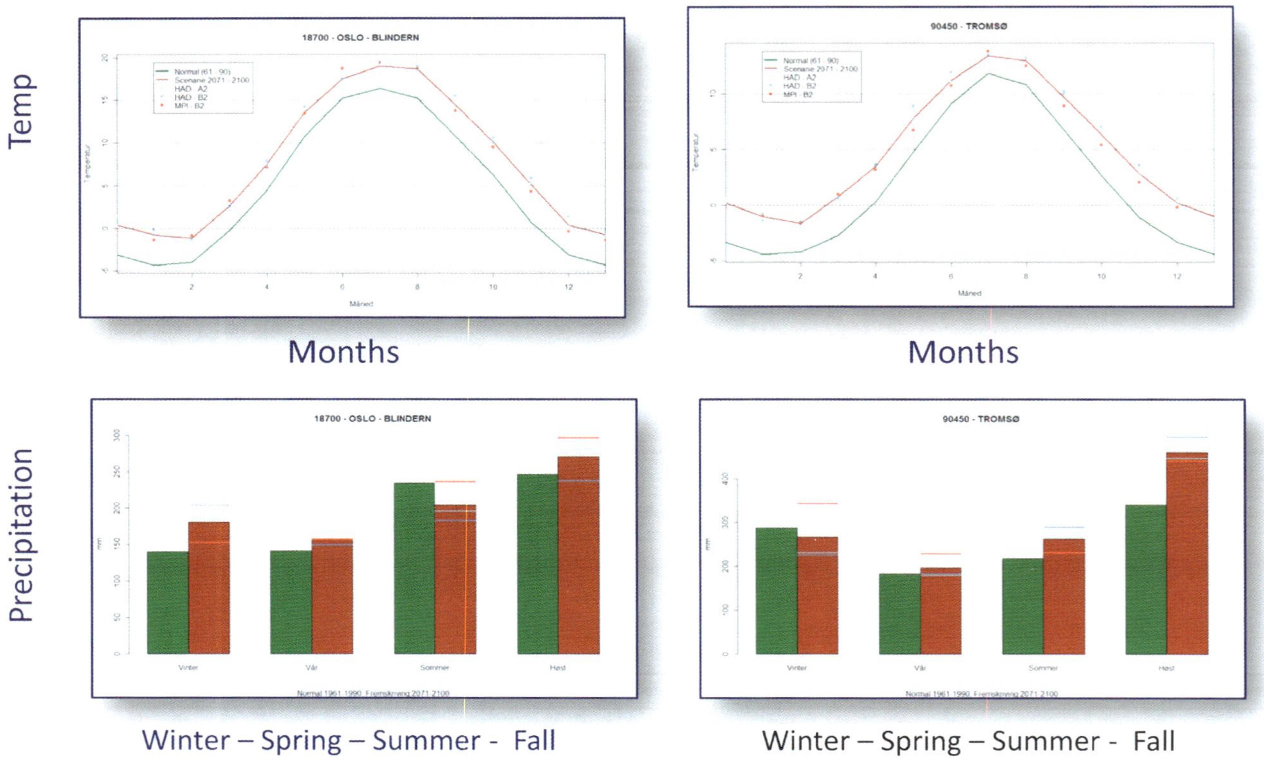

Fig. 1 Temperature and precipitation projections for 2071–2100 (*red*) compared to 1961–1990, Oslo (*left*) and Tromsø (*right*). The Norwegian. (Source: Meteorological Institute 2012. http://met.no)

Table 1 Cardinal temperatures for the four main snow mould fungi in Scandinavia and their growth rate at 0 °C in % of growth rate at optimum temperature. (From Årsvoll 1975)

Species	Cardinal temperatures, in °C			Growth rate at 0°C, in % of growth rate at the optimum temp
	Min	Max	Opt	
M. nivale	-6	28	21	8%
T. incarnata	-6	21	9-12	27%
T. ishikariensis	-6	18	9	61%
S. borealis	-6	18	3-6	80%

be higher than today, we fear the cold hardening of plants in fall will be counteracted by warmer temperatures and more overcast weather, and thereby render current cultivars more vulnerable to snow mould attack as well as other winter stress factors. A goal for our research is therefore to develop knowledge and tools that will enable us to mitigate crop loss due to snow moulds in a changing environment.

Snow Moulds/Snow Mould Pathogens

Snow moulds are fungal plant pathogens that have the ability to grow at low temperatures (sub-zero); they are either psychrophilic or mesophilic organisms with the ability to grow below zero temperatures. The duration of the snow cover affects how severely the crop or plant is injured by the pathogen; the longer the snow cover, the more severe the damage (Gaudet et al. 1989). The first European report on biotic winter injury was by the Swedish plant pathologist Jakob Eriksson in 1879 (Røed 1969), when he described injury caused by *Typhula graminum* to winter wheat. During the following 30 years, all the main snow mould species were recorded, described and identified in the Scandinavian countries. In spite of this knowledge, it was long assumed among breeders, plant culturists and farmers that the primary winter injury was freezing and that the fungi were secondary saprophytes.

Altogether, 33 species of pathogenic fungi have been isolated from grasses with winter injury in Norway (Årsvoll 1975). Most of these are of minor practical importance. The important species on grasses and cereals in Scandinavia are *Sclerotinia borealis*, *Typhula ishikariensis*, *T. incarnata* and *Microdochium nivale*. The most important characteristic of these fungi is the ability to grow at sub-zero temperatures and for two of the species, growth at 0 °C is approaching growth rate at the optimum temperature (Table 1).

Sclerotinia borealis

Sclerotinia borealis (Bubak and Vleugel) Schumacher and Kohn is the most extreme psychrophilic species and is restricted to regions with very stable, cold winter climate and long snow cover (more than 5–6 months). It causes snow scald or *Sclerotinia* snow mould on cereals and grasses. In 1906, Ulander observed winter injury on cocksfoot caused by a sclerotia-forming fungus (Ulander 1910). The fungus was later described by Vleugel. Early snow in autumn on unfrozen ground favours very severe injury (Jamalainen 1949, 1956; Ekstrand 1955; Røed 1960; Årsvoll 1975).

After snowmelt in spring, the disease appears as bleached patches of dead plants covered with greyish white mycelium. Large, greyish white sclerotia, soon turning black, are formed on or within infected tissue. Sclerotia remain dormant on the soil during summer, or if buried in soil, for several years. Late in autumn, sclerotia may germinate and form 1–6 apothecia. Ascospores spread to surrounding plants before snowfall and they serve as main source of inoculum. Infection may also occur through direct mycelial growth. In culture, the fungus is able to grow between −6 and 18 °C with optimum around 5 °C (Årsvoll 1975).

Typhula ishikariensis

Typhula ishikariensis Imai, is the main snow mould species in regions with a stable, cold winter climate and prolonged deep snow cover. It causes grey or speckled snow mould on grasses, cereals as well as some dicots. Overwintering grasses and forage legumes can suffer very severe injury (Ekstrand 1955; Jamalainen 1957; Ylimäki 1969). At localities with less than 4 months' snow cover, it is not common, except on golf green. *T. ishikariensis* is much more aggressive than the closely related *T. incarnata*.

The disease appears after snowmelt in spring as bleached patches of dead leaves, covered sparsely with a greyish-white mycelium and numerous dark brown to black, almost globose sclerotia with 0.5–2 mm diameter. Very young sclerotia may be light amber to cream coloured. Sclerotia on leaves are readily detached from the host and remain dormant during summer. They are the main source of dispersal and inoculum. New plants are mainly infected by direct growth of dicaryotic mycelium from sclerotia, but distribution and infection may also occur by basidiospores that are produced in autumn on fruiting bodies developed from a sclerotium. In culture, the fungus is able to grow between −6 and +18 °C with optimum at 9 °C (Årsvoll 1975).

Typhula incarnata

Typhula incarnata Lasch ex Fr. is widespread in temperate and cooler regions in Europe, North America, USSR and Japan and causes grey snow mould on cereals and grasses. Severe damage rarely occurs at localities with less than 3 months' snow cover, but mild attack has been observed even without a snow cover. It is most prevalent in areas with relative mild wet winter climate, and according to Scandinavian observations, it causes only relatively slight injury (Jamalainen 1957; Årsvoll 1975). The only exceptions are golf greens and winter barley (personal observations). *T. incarnata* is assumed to be the same species as Eriksson studied in 1879 and called *T. graminum* Karsten sensu Eriksson (Jamalainen 1957; Røed 1969).

The disease appears as small patches, first light yellowish brown, changing to greyish white, and progresses under prolonged snow cover. A sparse greyish-white mycelium develops on the surface of infected plants. Sclerotia are formed on or within infected tissues, particularly between the leaf sheets of the crown. Mature sclerotia are reddish brown to dark brown and are more or less elongated and flattened or irregularly shaped. They remain dormant during summer and serve as inoculum source in autumn, mainly through direct growth of dicaryotic mycelium (Årsvoll 1975). The basidiospores may also infect new plants, but are assumed to be of minor importance for disease development. The sclerotia may survive for several years when buried in soil. In culture, the fungus is able to grow between −6 and +21 °C. *T. incarnata* has good competitive saprophyte ability even at 10 °C (Ekstrand 1955; Jamalainen 1957; Domsch et al. 1980; Matsumoto and Sato 1982).

Microdochium nivale

Microdochium nivale (Fr.) Samuels & IC Hallett. is regarded as a serious pathogen in cold to temperate region of the northern hemisphere and also in Australia and New Zealand. It causes pink snow mould on cereals and grasses. Heavy attack on fodder grass and winter cereals is most often seen after 2–3 months of snow cover. Without seed dressing or late fall fungicide spraying, it may cause total loss of winter cereals (Jamalainen 1956, 1957; Årsvoll 1975).

The fungus was first described as *Lanosa nivalis* by Fries in 1825 (Jamalainen 1943). The nomenclature has since then changed several times. With its longest lasting name on the anomorph, *Fusarium nivale*, the species was described with two distinct varieties, var *nivale* and var *majus* (Booth 1971). The anamorph kept the varieties through two more name changes, to *Gerlachia nivalis* (Gams and Müller 1980) and to *Microdochium nivale* (Samuels and Hallet 1983). However, Glynn et al. (2005) have claimed the two varieties (var *nivale* and var *majus*) to be two distinct species, *M. nivale* and *M. majus*.

Immediately after snowmelt, the fungus shows a pale pinkish mycelium in patches and/or orange sporodochia. After drying, the dead leaves form a compressed paper-like layer (Fig. 2). Lawn turfs and golf greens may be attacked in autumn and spring when the weather is cold and humid. The disease symptoms then appear as small patches, first water-soaked and then yellow to orange-brown, sometimes with a fringe of pale pinkish mycelium. In some cases, leaf lesions with straw-coloured centres and dark margins may be found on leaves. Conidia are formed in aerial mycelium and in sporodochia. *M. nivale* does not produce chlamydospores and is

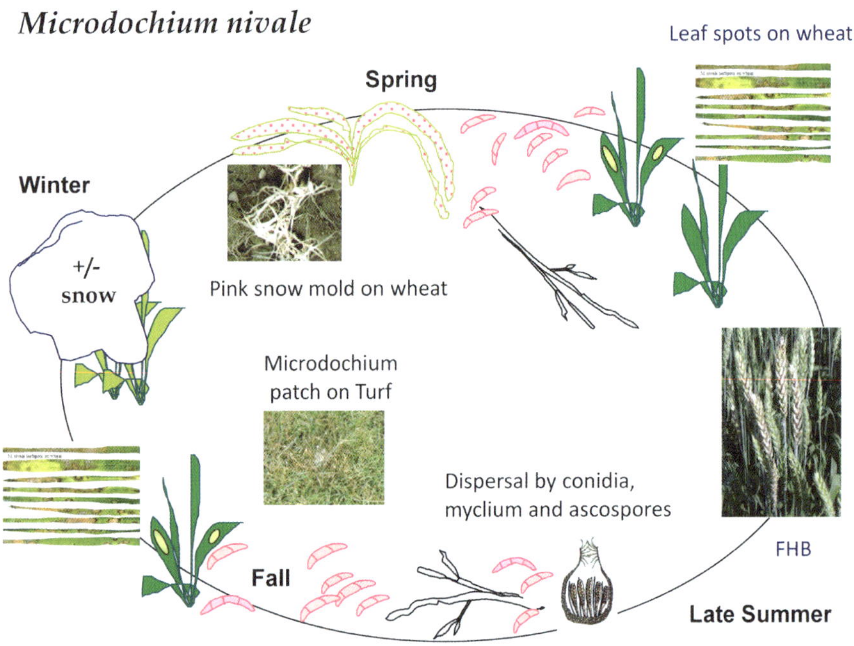

Fig. 2 The life cycle for *Microdochium nivale*

spread by infected seed, by mycelia and conidia from infected plants, and maybe also from soil (Domsch et al. 1980).

In culture, *M. nivale* is able to grow from −6 to +28 °C (Årsvoll 1975). Accordingly, summer survival in the Nordic countries should be no major problem. This fungus has, in fact, both a psychrophilic and a mesophilic phase, which permits its development both during the winter under snow and during the warmer conditions of summer (Fig. 2). During the summer growing season, it is the cause of Fusarium patch on turf grasses and stem rot, leaf blotch and is a part of the fungal complex causing "Fusarium Head Blight" (FHB) in cereals (Tronsmo et al. 2001; Pettitt et al. 2003; Simpson et al. 2004).

Projected Climatic Changes and Effect on Snow Mould Pathogens

Boland et al. (2004) and Roos et al. (2011) suggested that climate change will result in a significant decrease in the prevalence and severity of snow moulds. This seems plausible because reduced snow cover duration brought upon by a warmer climate in the Northern Hemisphere will ultimately result in shorter incubation period for the fungi. Under this scenario, snow moulds may become extinct in some regions because of the lack of persistent snow. However, other scenarios predict episodic extremes of weather, including years of heavy snowfall and colder winters (Rosenzweig et al. 2001; Anderson et al. 2004).

Projected changes in temperature and precipitation for Oslo (60° N and Tromsø (69° N) is shown in Fig. 1 (The Norwegian Meteorological Institute 2012. http://met.no) and are based on the average of three projections. From a stable winter temperature in Oslo around −5 °C, the coldest months will be around 0 °C. The projections show increased precipitation in autumn and winter, but somewhat drier summers. The temperature projections for Tromsø are quite similar to those for Oslo, with a change from a relatively stable temperature around −5 °C to around 0 °C in the coldest months. Even though the details in the three projections for precipitation vary, the tendency is for more precipitation, particularly in the autumn. The projected increase in average temperature represents a potential for increased crop production in the North. However, the quite unstable weather conditions, both in autumn and winter, will be a challenge for overwintering crops.

Since some snow mould fungi can live saprophytically in the soil for years and the inoculum of sclerotia-forming species can survive long periods in the soil in the absence of conducive environmental conditions (Hsiang et al. 1999), inoculum will be available to attack crops despite a long absence of snow. In this scenario, less frequent but more severe overall attacks by snow moulds will likely occur as a result of climate change. Moreover, the increased occurrence of weather extremes during the autumn such as increased rain or drought, warmer temperatures and early frosts will counteract the natural cold hardening of plants. Plants that are adapted to northern climates respond to decreasing temperatures in the autumn by several physiological changes in their metabolism, i.e. cold hardening. This process is essential for development of maximal resistance to low temperatures as well as snow moulds in grasses and cereals (Årsvoll 1977; Tronsmo 1984a; Gaudet and Chen 1988) and to other fungal plant pathogens (Tronsmo 1984b; Gaudet et al. 2010). Cold hardening also requires photosynthetic daylight. Unhardened or insufficiently hardened plants are more susceptible to infection by snow moulds than fully cold-hardened plants (Tronsmo 1984a; Gaudet and Chen 1988; Tronsmo 1994).

The projected future climate (Fig. 1) for Northern Scandinavia, with warmer autumns combined with increased precipitation and consequently reduced irradiance, will not provide environmental conditions for development of maximal cold hardening of over-wintering crop plants as grasses and cereals. Moreover, with warmer winter temperatures, plants will be exposed to winters with fluctuating temperatures whereby any levels of cold hardiness may be quickly lost. Under these conditions, our current cultivars of crop plants may suffer severe damage due to snow moulds.

The climatic change is, however, expected to affect the various snow mould species differently since different taxa of snow moulds utilise different strategies for their adaptation to the cold environments (Bruehl and Cunfer 1971; Hoshino et al. 2009). In Hokkaido, Japan, the distribution patterns of snow mould species have recently changed (Tamotsu Hoshino, Hokkaido, 2009, personal communication). In some regions, *T. ishikariensis* seems to have replaced *S. borealis* as the main snow mould species indicating that prevailing environmental conditions have changed for sufficiently long periods to permit this shift in species distribution. However, the psychrotrophic fungi *M. nivale* do not require snow cover for the mesophilic stage and this may permit accumulation of sufficient inoculum in years that are not conducive to snow mould development to replace other strictly psychrophilic fungi. Moreover, there are reports from Poland (Maria Wedzony, Krakow, 2010, personal communication) and The Netherlands (Gerard van't Klooster, Nijmegen, personal communication, 2007) that indicate that this species actually seems to have become more abundant during recent years in regions with much milder winter than Scandinavia. Its ability to grow at temperatures from sub-zero to 28 °C may be a contributing factor in this fungus's apparent increase in its area of adaption, spreading to "warmer" areas where it has not been recorded before, including golf courses in Hawaii, and California with abnormal high rainfall and cooler growing season temperatures (Wong 2006).

Environmental Adaptation Among Fungi

Most fungal species have a great capacity for environmental adaptation. The effects of a projected increase of 2–5 °C in average temperatures in the northern hemisphere on diseases such as *Microdochium* species will likely translate to the production of additional generations of asexual spores (conidia) during spring and late summer/autumn. Increased inoculum production facilitates the evolution of the parasites towards environmental adaptation and overcoming host resistance (Zhan et al. 2007; Garett et al. 2009). The sexual stage of *M. nivale* var *nivale* is not commonly found. Environmental or climatic stress is likely to promote sexual reproduction in such fungi (Grishkan et al. 2003). Microorganisms that can reproduce either sexually or asexually seem to preferentially express the sexual stage when exposed to environmental stresses (Taylor et al. 1999). A recent study by Schoustra et al. (2010) showed that genotypes of *Aspergillus nidulans* reverted to sexual reproduction in stress environments in which their fitness was lowest. Even in the asexual stage, isolates of *Fusarium* species display a great variability, in morphology, spore production, pathogenicity, temperature requirement as well as mycotoxin production (Summerell et al. 2010). Similar variability is displayed by *M. nivale* (Hofgaard et al. 2006).Variability within species of micro-organisms has been attributed to the presence of transposable elements, specifically retrotransposons that are stimulated by environmental stresses (Anaya and Roncero 1996; Gregory et al. 2009). Adaptive mutations in oomycete and fungal plant pathogens as a route for enhancing genome plasticity and rapid response to environmental changes, as well as the contribution of post-translational mechanisms are discussed by Gregory et al. (2009). These studies suggest that fungi react to environmental stresses by boosting their capacity for sexual reproduction. High reproductive capacity, as most plant pathogens possess, leads to large numbers of mutations that accumulate in the population. Sexual reproduction under stress environments may lead to recombination of advantageous mutations (i.e. new virulence combinations, higher aggressiveness, increased resistance to fungicides) that can lead to newer, fit genotypes. Climatic change, by augmenting the stress levels encountered by plant pathogens, may ultimately expand the opportunities for spread and exploitation of new habitats for some pathogens, including, *M. nivale*. In such a scenario, the likeliness for snow mould species to be extinct does not seem highly probable. Highly opportunistic micro-organisms will rather predominate with the resulting negative impacts on crop species.

Resistance to Snow Mould Fungi

Screening for resistance after artificial inoculation has revealed significant genetic variation in resistance to *M. nivale* in both forage grasses and winter rye (Miedaner et al. 1993; Tronsmo 1984a, 1992, 1993). Since the broad-sense heritability of resistance to *M. nivale* was reported to be sufficiently high (0.79 in winter rye and 0.49 in cocksfoot; Miedaner et al. 1993; Tronsmo 1993, respectively), improvement of resistance by selection should be feasible.

Current cultivars of perennial grasses and winter cereals develop maximal snow mould resistance only after cold hardening. Cold hardening enhances resistance both to freezing and to snow mould injury (Tronsmo 1984a, 1994). The ability to cold harden, however, varies among genotypes, as well as the ability to develop "cold-induced" snow mould resistance (Tronsmo 1994). Hömmö (1994, 1996) found that snow mould resistance in a detached leaf test was unaffected or repressed by cold hardening, and proposed the existence of different types of snow mould resistance. This finding indicates the possibilities for identifying and enhancing snow mould resistance that is independent of cold hardening in the fall. Moreover, there is also evidence for host specialisation and cultivar x isolate interactions in the wheat/*M. nivale* pathosystem (Diamond and Cooke 1997), and host preference for different species of grasses and cereals among isolates of *M. nivale* (Hageskal 2000; Simpson et al. 2000). For the development of cultivars with resistance to *M. nivale*, under a changing environment, it will be necessary to search for host-specific genes for resistance to the fungus that are expressed without cold hardening. However, genes encoding the ability to develop "cold-induced" snow mould resistance should not be discarded, since they may harbour the crucial trait for survival.

Mitigation of Crop Loss

The most common approaches to mitigate crop loss due to snow mould fungi has been the use of fungicides and cultivars with the best snow mould resistance available. An efficient control by fungicides is most often unreliable since it is difficult to target the right application time. There is also an increasing problem with development of fungicide resistance and restrictions on use. Restrictions are based both on environmental considerations and for minimising the risk for development of fungicide resistance.

As previously mentioned, current crop cultivar is not expected to give optimum yield under changing climatic conditions. There is therefore a need to develop high-yielding cultivars adapted to new environmental conditions that harbour resistance to snow mould fungi independent of cold hardening in the autumn. To achieve such a goal, we need to identify cold hardening independent snow mould resistance in genotypes of desired species and incorporate these in further classical breeding programs. However, we also need to identify traits (genes) governing the cold-induced snow mould resistance to facilitate breeding of cultivars robust enough for an *unpredictable* climate.

Another approach in the future may be application of suitable defence activators (elicitors). Experiments in controlled environment have shown interesting results, but from there to affordable and reliable use under field conditions is a long way (Eikemo et al. 2003; Hofgaard et al. 2005, 2010; Walters et al. 2009).

Concluding Remarks

This chapter has focused on the Scandinavian situation in a changing climate. The challenges we are facing are not very different from those in many other parts of the world. But for large parts of Scandinavia, the projected climate change will also give an opportunity for increased biomass production due to prolonged growing season. However, since crop productions are and will be carried out under temperature and day length conditions found at no other places on the earth, there is an urgent need for development of crop cultivars with potential to exploit the expected future conditions.

For all plants, as well as pathogens, the changing environment will pose a stress that eventually will lead to adaptation to the new environment. To be able to sustain crop production through development of new cultivars, we must know how the pathogens are expected to behave in a new environment. A better understanding of the biology and genetics of the snow mould fungi and other plant pathogens are therefore crucial.

References

Anaya N, Roncero MIG (1996) Stress-induced rearrangement of *Fusarium* retrotransposon sequences. Mol Gen Genet 253:89–94

Anderson PK, Cunningham AA, Patel NG, Morales FJ, Epstein PR, Daszak P (2004) Emerging infectious diseases of plants: pathogen pollution, climate change and agrotechnology drivers. Trends Ecol Evol 19:535–544

Årsvoll K (1975) Fungi causing winter damage on cultivated grasses in Norway. Meld Norg LandbrHøgsk 54(3):21

Årsvoll K (1977) Effects of hardening, plant age, and development in *Phleum pratense* and *Festuca pratensis* on resistance to snow mould fungi. Meld Norg LandbrHøgsk 56:1–14

Boland GJ, Melzer MS, Hopkin A, Higgins V, Nassuth A (2004) Climate change and plant diseases in Ontario. Can J Plant Pathol 26:335–350

Booth C (1971) The genus *Fusarium*. CMI, Kew

Bruehl GW, Cunfer B (1971) Physiologic and environmental factors that affect the severity of snow mold of wheat. Phytopathology 61:792–799

Diamond H, Cooke BM (1997) Host specialisation in *Microdochium nivale* on cereals. Proceedings of the Fifth European Fusarium Seminar. Cereal Res Commun 25(3/2):533–538

Domsch KH, Gams W, Anderson TH (1980) Compendium of soil fungi. Academic Press, London ISBN 0-12-220401-8

Eikemo H, Stensvand A, Tronsmo AM (2003) Induced resistance as a possible means to control diseases of strawberry caused by *Phytophthora* spp. Plant Disease 87:345–350

Ekstrand H (1955) Höstsädens og vallgräsens övervintring. Statens växtskyddsanstalt. Medd Nr 67(Stockholm), pp 125

Flæte O, Bardalen A, Dalen L, Drange H, Gjærum I, Hanssen-Bauer I, Hisdal H, Hovelsrud GK, Karlsen J, Larssen SA, Nyeggen E, Ottesen P, Pedersen S, Petkovic G, Sundby S, Vennemo H, Aanestad J (2010) Tilpassing til eit klima i endring. NOU 2012:10

Garett KA, Nita M, De Wolf ED, Gomez L, Sparks AH (2009) Plant pathogens as indicators of climate change. In: Letcher TM (ed) Climate change: Observed impact on planet earth. Elsevier, Amsterdam, The Netherlands, pp 425–437

Gams W, Müller E (1980) Conidiogenesis of *Fusarium nivale* and *Rhynchosporium oryzae* and its taxonomic implications. Neth J Plant Pathol 86:45–53

Gaudet DA, Chen THH (1988) The effect of freezing resistance and low temperature stress on development of cottony snow mold (*Coprinus psychromorbidus*) in winter wheat. Can J Bot 66:1610–1615

Gaudet DA, Bhalla MK, Clayton GC, Chen THH (1989) Effect of cottony snow mold and low temperatures on winter wheat survival in central and northern Alberta. Can J Plant Pathol 11:291–296

Gaudet DA, Wang Y, Frick M, Puchalski B, Penniket C, Ouellet T, Robert L, Singh J, Laroche A (2010) Low temperature induced defence gene expression in winter wheat in relation to resistance to snow moulds and other wheat diseases. Plant Sci 180:99–110

Gaudet DA, Tronsmo AM, Laroche A (2012) Climate change and plant diseases. In temperature adaptation in a changing climate: Nature at risk (Eds KB Storey and KK Tanino). CAB International. Chapter 10, p 144–158. ISBN-13 9781845938222

Glynn NC, Hare MC, Parry DW, Edwards SG (2005) Phylogenetic analysis of EF-1 alpha gene sequences from isolates of *Microdochium nivale* leads to elevation of varieties *majus* and *nivale* to species status. Mycol Res 109:872–880

Gregory PJ, Johnson SN, Newton AC, Ingram JSI (2009) Integrating pests and pathogens into the climate change/food security debate. J Exp Bot 60:2827–2838

Grishkan I, Korol AB, Nevo E, Wasser SP (2003) Ecological stress and sex evolution in soil microfungi. Proceedings of the Royal Society of London. Ser B: Biol Sci 270:13–18

Hageskal G (2000) Aggressivitet og vertsspesifisitet i snømuggsoppen *Microdochium nivale*. Master thesis, Norwegian Agricultural University (NLH) (Present: Norwegian University of Life Sciences)

Hofgaard IS, Ergon Å, Wanner LA, Tronsmo AM (2005) The effect of chitosan and Bion on resistance to pink snow mould in perennial ryegass and winter wheat. J Phytopathol 153:108–119

Hofgaard IS, Wanner LA, Hageskal G, Henriksen B, Klemsdal S, Tronsmo AM (2006) Isolates of *Microdochium nivale* and *M. majus* differentiated by pathogenicity on perennial ryegrass (*Lolium perenne* L.) and *in vitro* growth at low temperature. J Phytopathol 154:267–274

Hofgaard IS, Ergon Å, Henriksen B, Tronsmo AM (2010) The effect of potential resistance inducers on development of *Microdochium majus* and *Fusarium culmorum* in winter wheat. Eur J Plant Pathol 128(2):269–281

Hoshino T, Xiao N, Tkachenko O (2009) Cold adaptation in the phytopathogenic fungi causing snow molds. Mycoscience 50:26–38

Hsiang T, Matsumoto N, Millett SM (1999) Biology and management of *Typhula* snow molds of turfgrass. Plant Dis 83:88–798

Hömmö LM (1994) Screening winter rye cultivars for snow mould (*Microdochium nivale*) resistance. Plant Pathol 43:740–750

Hömmö LM (1996) Effect of hardening and dehardening on snow mould (*Microdochium nivale*) resistance. Nord Jordbruksforskning 78(2):89

Jamalainen EA (1943) Über die *Fusarien* Finnlands. I. Die Staatliche Landwirtschaftliche Versüchstatigkeit. Veröffentlichung No. 122

Jamalainen EA (1949) Overwintering of Gramineae plants and parasitic fungi I. *Sclerotinia borealis* Bubak & Vleugel. J Sci Agric Soc Finl 21:125–142

Jamalainen EA (1956) Overwintering of plants in Finland with respect to damage caused by low-temperature pathogens. Finn State Agric Res. Board Publ 148:5–30

Jamalainen EA (1957) Overwintering of *Gramineae* plants and parasitic fungi II. On the *Typhula sp*. fungi in Finland. J Sci Agric Soc Finl 29:75–81

Matsumoto N, Sato F (1982) The competitive saprophytic abilities of *Typhula incarnata* and *T. ishikariensis*. Ann Phytopatholog Soc Jpn 48:419–424

Miedaner T, Höxter H, Geiger HH (1993) Development of a resistance test for winter rye to snow mold (*Microdochium nivale*) under controlled environment conditions in regard to field inoculations. Can J Bot 71(1):136–144

Pettitt T, Xu X, Parry D (2003) Association of Fusarium species in the wheat stem rot complex. Eur J Plant Pathol 109:769–774

Roos J, Hopkins R, Kvarnheden A, Dixelius C (2011) The impact of global warming on plant diseases and insect vectors in Sweden. Eur J Plant Pathol 129:9–19

Rosenzweig C, Iglesias A, Yang XB, Epstein PR, Chivian E (2001) Climate change and extreme weather events; Implications for food production, plant diseases, and pests. Glob Chang Hum Health 2:90–104

Røed H (1960) *Sclerotinia borealis* Bub. & Vleug., a cause of winter injuries to winter cereals and grasses in Norway. Acta Agric Scand 10(1):74–82

Røed H (1969) Et bidrag til oppklaring av forholdet mellom *Typhula graminum* Karst. og *Typhula incarnata* Lasch ex Fr. Friesia IX:219–225

Samuels GJ, Hallett IC (1983) *Microdochium stoveri* and *Monographella stoveri*, new combinations for *Fusarium stoveri* and *Micronectriella stoveri*. Trans Br mycol Soc 81(3):473–483

Schoustra S, Rundle HD, Dali R, Kassen R (2010) Fitness-associated sexual reproduction in a filamentous fungus. Curr Biol 20:1350–1355

Simpson DR, Rezanoor HN, Parry DW, Nicholson P (2000) Evidence for differential host preference in *Microdochium nivale* var. *majus* and *Microdochium nivale* var. *nivale*. Plant Pathol 49:261–268

Simpson DR, Thomsett MA, Nicholson P (2004) Competitive interactions between *Microdochium nivale* var. *majus*, *M. nivale* var. *nivale* and *Fusarium culmorum in planta* and *in vitro*. Environ Microbiol 6(1):79–87

Summerell B, Laurence M, Liew E, Leslie J (2010) Biogeography and phylogeography of *Fusarium*: a review. Fungal Divers 44:3–13

Taylor JW, Jacobson DJ, Fisher MC (1999) The evolution of asexual fungi: Reproduction speciation and classification. Annu Rev Phytopathol 37:197–246

Tronsmo AM (1984a) Predisposing effects of low temperature on resistance to winter stress factors in grasses. Acta Agric Scand 34:210–220

Tronsmo AM (1984b) Resistance to the rust fungus *Puccinia poae-nemoralis* in *Poa pratensis* is induced by low-temperature hardening. Can J Bot 62:2891–2892

Tronsmo AM (1992) Resistance to snow mould fungi in breeding material of grasses. Nor J Agric Sci 7(Suppl.):35–38

Tronsmo AM (1993) Resistance to winter stress factors in half sib families of *Dactylis glomerata*, tested in controlled environment. Acta Agric Scand., Section B, Soil Plant Sci 43:89–96

Tronsmo AM (1994) Effect of different cold hardening regimes on resistance to freezing and snow mould infection in timothy varieties of different origin. In: Dørffling K, Brettschneider B, Tantau H, Pithan K (eds) Crop adaptation to cool climates cost 814 Workshop. ECSP-EEC-EAEC, pp 83–89

Tronsmo AM, Hsiang T, Okuyama H, Nakajima T (2001) Low temperature diseases caused by *Microdochium nivale*. In: Iriki N, Gaudet DA, Tronsmo AM, Matsumoto N, Yoshida M, Nishimune A (eds) Low temperature plant microbe interactions under snow. Hokkaido National Experiment Station, Sapporo, pp 75–86

Ulander A (1910) Redogörelse for verksamheten vid Sveriges Utsädesföreningsfilial i Lulea år 1906–1909. Sver Utsädesförenings Tidskri 20:33–53

Walters DR, Paterson L, Walsh DJ, Havis ND (2009) Priming for plant defense in barley provides benefits only under high disease pressure. Physiol Mol Plant Pathol 73:95–100

Wong FP (2006) The 2006 California disease round up. USGA Green Section Meeting. http://faculty.ucr.edu/~frankw/Downloads/070108%20USGA.ppt

Ylimäki A (1969) Typhula blight of clovers. Ann Agric Fenni 8:30–37

Zhan J, Mundt CC, McDonald BA (2007) Sexual reproduction facilitates the adaptation of parasites to antagonistic host environments: Evidence from empirical study in the wheat-*Mycosphaerella graminicola* system. Int J Parasitol 37:861–870

Mechanisms of Snow Mold Resistance in Wheat

Denis Gaudet and André Laroche

Introduction

In the deep snow regions of temperate, boreal, and subarctic zones, winter cereals must survive harsh winter conditions in a live state under a deep, persistent snow cover that insulates the root and crown zone. Snow cover maintains soil temperatures at crown level between 0 and −10 °C despite much lower air ambient temperatures, thereby protecting plants against low-temperature injury (Bruehl 1982; Gaudet et al. 1989). The protracted snow cover creates a dark, humid environment with constant temperatures, which prevents photosynthesis and reduces plant metabolism. These conditions favor development of snow molds, which can cause extensive damage to agricultural, ornamental, and native plants (Bruehl 1982).

In northern Canada, where the snow cover can last as long as 200 days, cottony snow mold (sterile Low Temperature Basidiomycete [LTB]), snow scald (*Sclerotinia borealis* Bub. & Vleug.), and speckled snow mold (*Typhula ishikariensis* Imai; *T. idahoensis* Remsberg) are the most important snow molds on winter cereals (Gaudet 1994). Beneath the snow on frozen soil, the optimum temperature for development of the LTB fungus is −3 °C (Gaudet 1986; Gaudet 1994) and of *Sclerotinia borealis* is −2.0 °C (Årsvoll 1976). In the Pacific Northwestern United States, Japan, and Scandinavia, where temperate or northern regions are modified by a maritime climate and milder winter temperatures and snow cover in excess of 70 days on unfrozen soil prevail, *T. isikariensis* and pink snow mold (*Microdochium nivale* [Fr.] Samuels and Hallet) are the most important snow molds (Bruehl 1982; Nissinen 1996). Gray snow mold (*Typhula incarnata* Lasch ex Fr.) and snow rot (*Pythium iwayami* S. Ito. Hirane) also damage crops but are considered of minor importance in most years. These snow mold fungi develop under snow at temperatures slightly above 0°C (Bruehl 1982).

D. Gaudet (✉) · A. Laroche
Agriculture and Agri-Food Canada, Lethbridge Research Centre, 5403 1st Avenue South, Lethbridge, Alberta T1J 4B1, Canada
e-mail: denis.gaudet@agr.gc.ca

R. Imai et al. (eds.), *Plant and Microbe Adaptations to Cold in a Changing World*, 319
DOI 10.1007/978-1-4614-8253-6_27, © Springer Science+Business Media New York 2013

Infection of wheat by the LTB and pink snow mold fungi occurs in the leaves by mechanical penetration of the stomata and cuticle throughout the winter under a continuous blanket of snow (Gaudet and Kokko 1985; Gaudet et al. 1989; Oshiman et al. 1995; Kuwabara and Imai 2009). Conversely, *Typhula spp*. and *S. borealis* spores are released from fruiting structures and infect leaves in the autumn but further development in leaves requires a persistent snow cover (Arsvoll 1976; Bruehl 1982). Upon entering the substomatal cavity, hyphal development of the LTB fungus becomes stalled, possibly due to the action of preformed resistance responses (described below; Gaudet and Kokko 1985). Eventually, fungal development becomes widespread throughout the leaves and hyphae spread down to the crowns and roots (Gaudet et al. 1989). If crown tissues are unable to regenerate new growth in the spring, plants die. Sublethal injury to crowns will both reduce and delay development in the spring and result in yield losses (Bruehl 1982).

Genetic resistance to snow molds among winter wheat cultivars has been reported (Bruehl 1982; Amano and Osanai 1983; Gaudet and Kozub 1991). A narrow continuum in variation exists among winter wheat cultivars for resistance to snow molds (Bruehl 1982) that is conditioned by the additive effect of two or three loci that are relatively highly heritable (Iriki and Kuwabara 1993). Snow mold resistance differs from typical hypersensitive resistance responses in that it is expressed quantitatively and resistant cultivars exhibit greater survival and yield than susceptible cultivars; however, given a sufficiently severe snow mold attack, even the most resistant cultivar will eventually succumb (Gaudet and Kozub 1991; Nissinen 1996). In general, the level of freezing tolerance among winter wheat cultivars appears negatively correlated with resistance to snow molds (Gaudet and Kozub 1991). An exception is resistance to *S. borealis* where a positive correlation between freezing tolerance and snow mold resistance has been identified (Amano and Osanai 1983).

A second form of resistance is developmentally based resistance (Bruehl 1982; Gaudet and Chen 1987; Kuwabara and Imai 2009). Early seeding is essential for full expression of genetic resistance in winter cereals and even moderately susceptible varieties can develop substantial levels of snow mold resistance if seeded early (Bruehl and Cunfer 1971; Bruehl 1982). Unfortunately, seeding sufficiently early to attain full developmental resistance is often not possible or impractical and may be associated with increased incidence of freezing damage (Fowler 1982, 2012) and severity of other diseases.

Factors Affecting the Expression of Snow Mold Resistance

Before snow mold resistance is expressed, cereals and grasses must undergo cold acclimation or hardening, a process that normally occurs during the autumn and early winter (Årsvoll 1977; Tronsmo 1984a; Tronsmo et al. 1993; Gaudet 1994; Gaudet et al. 1999; Kuwabara and Imai 2009). Cold acclimation is a process by which plants acquire freezing tolerance following exposure to low nonfreezing temperatures and the same conditions that induce freezing tolerance also induce snow mold resistance (Tronsmo 1984a; Gaudet and Chen 1987; Chinnusamy et al. 2007; Kuwabara and Imai 2009). Cold-induced resistance has also been demonstrated to be effective against rust in grasses (Tronsmo 1984b) and powdery mildew in wheat (Gaudet et al. 2010) suggesting that this form of resistance is effective against a broad range of plant pathogens.

While freezing resistance and snow mold resistance are both cold-induced, important distinctions between the two exist. In addition to differences that exist in the underlying physiological basis between the two, freezing resistance along with the expression of a series of low-temperature-associated genes increases rapidly in winter wheat within a few days following exposure to low acclimating temperatures (Houde et al. 1992; Thomashow 2003; Sung and Amasino 2005) whereas, snow mold resistance increases gradually following exposure to low acclimating temperatures, requiring up to 10 weeks acclimation prior to achieving maximum development (Gaudet and Chen 1987). Other abiotic stresses such as pollution and drought (Pääkkönen et al. 1998), wounding (Orozco-Cardenas and Ryan 1999), and UV light (Wingender et al. 1989) have also been implicated disease defense responses. It therefore appears that snow mold resistance and freezing resistance are two distinct physiological reactions that are both induced by cold above-zero temperatures. However, the two processes are indirectly linked because subzero low-temperature stress increases the susceptibility of plants to snow mold and snow mold stress reduces the plant's tolerance to freezing temperatures (Tronsmo 1984a; Gaudet and Chen 1988).

Current Models of Host–Parasite Interactions in Plants

The nature of snow mold resistance remains largely unknown but considerable progress in understanding the physiological and molecular nature of resistance to many plant diseases, particularly to biotropic and some necrotropic plant pathogens, has been made during the past 20 years. At a molecular level, pathogen-produced molecules, referred to as Pathogen or Microbe-Associated Molecular Patterns, PAMPs or MAMPs, are recognized by the host as foreign or "nonself" and initiate PAMP-triggered immunity (PTI) (Jones and Dangl 2006). For nonadapted pathogens, which refer to pathogens that do not normally attack a particular plant species (Heath 2000b; Thordal-Christensen 2003; Abramovitch et al. 2006), PTI is effective because the host is able to rapidly initiate and deliver a series of defense responses that terminate infection. This resistance is also termed nonhost resistance or basal or innate immunity.

For adapted pathogens, which normally attack the particular plant species, a battery of pathogen effectors are produced that globally suppress the defense responses, rendering the plant susceptible. However, some members of a particular plant species may have individuals that are resistant to one or more of the pathogen biotypes and this resistance is known as host resistance. Genetically, these interactions are controlled by resistance (*R*) genes that detect a specific effector that is the product of a pathogen avirulence (*Avr*) gene; this resistance is referred to as Effector-Triggered Immunity (Belkhadir et al. 2004; Abramovitch et al. 2006; Chisholm et al. 2006; Jones and Dangl 2006; Bent and Mackey 2007). In molecular terms, *R* gene products frequently belong to a class of nucleotide-binding leucine-rich repeat receptors that interact either directly or indirectly with a specific effector delivered into hosts (Jones and Dangl 2006).

In addition to these specific types of pathogen-based immunity, plants can also activate general defense responses that are induced in response to different physiological and mechanical stresses. Induced resistance (IR) is typically systemic and effective against a broad spectrum of pathogens and insects (Koornneef and Pieterse 2008). Examples of IR include systemic acquired resistance (SAR), rhizobacteria-induced systemic resistance, and wound-induced resistance in response to mechanical injury during insect feeding. This type of resistance often acts systemically throughout the plant. These various forms of resistances appear to be branches of the same plant-immune system (Abramovitch et al. 2006; Chisholm et al. 2006; Jones and Dangl 2006; Bent and Mackey 2007).

Once detection of biotic or abiotic stresses has been made, mitogen-activated protein kinase cascades transmit these extracellular stimuli into adaptive responses. The hormones salicylic acid (SA), jasmonic acid (JA) and ethylene (ET) have major roles in the regulation of plant defense signaling networks involved in host–parasite interactions in plants (Jones and Dangl 2006; Bent and Mackey 2007). In general, SA is induced in resistance reactions involving biotrophic pathogens while JA/ET is induced when necrotrophic plant pathogens are involved (Grant and Jones 2009). Furthermore, other plant hormones, such as auxins, gibberellins and abscisic acid (ABA), can interact with the defense signaling networks to augment or decrease the resistance response (Robert-Seilaniantz et al. 2011). These resistance responses include the hypersensitive response (Heath 2000a), callose and lignin strengthening of the cell wall (Hématy et al. 2009), upregulation of *phenylalanine ammonium lyase* (PAL), synthesis of antimicrobial phenolic compounds of secondary metabolites, such as phytoalexins (Pääkkönen et al. 1998), small defense peptides, including defensins, thionins, and lipid transfer proteins (LTPs) (García-Olmedo et al. 1998), and the accumulation of pathogenesis-related (PR) proteins, such as PR-1, chitinases, and glucanases (Van Loon et al. 2006). Mechanical barriers posed by deposits of callose and increased lignification at the infection site may also occur (Jacobs et al. 2003; Bhuiyan et al. 2009). Collectively, these responses attenuate or terminate infection.

Molecular Responses to Cold Hardening and Snow Mold Infection

Microarray studies for gene expression following exposure of winter wheat to different hardening regimes showed that the expression of many genes associated with freezing resistance are upregulated during the first few days following exposure to hardening conditions (Fowler and Thomashow 2002; Houde et al. 2006; Gaudet et al. 2010). This contrasted with the upregulation of genes associated with the JA/ET signaling pathway, which requires exposure of 3 to 6 weeks at hardening temperatures (Houde et al. 2006; Gaudet et al. 2010). These included LTPs, defensins, lipoxygenases, NPR3, ET-responsive genes, and numerous stress-related transcription factors including WRKY, zinc finger, and other transcription factors and could be associated with snow mold resistance (Gaudet et al. 2010). A second increase in defense gene expression has been observed in response to snow mold infection under controlled conditions in wheat (Ergon et al. 1998; Gaudet et al. 2010) and grasses (Pociecha et al. 2008a). The basis for this second burst in plant defense responses is unknown but may be due directly to attack by snow mold fungi (Ergon et al. 1998) or to additional stresses resulting from the transfer of plants from above-zero growth and hardening temperatures in the light to prolonged dark conditions, near or below freezing temperatures that are required to obtain snow mold infection (Gaudet 1986).

Another phytohormone implicated in snow mold resistance is ABA. Higher levels of ABA were associated with resistance to *M. nivale* in meadow fescue (Pociecha et al. 2008b; Pociecha et al. 2009) but not in winter rye (PŁażek et al. 2011). ABA is induced during hardening (Dörffling et al. 1990; Galiba et al. 1993; Veisz et al. 1996) and is involved in several aspects in the development of resistance to abiotic stresses including to freezing temperatures (Fan et al. 2009; Peleg and Blumwald 2011). However, activation of ABA biosynthetic and signaling pathways has been shown to promote disease susceptibility to several plant pathogens by inhibiting the SA and JA pathways (Robert-Seilaniantz et al. 2011). For interactions involving necrotropic fungi, the key may lie in the apparent ability of ABA to specifically induce the JA pathway thereby leading activation of downstream defense genes and increased callose production (Ton et al. 2009).

During infection of winter wheat by the snow mold fungi, some PR proteins and defense-related signaling pathway genes were upregulated earlier and at a higher level of response in plants that had been hardened prior to inoculation compared to unhardened plants (Ergon et al. 1998; Gaudet et al. 2010). This suggests that hardening has a priming effect on cold-induced resistance. Priming is a phenomenon in plants that is induced following plant detection of pathogens, beneficial microorganisms, or application of specific chemicals; the defense responses in primed plants are expressed more rapidly and at higher levels when subsequently challenged by biotic or abiotic stresses (Conrath 2009). SAR and IR are both involved in priming downstream defense responses that impart resistance to a broad range of plants and implicate a number of signaling molecules including SA and JA (Durrant and Dong 2004; Conrath 2009). Cold-induced priming would be particularly relevant in plants under snow during the winter where they are in a dormant state. The large number and diversity of defense-related genes upregulated during hardening (Houde et al. 2006; Gaudet et al. 2010) may contribute to the priming of resistance to snow molds.

Defence responses downstream of primed signal responses activate expression of PR proteins, PAL and phenolic compounds, which have been associated with resistance to snow molds in wheat (Ergon et al. 1998; Gaudet et al. 2003a, b) and ryegrass (Pociecha et al. 2008a). Cold-induced members of the PR-14 class of non-specific LTPs have been shown to be upregulated during cold hardening and associated with resistance to snow molds (Gaudet et al. 2003b). When expressed in *Pichia pastoris*, LTPs were shown to exhibit *in vitro* toxicity to the LTB snow mold fungus (Sun et al. 2008). Interestingly, some PR proteins such as chitinases and β-1,3-glucanases also exhibit antifreeze properties and may serve a duel function (Hon et al. 1995; Kuwabara and Imai 2009).

During the course of snow mold infection under controlled conditions, primed transcripts for the preformed defense responses decrease suggesting that plants are unable to maintain these expressed transcripts throughout the course of the snow mold infection; this could explain why inoculated plants under controlled environment conditions and in the field eventually succumb to snow molds given a sufficiently long incubation period (Gaudet and Chen 1987; Gaudet et al. 1989). This coincides with widespread growth of snow mold fungi throughout leaf tissues and spread to the crown where it eventually kills the plant. Under field conditions, spread and eventual death of plants occurs in the late winter and early spring, often following long exposure to snow molds under snow (Nissinen 1996).

Fructans and Snow Mold Resistance

Carbohydrate accumulation during the autumn and early winter in the form of simple sugars and complex carbohydrates known as fructans, are important components of winter survival in winter wheat. Fructans are water soluble, complex linear or branched carbohydrates, which provide an energy source during the winter when plants are photosynthetically inactive. Early work by G. W. Bruehl and coworkers in the Pacific Northwest of the United States (Bruehl and Cunfer 1971; Kiyomoto and Bruehl 1977; Bruehl 1982) demonstrated that snow mold-resistant winter wheat varieties accumulated higher fructan content in crowns during autumn and elevated fructan content remaining in the spring, even after snow mold infection (Kiyomoto and Bruehl 1977; Yoshida et al. 1998; Gaudet et al. 2001). Bertrand et al. (2011) similarly demonstrated that fructan content, particularly those that possessed a high degree of polymerization, was associated with both freezing resistance and snow mold resistance in grasses. Developmental stage of plants also affects fructan content with older, larger plants accumulating higher levels of fructan than younger plants (Gaudet et al. 2001; Østrem et al. 2011). The implication of fructans in snow mold resistance is further supported by the research demonstrating that early seeding of winter wheat, which results in development of larger plants in the autumn, is necessary for expression of snow mold resistance (Kiyomoto and Bruehl 1977; Bruehl 1982; Gaudet et al. 1989). It has also been suggested that fructans have a role in protecting plants against desiccation injury caused by low temperatures and drought by helping to maintain membrane stability (Hincha et al. 2007).

The direct role of fructans in snow mold resistance remains uncertain. Some evidence suggests that complex carbohydrates such as fructan are less desirable substrates for growth of snow molds and this feature could contribute to resistance (Kawakami and Yoshida 2012). Maintaining a positive energy balance in plants throughout the winter and early spring when plants are unable to manufacture carbohydrates via photosynthesis is also important to winter survival (Huner et al. 1998). Snow mold-resistant cultivars tend to accumulate carbohydrates earlier and at higher levels in the autumn and metabolize carbohydrates more slowly under snow during the winter and early spring compared to susceptible cultivars (Kiyomoto and Bruehl 1977; Yoshida et al. 1998; Gaudet et al. 2001).

Early-seeded treatments also accumulated higher fructan per unit dry weight of tissues compared to later-seeded treatments (Gaudet et al. 2001). Additionally, even susceptible cultivars, if seeded very early, accumulated high fructan levels and developed relatively high levels of resistance to snow molds (Kiyomoto and Bruehl 1977; Gaudet et al. 2001). The increased activities of sucrose:sucrose 1-fructosyltransferase and sucrose:fructan 6-fructosyltransferase, involved in fructan biosynthesis, and decreased activity of fructan exohydrolase, involved in fructan breakdown, has been associated with snow mold resistance in winter cereals (Kawakami and Yoshida 2002, 2005, 2012). Collectively, these results strongly implicate a key role for fructans in snow mold resistance in winter cereals. Conversely, snow mold-resistant winter wheat cultivars are among the most susceptible to freezing injury whereas cold, hardy varieties accumulate low levels of fructans and are susceptible to snow mold (Gaudet and Kozub 1991). Moreover, winter wheat cultivars seeded very early fail to develop high levels of freezing resistance (Fowler 1982). Thus, a negative pleotropic effect between freezing resistance and snow mold resistance content may exist among winter cereals. A similar pleotropic effect has not been observed in grasses (Pociecha et al. 2008b; Bertrand et al. 2011).

Another mechanism for snow mold resistance that involves the role of decreasing plant water potentials caused by carbohydrate accumulation has been proposed to reduce, if not prevent, the growth of some snow mold fungi (Bruehl and Cunfer 1971; Tronsmo 1986). Fungal plant pathogens have characteristic ranges of water potential values for optimum growth rates (Cook and Papendick 1972). Growth of *M. nivale*, *T. ishikariensis*, and *T. incarnata* is impaired at water potentials below −1.0 MPa while optimum growth of *S. borealis* occurred at water potentials between −1.0 and −2.0 MPa (Bruehl and Cunfer 1971; Tronsmo 1986). Water potentials in grasses decrease during the hardening process, possibly explaining the less extensive growth of snow mold fungi on hardened plants (Tronsmo 1986). Snow mold-resistant cultivars are also characterized by their low water content (Yoshida et al. 1997).

A Model for Snow Mold Resistance

Taking into account previous models (Gaudet et al. 1999; Kuwabara and Imai 2009), it is likely that snow mold resistance is a form of stress-induced resistance that is typically systemic and effective against a broad spectrum of pathogens and insects.

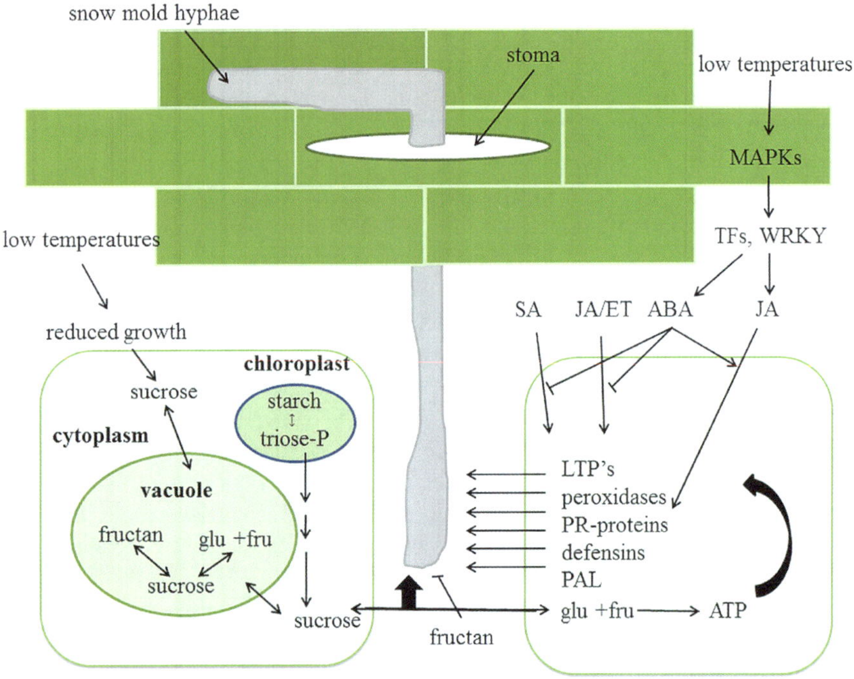

Fig. 1 A model for snow mold resistance in cereals. Low temperatures reduce growth while photosynthate production is maintained resulting in fructan accumulation that serves to maintain homeostasis during the winter. Cold-induced induction of the ABA-mediated JA pathway and suppression of SA and JA/ET pathway results in upregulation of defense genes specific to necrotropic fungi. Soluble fructan reserves maintain defense gene expression that retards growth of snow mold fungi during the winter. Fructans may also act as a poor nutritional substrate retarding fungal growth compared to simple sugars promoting growth. As the winter progresses, reserves diminish along with defense gene expression resulting in a progressive attack by snow mold fungi. Abbreviations: ABA, abscisic acid; ATP, adenosine triphosphate; ET, Ethylene; LTP, lipid transfer proteins; PR proteins, pathogenesis-related proteins; JA, jasmonic acid; MAPK, mitogen activated kinases; PAL, phenylalanine ammonium lyase; SA, salicylic acid; TFs, transcription factors

This resistance is an extension of basal resistance in plants involving predominantly the JA/ET pathway, a pathway that is suited to a resistance response involving necrotropic plant pathogens such as snow molds (Fig. 1). ABA produced during hardening may also promote resistance via the JA pathway. Infection through the stomata may become temporarily stopped or slowed in the substomatal cavity by action of constitutive, preformed defense responses that develop during hardening and during freezing conditions under snow. During early infection stages, snow mold fungi produce enzymes and other metabolites (PAMPS) that may be perceived by host plants to induce a second round of plant defense expression observed under artificial infection conditions. Eventually, a diminution of preformed defense responses throughout the course of the winter occurs and this coincides with the loss of fructan reserves in the leaves and crowns. The difference in levels of resistance observed among early-seeded resistant cultivars and later-seeded susceptible cultivars is likely attributable to ability of more resistant cultivars to accumulate and maintain higher levels of water-soluble carbohydrates and hence, energy levels, to induce defense responses sufficient to slow progression of fungi into the crowns.

Future Work

Given that climatic change models predict that crops will be subjected to a wide range of winter stresses (Gaudet et al. 2012), a thorough understanding of the genetic basis of resistance to low temperatures and snow molds and the environmental factors that induce resistance to these two stresses is critical. The silencing of defense-related candidate genes differentially expressed during hardening and infection by snow molds will be key to identifying crucial transcription factors and kinases involved in resistance to this group of low temperature plant pathogens.

References

Abramovitch RB, Anderson JC, Martin GB (2006) Bacterial elicitation and evasion of plant innate immunity. Nat Rev Mol Cell Biol 7:601–611

Amano Y, Osanai S-I (1983) Winter wheat breeding for resistance to snow mold and cold hardiness. III. Varietal differences of ecological characteristics on cold acclimation and relationships of them to resistance. Bull Hokkaido Natl Agric Exp Stn 50:83–95

Årsvoll K (1976) *Sclerotinia borealis*. Sporulation, spore germination and pathogenesis. Meld Norg LandbrHogsk 55:1–11

Årsvoll K (1977) Effects of hardening, plant age, and development in *Phleum pratense* and *Festuca pratensis* on resistance to snow mould fungi. Meld Norg LandbrHøgsk 56:1–14

Belkhadir Y, Subramaniam R, Dangl JL (2004) Plant disease resistance protein signaling: NBS-LRR proteins and their partners. Curr Opin Plant Biol 7:391–399

Bent AF, Mackey D (2007) Elicitors, effectors, and R genes: the new paradigm and a lifetime supply of questions. Ann Rev Phytopathol 45:399–436

Bertrand A, Castonguay Y, Azaiez A, Hsiang T, Dionne J (2011) Cold-induced responses in annual bluegrass genotypes with differential resistance to pink snow mold (*Microdochium nivale*). Plant Sci 180:111–119

Bhuiyan NH, Selvaraj G, Wei Y, King J (2009) Role of lignification in plant defense. Plant Signal Behav 4:158–159

Bruehl GW (1982) Developing wheats resistant to snow mold. Plant Dis 66:1190–1093

Bruehl GW, Cunfer B (1971) Physiologic and environmental factors that affect the severity of snow mold of wheat. Phytopathology 61:792–799

Chinnusamy V, Zhu J, Zhu J-K (2007) Cold stress regulation of gene expression in plants. Trends Plant Sci 12:444–451

Chisholm ST, Coaker G, Day B, Staskawicz BJ (2006) Host-microbe interactions: shaping the evolution of the plant immune response. Cell 124:803–814

Conrath U (2009) Priming of induced plant defense responses. Adv Bot Res 51:361–395

Cook RJ, Papendick RI (1972) Influence of water potential of soils and plants on root disease. Ann Rev Phytopathol 10:349–374

Dörffling K, Schulenburg S, Lesselich G, Dörffling H (1990) Abscisic acid and proline levels in cold hardened winter wheat leaves in relation to variety-specific differences in freezing resistance. J Agron Crop Sci 165:230–239
Durrant WE, Dong X (2004) Systemic acquired resistance. Ann Rev Phytopathol 42:185–209
Ergon Å, Klemsdal SS, Tronsmo AM (1998) Interactions between cold hardening and *Microdochium nivale* infection on expression of pathogenesis-related genes in winter wheat. Physiol Mol Plant Pathol 53:301–310
Fan J, Hill L, Crooks C, Doerner P, Lamb C (2009) Abscisic acid has a key role in modulating diverse plant-pathogen interactions. Plant Physiol 150:1750–1761
Fowler DB (1982) Date of seeding, fall growth, and winter survival of winter wheat and rye. Agron J 74:1060–1063
Fowler DB (2012) Wheat production in the high winter stress climate of the great plains of North America—An experiment in crop adaptation. Crop Sci 52:11–20
Fowler S, Thomashow MF (2002) Arabidopsis transcriptome profiling indicates that multiple regulatory pathways are activated during cold acclimation in addition to the CBF cold response pathway. Plant Cell 14:1675–1690
Galiba G, Tuberosa R, Kocsy G, Sutka J (1993) Involvement of chromosomes 5 A and 5D in cold-induced abscisic acid accumulation in and frost tolerance of wheat calli. Plant Breed 110:237–242
Garcàa-Olmedo F, Molina A, Alamillo JM, Rodràguez-Palenzuéla P (1998) Plant defense peptides. Biopolymers 47:479–491
Gaudet DA (1986) Effect of temperature on pathogenicity of sclerotial and non-sclerotial isolates of *Coprinus psychromorbidus* under controlled conditions. Can J Plant Pathol 8:394–399
Gaudet DA (1994) Progress towards understanding interactions between cold hardiness and snow mold resistance and development of resistant cultivars. Can J Plant Pathol 16:241–246
Gaudet DA, Bhalla MK, Clayton GC, Chen THH (1989) Effect of cottony snow mold and low temperatures on winter wheat survival in central and northern Alberta. Can J Plant Pathol 11:291–296
Gaudet DA, Chen THH (1987) Effects of hardening and plant age on development of resistance to cottony snow mold (*Coprinus psychromorbidus*) in winter wheat under controlled conditions. Can J Bot 65:1152–1156
Gaudet DA, Chen THH (1988) The effect of freezing resistance and low temperature stress on development of cottony snow mold (*Coprinus psychromorbidus*) in winter wheat. Can J Bot 66:1610–1615
Gaudet DA, Kokko EG (1985) Penetration and infection of winter wheat leaves by *Coprinus psychromorbidus* under controlled conditions. Can J Bot 62:955–960
Gaudet DA, Kozub GC (1991) Screening winter wheat for resistance to cottony snow mold under controlled conditions. Can J Plant Sci 71:957–965
Gaudet DA, Laroche A, Frick M, Huel R, Puchalski B (2003a) Cold induced expression of plant defensin and lipid transfer protein transcripts in winter wheat. Physiol Plant 117:195–205
Gaudet DA, Laroche A, Frick M, Huel R, Puchalski B (2003b) Plant development affects the cold-induced expression of plant defence-related transcripts in winter wheat. Physiol Mol Plant Pathol 62:175–184
Gaudet DA, Laroche A, Puchalski B (2001) Seeding date alters carbohydrate accumulation in winter wheat. Crop Sci 41:728–738
Gaudet DA, Laroche A, Yoshida M (1999) Low temperature-wheat-fungal interactions: a carbohydrate connection. Physiol Plant 106:437–444
Gaudet DA, Tronsmo AM, Laroche A (2012) Climate change and plant diseases. Temperature adaptation in a changing climate. Nature at risk. KB Story and KK Tanino. Wallingford, UK, CABI:144–159
Gaudet DA, Wang Y, Frick M, Puchalski B, Penniket C, Ouellet T, Robert L, Singh J, Laroche A (2010) Low temperature induced defence gene expression in winter wheat in relation to resistance to snow moulds and other wheat diseases. Plant Sci 180:99–110

Grant MR, Jones JDG (2009) Hormone (dis)harmony moulds plant health and disease. Science 324:750–752
Heath MC (2000a) Hypersensitive response-related death. Plant Mol Biol 44:321–334
Heath MC (2000b) Nonhost resistance and nonspecific plant defenses. Curr Opin Plant Biol 3:315–319
Hématy K, Cherk C, Somerville S (2009) Host–pathogen warfare at the plant cell wall. Curr Opin Plant Biol 12:406–413
Hincha DK, Livingston III DP, Premakumar R, Zuther E, Obel N, Cacela C, Heyer AG (2007) Fructans from oat and rye: Composition and effects on membrane stability during drying. Biochim Biophys Acta (BBA)—Biomembr 1768:1611–1619
Hon WC, Griffith M, Mlynarz A, Kwok YC, Yang D (1995) Antifreeze proteins in winter rye sre similar to pathogenesis-related proteins. Plant Physiol 109:879–889
Houde M, Belcaid M, Ouellet F, Danyluk J, Monroy A, Dryanova A, Gulick P, Bergeron A, Laroche A, Links M, MacCarthy L, Crosby W, Sarhan F (2006) Wheat EST resources for functional genomics of abiotic stress. BMC Genomics 7:149
Houde M, Danyluk J, Laliberte J-F, Rassart E, Dhindsa RS, Sarhan F (1992) Cloning, characterization, and expression of a cDNA encoding a 50-Kilodalton protein specifically induced by cold acclimation in wheat. Plant Physiol 99:1381–1387
Huner NPA, Öquist G, Sarhan F (1998) Energy balance and acclimation to light and cold. Trends Plant Sci 3:224–230
Iriki N, Kuwabara T (1993) Half diallel analysis of field resistance of winter wheat to *Typhula ishikariensis* Biotype A in artificially infested plots. Japan J Breed 43:495–501
Jacobs AK, Lipka V, Burton RA, Panstruga R, Strizhov N, Schulze-Lefert P, Fincher GB (2003) An Arabidopsis callose synthase, *GSL5*, is required for wound and papillary callose formation. Plant Cell 15:2503–2513
Jones JDG, Dangl JL (2006) The plant immune system. Nature 444:323–329
Kawakami A, Yoshida M (2002) Molecular characterization of sucrose:sucrose 1-fructosyltransferase and sucrose:fructan 6-fructosyltransferase associated with fructan accumulation in winter wheat during cold hardening. Biosci Biotechnol Biochem 66:2297–2305
Kawakami A, Yoshida M (2005) Fructan:fructan 1-fructosyltransferase, a key enzyme for biosynthesis of graminan oligomers in hardened wheat. Planta 223:90–104
Kawakami A, Yoshida M (2012) Graminan breakdown by fructan exohydrolase induced in winter wheat inoculated with snow mold. J Plant Physiol 169:294–302
Kiyomoto RK, Bruehl GW (1977) Carbohydrate accumulation and depletion by winter cereals differing in resistance to *Typhula idahoensis*. Phytopathology 67:206–211
Koornneef A, Pieterse CMJ (2008) Cross talk in defense signaling. Plant Physiol 146:839–844
Kuwabara C, Imai R (2009) Molecular basis of disease resistance acquired through cold acclimation in overwintering plants. J Plant Biol 52:19–26
Nissinen O (1996) Analysis of climatic factors affecting snow mould injury in first-year timothy (*Phleum pratence* L.) with special reference to *Sclerotinia borealis*. Acta Univ Oulu A 289:1–115
Orozco-Cardenas M, Ryan CA (1999) Hydrogen peroxide is generated systemically in plant leaves by wounding and system in via the octadecanoid pathway. Proc Nat Acad Sci 96:6553–6557
Oshiman K, Kobayashi I, Shigemitsu H, Kunoh H (1995) Studies on turfgrass snow mold caused by Typhula ishikariensis. II. Microscopical observation of infected bentgrass leaves. Mycoscience 36:179–185
Østrem L, Rapacz M, Jørgensen M, Höglind M (2011) Effect of developmental stage on carbohydrate accumulation patterns during winter of timothy and perennial ryegrass. Acta Agric Scand, Section B – Soil Plant Sci 61:153–163
Pääkkönen E, Seppänen S, Holopainen T, Kokko H, Kärenlampi S, Kärenlampi L, Kangasjärvi J (1998) Induction of genes for the stress proteins PR-10 and PAL in relation to growth, visible injuries and stomatal conductance in birch (*Betula pendula*) clones exposed to ozone and/or drought. New Phytol 138:295–305
Peleg Z, Blumwald E (2011) Hormone balance and abiotic stress tolerance in crop plants. Curr Opin Plant Biol 14:290–295

PŁażek A, Dubert F, Pociecha E, Janowiak F, Kolasińska I, Maciejewski M (2011) Resistance of winter rye (*Secale cereale* L.) to *Microdochium nivale* depends on soluble carbohydrate content but not on abscisic acid level. J Phytopathol 159:751–758
Pociecha E, Płażek A, Janowiak F, Janeczko A, Zwierzykowski Z (2008a) Physiological basis for differences in resistance to *Microdochium nivale* (Fr.) Samuels and Hallett in two androgenic genotypes of Festulolium derived from tetraploid F1 hybrids of *Festuca pratensis* and *Lolium multiflorum* (Festulolium). J Phytopath 156:155–163
Pociecha E, Płażek A, Janowiak F, Zwierzykowski Z (2008b) ABA level, proline and phenolic concentration, and PAL activity induced during cold acclimation in androgenic Festulolium forms with contrasting resistance to frost and pink snow mould (*Microdochium nivale*). Physiol Mol Plant Pathol 73:126–132
Pociecha E, Płażek A, Janowiak F, Waligórski P, Zwierzykowski Z (2009) Changes in abscisic acid, salicylic acid and phenylpropanoid concentrations during cold acclimation of androgenic forms of Festulolium (*Festuca pratensis*, *Lolium multiflorum*) in relation to resistance to pink snow mould (*Microdochium nivale*). Plant Breed 128:397–403
Robert-Seilaniantz A, Grant M, Jones JDG (2011) Hormone crosstalk in plant disease and defense: more than just jasmonate-salicytate antagonism. Ann Rev Phytopathol 49:317–343
Sun J-Y, Gaudet DA, Lu Z-X, Frick M, Puchalski B, Laroche A (2008) Characterization and antifungal properties of wheat nonspecific lipid transfer proteins. Mol Plant-Microbe Interact 21:346–360
Sung S, Amasino RM (2005) Remenbering winter: toward a molecular understanding of vernalization. Ann Rev Plant Biol 56:491–508
Thomashow MF (2003) Plant cold acclimation: freezing tolerance genes and regulatory mechanisms. Annu Rev Plant Physiol Plant Mol Biol 50:571–599
Thordal-Christensen H (2003) Fresh insights into processes of nonhost resistance. Curr Opin Plant Biol 6:351–357
Ton J, Flors V, Mauch-Mani B (2009) The multifaceted role of ABA in disease resistance. Trends Plant Sci 14:310–317
Tronsmo A, Gregersen PL, Hjelsjord J, Sandal T, Bryngelsson T, Collinge DB (1993) Cold induced disease resistance. Mechanisms of plant defense responses. B Fritig and M Legrand, Kluwer Academic Publishers 369
Tronsmo AM (1984a) Predisposing effects of low temperature on resistance to winter stress factors in grasses. Acta Agric Scand 34:210–220
Tronsmo AM (1984b) Resistance to the rust fungus *Puccinia poae-nemoralis* in *Poa pratensis* is induced by low-temperature hardening. Can J Bot 62:2891–2892
Tronsmo AM (1986) Host water potentials may restrict development of snow mold fungi in low temperature hardened grasses. Physiol Plant 68:175–179
Van Loon LC, Rep M, Pieterse CMJ (2006) Significance of inducible defense-related proteins in infected plants. Ann Rev Phytopathol 44:135–162
Veisz O, Galiba G, Sutka J (1996) Effect of abscisic acid on the cold hardiness of wheat seedlings. J Plant Physiol 149:439–443
Wingender R, Röhrig H, Höricke C, Wing D, Schell, J (1989) Differential regulation of soybean chalcone synthase genes in plant defence, symbiosis and upon environmental stimuli. Mol Gen Genet 218:315–322
Yoshida M, Abe J, Moriyama M, Kuwabara T (1998) Carbohydrate levels among winter wheat cultivars varying in freezing tolerance and snow mold resistance during autumn and winter. Physiol Plant 103:8–16
Yoshida M, Abe J, Moriyama M, Shimokawa S, Nakamura Y (1997) Seasonal changes in the physical state of crown water associated with freezing tolerance in winter wheat. Physiol Plant 99:363–370

Activating Disease Resistance in Turfgrasses Against Fungal Pathogens: Civitas and Harmonizer

Tom Hsiang, Paul H. Goodwin, Alejandra M. Cortes-Barco, Brady T. Nash, and Jonathan Tung

Introduction

Plants possess many resistance mechanisms against diseases that are triggered following pathogen attack. Under intensive plant maintenance systems and stressful growth conditions, these mechanisms may be insufficient to guard against disease outbreaks without significant economic loss. However, there are compounds that can pretrigger resistance against pathogens of plants, although very little of this research has been done on monocots compared with dicots. These compounds generally have little or no direct antifungal activity, but activate defense mechanisms of the plant prior to pathogen attack, resulting in greater subsequent resistance. We have been investigating the activity and mode of action of several new plant defense-activating compounds, particularly in the turfgrass species, creeping bentgrass (*Agrostis stolonifera*), and present here an overview of some of our recent research findings.

Activated Resistance Against Plant Diseases

Activated, acquired, or induced resistance is a physiological state in which environmental, chemical, or biological stimuli pretrigger a part of the plant's defenses against subsequent pathogen attack. During induced resistance, a plant defense activator is recognized by the plant or stimulates a part of a recognition-signaling pathway. The signaling pathway(s) eventually promote the expression of defense genes resulting in the expression of defense mechanisms such as antimicrobial proteins (Edreva 2004). The enhanced resistance is expressed locally at the site of infection and, in some cases, systemically throughout the plant (Métraux et al. 2002).

T. Hsiang (✉) · P. H. Goodwin · A. M. Cortes-Barco · B. T. Nash · J. Tung
School of Environmental Sciences, University of Guelph,
50 Stone Road E, Guelph, ON, N1G 2W1, Canada
e-mail: thsiang@uoguelph.ca

R. Imai et al. (eds.), *Plant and Microbe Adaptations to Cold in a Changing World*,
DOI 10.1007/978-1-4614-8253-6_28, © Springer Science+Business Media New York 2013

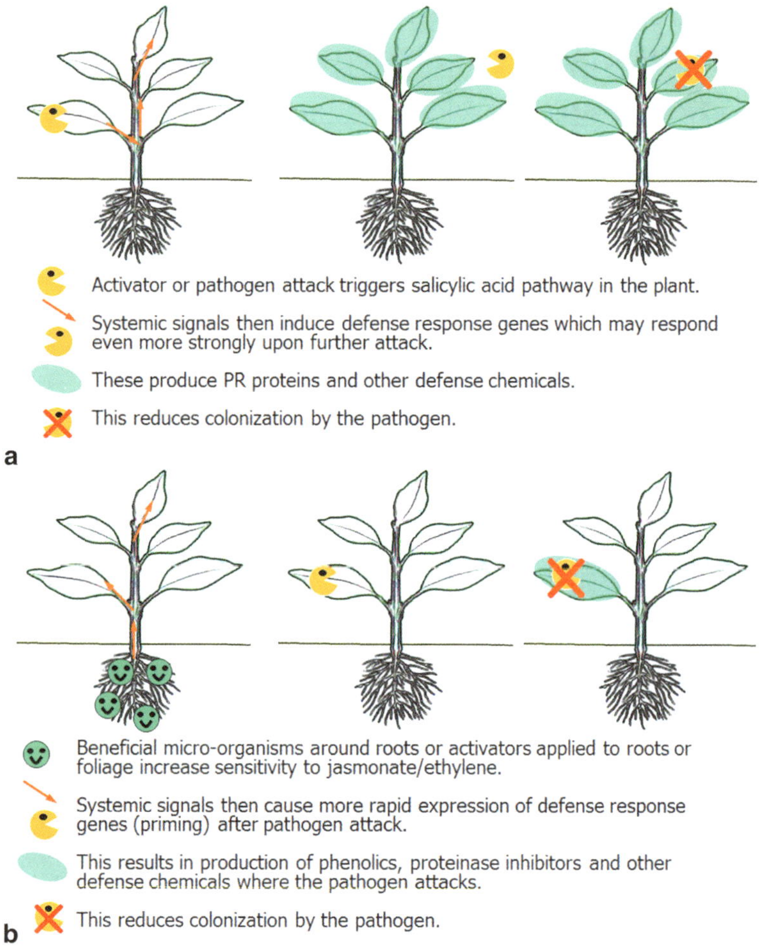

Fig. 1 Diagrammatic representation of (**a**) SAR and (**b**) ISR

The two major recognized forms of pretriggered systemic resistance are called systemic acquired resistance (SAR; Fig. 1a) and induced systemic resistance (ISR; Fig. 1b). In SAR, resistance can be activated by natural or synthetic chemicals and avirulent or attenuated pathogens (Lyon 2007; Vallad and Goodman 2004). These activators will locally induce resistance mechanisms that result in a signal that is translocated systemically throughout the plant. As a result, the whole plant may become more resistant to subsequent pathogen attack (Sticher et al. 1997). Salicylic acid (SA) is an important signaling molecule involved in this type of resistance (Métraux et al. 2002). Although SAR is associated with the direct stimulation of expression of a set of defense-related genes regulated by SA, there is evidence that some defense-related genes can also show potentiation (priming), which means that some elements in defense triggering, such as production of nonactivated protein kinase, are affected so that expression of defense genes can respond stronger and faster upon later pathogen attack (Conrath et al. 2002).

In contrast to SAR, ISR is normally induced in roots by beneficial soil-borne plant growth-promoting rhizobacteria (PGPR) and plant growth-promoting fungi. The signal molecules are ethylene and jasmonic acid, which regulate the response of a set of plant genes that are typically also expressed in response to wounding (Métraux et al. 2002). In ISR, PGPRs in the soil remain localized at the root surface or invade the root without causing apparent injury, but still induce resistance in the leaves or stems, demonstrating that ISR is systemic similar to SAR (Sticher et al. 1997). Although some genes may be induced directly during ISR, more genes show priming of expression with ISR compared with SAR (Besser et al. 2000; Verhagen et al. 2004).

The purpose of this work was to examine the activity of the products Civitas and Harmonizer directly on fungal pathogens and indirectly on pathogens attacking turfgrass in the laboratory and the field, and to investigate the mode of action by assessing expression of ISR- or SAR-marker genes after exposure to these compounds.

Understanding New Defense Activators

New plant defense activators are continually being developed for plants. Some important defining characteristics of such compounds are as follows: (1) the plant is directly affected by the compound; (2) the compound may have weak or no effect on the pathogen; and (3) the compound causes systemic activation of resistance. In addition to determining their effectiveness, it is important to also understand their mode of action (SAR, ISR, or another form of induced resistance), as this is important for understanding the compatibility of different defense activators, potential demands on plant resources, and the duration of the protection. A common way to address the mode of action of defense activators is to examine the effects of defense activators on plant gene expression.

Mode of Action of Civitas

In the last few years, we have been testing a new compound produced by Petro-Canada called Civitas. It is a mixture of food-grade synthetic isoparaffins and a food-grade emulsifier. It is a clear, colorless liquid at room temperature, composed of molecules with between 16 and 36 carbons. The formulation has been registered in the United States (February 2009) and Canada (January 2011), and is being investigated in other parts of the world. Label rates are 1.2–25 % in 4–20 l water per 100 m^2 of grass surface.

One of the features of a resistance-inducing compound is that it should have a weak or no direct effect on the pathogen. We tested 16 different turfgrass pathogens on culture media amended with 0–20 % Civitas and found that there was only slight

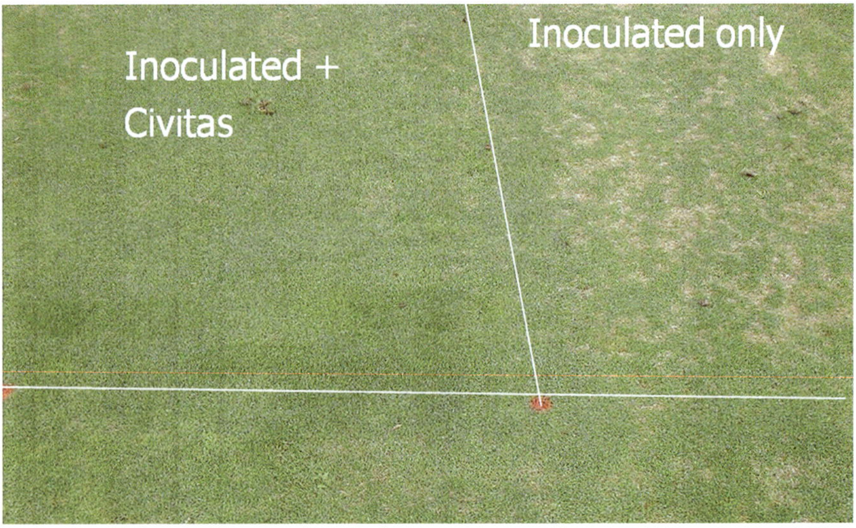

Fig. 2 Field trial with 5 % Civitas applied in water at 10 l per 100 m^2. The plots were inoculated in July 2006 with a grain inoculum of *Sclerotinia homoeocarpa*, which causes dollar spot disease. This photo was taken a month later

Fig. 3 Civitas was applied to soil of pots containing 2-week-old creeping bentgrass plants and inoculated 7 days later. This photo was taken 5 days after the inoculation showing much more fungal activity and disease in the inoculated control

inhibition. However, even this inhibition disappeared after 10 days, since growth rates after that time were similar between amended media and nonamended controls (Cortes et al. 2010b).

Civitas was sprayed onto turfgrass in the field and found to have significant activity against various turfgrass diseases with almost full disease suppression in some cases (Fig. 2). We also conducted tests on turfgrass grown in the laboratory in plastic containers and found that even when the compound was directly applied to soil without contact to leaves, the leaves became more resistant to several fungal pathogens (Fig. 3), and could reduce diseased area by 20–40 % (Cortes et al. 2010b). This activity gave us the idea that the compound was activating a form of systemic resistance in the plant.

To obtain more definitive evidence that the plant was being directly affected by the Civitas treatment, we examined gene expression changes in treated plants. First, we needed to find marker genes as indicators of SAR or ISR. Many marker genes for SAR or ISR have been identified in dicotyledonous plants, but there have been relatively few studies on monocotyledonous plants such as rice, wheat, maize, and barley, compared with dicotyledonous plants such as Arabidopsis and tobacco. We needed to find the corresponding genes in *A. stolonifera*, but this was challenging since genomic sequences or expressed sequence tag libraries were limited in availability for this species and other turfgrasses. Using PCR with primers in conserved regions of known SAR- and ISR-related genes in various cereals, we found the corresponding genes in *A. stolonifera* and examined whether they responded to known SAR or ISR defense activators. We then examined the expression of the genes at set time intervals, starting before the application of Civitas to the roots, both before and after inoculation of the plants with the turfgrass pathogen *Microdochium nivale*, and throughout the infection period.

Figure 4 (Cortes et al. 2010a) shows examples of the response of three marker genes for ISR (*AsOPR4*, *AsGNS5*, and *AsOS1*) in *A. stolonifera*, in comparison (2R, 3R)-butanediol that is known to cause ISR in other plants. For *AsOPR4*, treatment with water resulted in no significant change in expression, whereas (2R, 3R)-butanediol caused induction before inoculation and then exhibited priming after inoculation. In contrast, Civitas did not cause any significant induction of *AsOPR4* prior to inoculation, but it also led to priming after inoculation. *AsGNS5* showed a very similar pattern with (2R, 3R)-butanediol, but not Civitas, causing induction and both compounds showing strong priming. The third gene, *AsOS1*, showed induction with Civitas but not with (2R, 3R)-butanediol. As with the two other genes, high levels of priming for *AsOS1* were observed with both (2R, 3R)-butanediol and Civitas.

Our conclusions on the activity of the defense activator Civitas are as follows: (1) it has minor direct effects on fungi, (2) it has suppressive effects against diseases in the laboratory and field, (3) it primes defense-response genes for greater and faster expression after infection, and (4) its mode of action is ISR based on gene expression analysis when compared with gene expression after application of a known ISR or SAR activator (Cortes et al. 2010a, b).

Mode of Action of Harmonizer

Civitas can also be combined with Harmonizer, which is a green pigment containing polychlorinated copper (Cu) II phthalocyanine (Nash 2011). Harmonizer is also produced by Petro-Canada and metal phthalocyanines, such as copper phthalocyanine, have been used on turfgrasses for more than 50 years as colorants (Liu et al. 2007; Ostmeyer 1994). Harmonizer when applied to grass gives it a distinct deep green color (Fig. 5). According to the manufacturer, Petro-Canada, the addition of Harmonizer to Civitas can allow for increased efficacy in controlling turfgrass diseases compared with applying Civitas alone.

Although the mode of action for Harmonizer is not known, the active ingredient may be the metal, copper. Treatment of roots with copper sulfate has been found

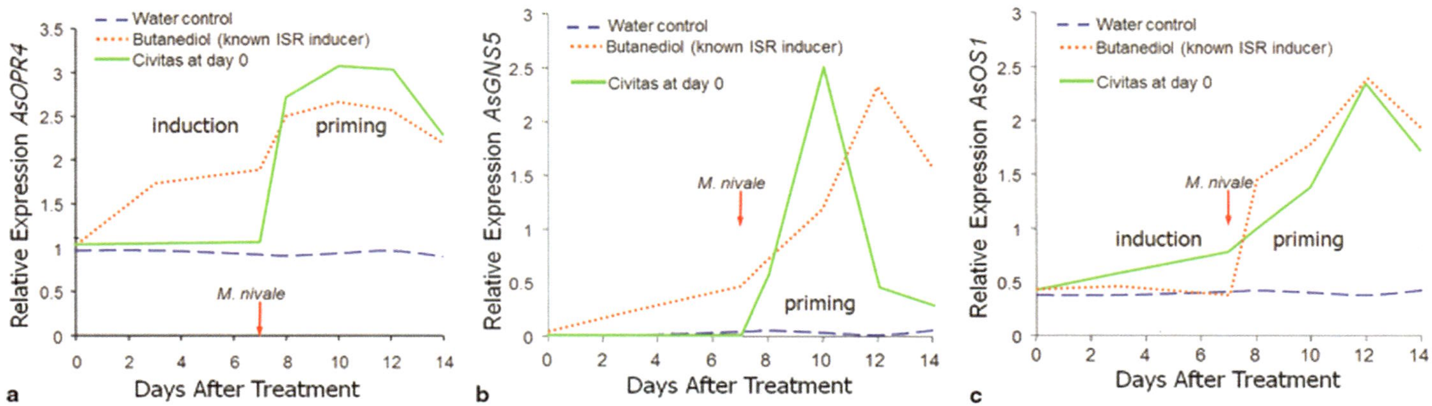

Fig. 4 Expression of *Agrostis stolonifera* genes (**a**) *AsOPR4* (**b**) *AsGNS5*, or (**c**) *AsOS1* after treatment with water, ISR activator butanediol, or 10 % PC1. Gene expression based on the amount of mRNA was determined relative to the expression of a constitutive ubiquitin gene. Treatments (water, butanediol, or Civitas) were applied at day 0 and plants were inoculated with a grain inoculum of *Microdochium nivale* 7 days later

Fig. 5 The use of Harmonizer in the field on *Agrostis stolonifera*. The grass acquires a distinctly *green* color after treatment

to induce resistance to root disease caused by *Verticillium dahliae* and increased expression of defense-related genes (Chmielowska et al. 2010). Although little information is available on the effects of metal phthalocyanines on disease reduction in plants, it is possible that metal phthalocyanines induce disease resistance through the production of ROS in plants (Vol'pin et al. 2000).

As with Civitas, Harmonizer was tested for direct activity against pathogens. Four turfgrass pathogens (*Rhizoctonia solani*, *Sclerotinia homoeocarpa*, *Microdochium nivale*, and *Colletotrichum cereale*) were grown on media amended with Harmonizer at concentrations up to 10 %, which is 20 times the normal field rate for application. *Colletotrichum cereale* showed the greatest sensitivity with a reduction of 50 % of radial growth estimated when grown on 0.1 % Harmonizer. The other three fungal pathogens showed much less sensitivity, with *Microdochium nivale* having a 50 % growth reduction at 8 % Harmonizer, while the remaining two were calculated to require more than 10 % Harmonizer for 50 % growth reduction.

In laboratory tests, Harmonizer applied at a rate of 5 ml of 5 % solution to foliage of *A. stolonifera* in 9-cm-diameter mason jars reduced disease caused by *S. homoeocarpa* by 93 % on average (Fig. 6a). However, since Harmonizer at 5 % has some direct effects on fungal growth, it was also directly applied to soil to avoid direct contact with the pathogen (Fig. 6b). An application of 10 ml of 5 % solution to 70 ml of soil in pots of *A. stolonifera* reduced disease caused by *S. homoeocarpa* by 40 % at 3 weeks after Harmonizer treatment and 2 weeks after inoculation of *A. stolonifera*. This demonstrated that Harmonizer is able to induce resistance in plants and the resistance is systemically transported from the roots to foliage.

In field tests, Harmonizer alone at 0.5 % applied at 10 l per 100 m^2 reduced disease caused by *S. homoeocarpa* by 45 % (Fig. 7). In the same trial, Civitas alone at 5 % reduced disease by 58 %, while Civitas and Harmonizer combined gave 69 % disease reduction. Thus, both Civitas and Harmonizer were effective, but the combination of Civitas and Harmonizer was more effective than either chemical alone. By comparison, the fungicide check, Daconil (containing 45 % chlorothalonil) reduced disease by 80 % (Fig. 7).

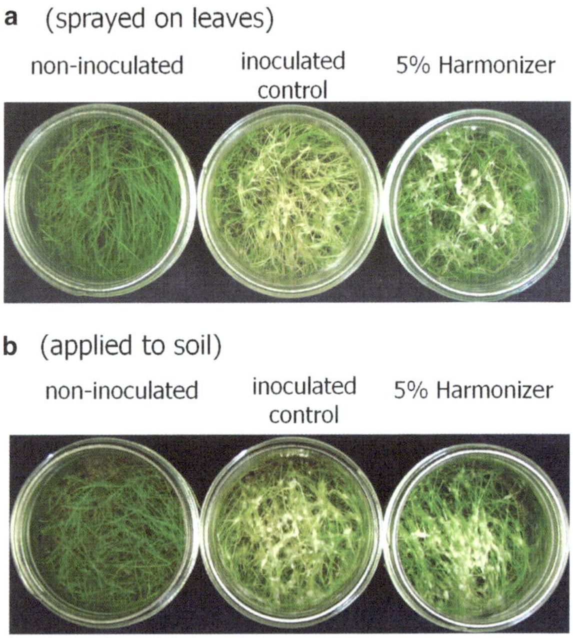

Fig. 6 Harmonizer applied (**a**) by spraying or (**b**) to soil and leaves inoculated 7 days later with the fungal pathogen *Sclerotinia homoeocarpa*. This photo was taken 7 days after inoculation, 14 days after Harmonizer treatment

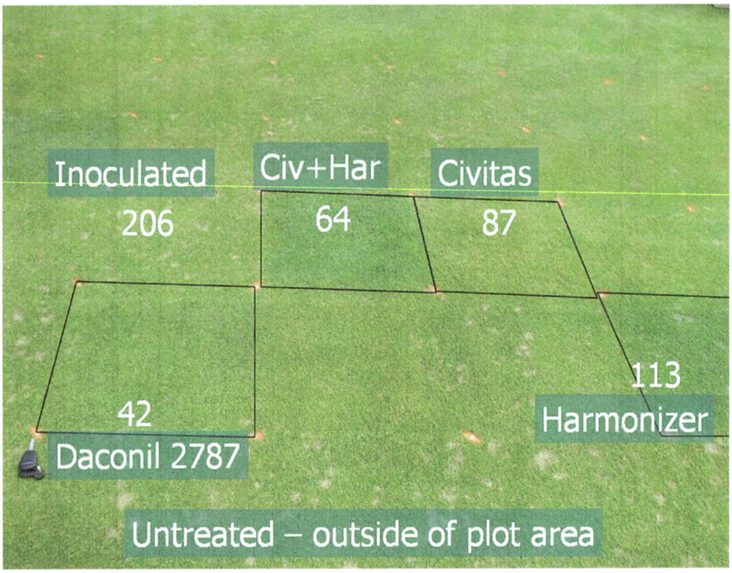

Fig. 7 Harmonizer, Civitas, and a standard fungicide (Daconil 2787 containing chlorothalonil) applied to field plots of *Agrostis stolonifera*. The fungus *Sclerotinia homoeocarpa* was inoculated onto the plots and rated through the season. This photo was taken 3 weeks after the last treatment in August 2010 and numbers reflect the amount of disease over the entire trial where low numbers represent lower disease

Table 1 Effect of Harmonizer on expression of SAR- and ISR-marker genes by RT-PCR in *Agrostis stolonifera*. These genes had previously been identified in *A. stolonifera* as being induced by benzothiadiazole (SAR) or butanediol (ISR)

Gene	Mode of action	Expression
Lipoxygenase	SAR	No change
Aspartic protease	SAR	No change
Hypersensitive-response-induced gene	SAR	No change
Glucanase 5	ISR	No change
12-oxo-phytodienoic acid reductase 7	ISR	No change
Allene oxide synthase 1	ISR	No change

To obtain more definitive evidence that the plant defenses were being directly affected by Harmonizer, we examined gene expression changes in treated plants. First, we tested three genes identified as marker genes for SAR in *A. stolonifera* (lipoxygenase, aspartic protease, and a hypersensitive-response-induced gene; Tung 2011), and found that there was no change in gene expression as measured by relative reverse transcriptase PCR after exposure to Harmonizer (Table 1). We also tested three marker genes for ISR in *A. stolonifera* (glucanase 5, 12-oxo-phytodienoic acid reductase 7, and allene oxide synthase 1; Cortes et al. 2010a) and they did not show any response to Harmonizer treatment (Table 1). These results implied that Harmonizer was not activating exhibiting classical ISR or SAR based on these limited number of genes. We then decided to use the power of next-generation sequencing to examine gene expression effects caused by Harmonizer on a much broader scale.

Soil in pots containing *A. stolonifera* was treated with 5 % Harmonizer or water and leaves were collected 7 days later. Total RNA was extracted from these leaves and sent for sequencing, which was done with an Illumina-Solexa GAIIx machine using a single lane to give 36 bp paired-end reads with a 200 bp insert size. After sequencing, there were more than 19 million 36 bp reads for the water control sample and more than 17 million reads for the Harmonizer-treated sample. These reads were assembled into contigs used the program Velvet (version 1.0.12, www.ebi.ac.uk/~zerbino/velvet; Zerbino and Birney 2008) over a range of Kmers, with Kmer 29 being selected as yielding the largest assembled contigs (highest N50 value).

Following assembly, the 36 bp reads from the Harmonizer data set were compared with the assembled contigs to determine read counts that were compared between Harmonizer- and water-treated samples to assess gene expression levels. The data were normalized by adjusting for read counts between the Harmonizer-treated and water-control samples. The adjusted counts were then subjected to a Chi-square test at $p=0.05$ to assess whether expression at twofold or greater in Harmonizer-treated versus water-control samples was significant. Among the matching contigs, 1,000 of them (referred to as nodes) were identified as having more than two times increased expression with Harmonizer. From these 1,000 nodes, 633 had matches with annotations in the NCBI Genbank NR database and, among these, 20 genes were identified from their annotations as potentially related to disease resistance.

Table 2 Twenty genes selected for their annotations potentially related to disease, out of a pool of 1,000 genes, showing greater than twofold overexpression (compared with water) in *Agrostis stolonifera* treated with Harmonizer in an RNAseq experiment. Three of these (PR-5, PR-10, and PR-1; basic) were chosen for further RT-PCR experiments

Gene	Expression
Cell wall-associated hydrolase	2.0×
PAL-4	2.0×
PAL-1	2.1×
PR class i	2.1×
WRKY-6	2.1×
Bacterial-induced peroxidase	2.2×
Harpin-induced protein	2.2×
Ethylene-responsive protein	2.3×
Hypersensitivity-related protein	2.5×
Disease-resistance protein	2.8×
PR-5	2.8×
Chitinase	3.0×
Jasmonate-induced protein	3.2×
Lipoxygenase	3.3×
Nematode-resistance protein	3.5×
Endo-beta glucanase	3.6×
Rapidly elicited protein	4.1×
4-coumarate coenzyme A ligase	4.5×
PR-10	4.5×
PR-1 (basic)	20×

Several of these genes were confirmed by RT-PCR to show induction by Harmonizer (Table 2). This indicates that Harmonizer does affect gene expression systemically (*A. stolonifera*-treated roots showed altered gene expression in leaves) and these changes include a number of defense-related genes.

A comparison to genes induced with a well-known activator of SAR in *A. stolonifera* (Actigard) showed that there was very little overlap in the changes in defense gene expression. This indicates that the impact of Harmonizer on *A. stolonifera* appears to be different from SAR or ISR. Further work is needed to characterize these differences and determine if the induced defense genes actually contribute to the reductions in disease observed with Harmonizer.

Our conclusions on the activity of the defense activator Harmonizer are as follows: (1) it has minor direct effects on fungi, except for *C. cereale*, which shows high sensitivity in laboratory tests, (2) it has mild suppressive effects against diseases in laboratory and field tests, (3) it induces defense-related genes in other parts of the plant after treatment, and (4) the mode of action of Harmonizer has some overlap with SAR, but does not exactly match that caused by well-known ISR or SAR activators based on gene expression analysis.

Acknowledgments This work was supported by the Natural Sciences and Engineering Research Council of Canada, the Ontario Turfgrass Research Foundation, and Petro-Canada.

References

Besser K, Jarosch B, Langen G, Kogel KH (2000) Expression analysis of genes induced in barley after chemical activation reveals distinct disease resistance pathways. Mol Plant Pathol 1:277–286

Chmielowska J, Veloso J, Gutiérrez J, Silvar C, Díaz J (2010) Cross-protection of pepper plants stressed by copper against a vascular pathogen is accompanied by the induction of a defence response. Plant Sci 178:176–182

Conrath U, Pieterse CM, Mauch-Mani B (2002) Priming in plant-pathogen interactions. Trends Plant Sci 7:210–216

Cortes-Barco AM, Hsiang T, Goodwin PH (2010a) Induced systemic resistance against three foliar diseases of *Agrostis stolonifera* by (2R, 3R)-butanediol or an isoparrafin mixture. Ann Appl Biol 157:179–189

Cortes-Barco AM, Goodwin PH, Hsiang T (2010b) Comparison of induced resistance activated by benzothiodiazole, (2R, 3R)-butanediol and an isoparaffin mixture against anthracnose of *Nicotiana benthamiana*. Plant Pathol 59:643–653

Edreva A (2004) A novel strategy for plant protection: induced resistance. J Cell Mol Biol 3:61–69

Liu H, McCarty B, Baldwin CM, Sarvis WG, Long SH (2007) Painting dormant bermudagrass putting greens. Golf Course Manage 75(11):86–91

Lyon G (2007) Agents that can elicit induced resistance. In: Walters D, Newton A, Lyon G (eds) Induced resistance for plant defence: a sustainable approach to crop protection. Blackwell Publishing, Oxford, pp 9–30

Métraux JP, Nawrath C, Genoud T (2002) Systemic acquired resistance. Euphytica 124:237–243

Nash BT (2011) Activation of disease resistance and defense gene expression in *Agrostis stolonifera* and *Nicotiana benthamiana* by a copper-containing pigment and a benzothiadiazole. M.Sc. Thesis, University of Guelph. 300 pp

Ostmeyer T (1994) The color green. Golf Course Manage 62(8):40–44

Sticher L, Mauch-Mani B, Métraux JP (1997) Systemic acquired resistance. Ann Rev Phytopathol 35:235–270

Tung J (2011) The effectiveness of induced plant disease resistance: genotypic variation and quantification by chlorophyll fluorescence. M.Sc. Thesis, University of Guelph. 192 pp

Vallad GE, Goodman RM (2004) Systemic acquired resistance and induced systemic resistance in conventional agriculture. Crop Sci 44:1920–1934

Verhagen B, Glazebrook J, Zhu T, Chang HS, van Loon L, Pieterse C (2004) The transcriptome of rhizobacteria-induced systemic resistance in Arabidopsis. Mol Plant Microbe Interact 17:895–908

Vol'pin ME, Novodarova GN, Krainova NY, Lapikova VP, Aver'yanov AA (2000) Redox and fungicidal properties of phthalocyanine metal complexes as related to active oxygen. J Inorg Biochem 81:285–292

Zerbino DR, Birney E (2008) Velvet: algorithms for de novo short read assembly using de Bruijn graphs. Genome Res 18:821–829

Index

R. Imai et al. (eds.), *Plant and Microbe Adaptations to Cold in a Changing World*,
DOI 10.1007/978-1-4614-8253-6, © Springer Science+Business Media New York 2013

L

M

Printed by Printforce, the Netherlands